Chemometric Monitoring: Product Quality Assessment, Process Fault Detection, and Applications

Chemometric Monitoring: Product Quality Assessment, Process Fault Detection, and Applications

By
Madhusree Kundu, Palash Kumar Kundu,
and Seshu Kumar Damarla

CRC Press is an imprint of the
Taylor & Francis Group, an **informa** business

CRC Press
Taylor & Francis Group
6000 Broken Sound Parkway NW, Suite 300
Boca Raton, FL 33487-2742

Printed on acid-free paper

International Standard Book Number-13: 978-1-1387-4621-3 (Paperback)
978-1-4987-8007-0 (Hardback)

Library of Congress Cataloging-in-Publication Data

Names: Kundu, Madhusree, author. | Kundu, Palash Kumar, author. | Damarla, Seshu K., author.
Title: Chemometric monitoring : product quality assessment, process fault detection and applications / Madhusree Kundu, Palash Kumar Kundu, Seshu K. Damarla.
Description: Boca Raton : CRC Press, 2018. | Includes bibliographical references and index.
Identifiers: LCCN 2017015762| ISBN 9781498780070 (hardback : acid-free paper) | ISBN 9781315155135 (ebook)
Subjects: LCSH: Chemometrics.
Classification: LCC QD75.4.C45 K86 2018 | DDC 543.01/5195–dc23
LC record available at https://lccn.loc.gov/2017015762

Visit the Taylor & Francis Web site at
http://www.taylorandfrancis.com

and the CRC Press Web site at
http://www.crcpress.com

Dedication

Dedicated in the fond memory of our beloved brother, Late Pradyumna Majumer

Dr. Madhusree Kundu

Dr. Palash Kundu

Dedicated to my father, Late Damarla Venugopalarao

Seshu Kumar Damarla

Contents

Preface

There have been specific ways of familiarizing oneself with chemometrics so far. Chemometrics to date is evolving not as a discipline on its own but due to combined advancements in multivariate statistical applications (biomedical signal processing specifically) fault detection and diagnosis leading to fault-tolerant system design, and machine learning–related research (accomplished by the computer science discipline; especially). In recent times, a new interface is simmering between bioinformatics and chemometrics, courtesy of analytical data provided by spectroscopy, chromatography, and gene expression profiles or protein sequences. All these various efforts are being consolidated under the single platform of chemometrics. We desired to contribute to this coveted transformation along with many others, which steered us toward writing this book.

We wanted to write a book in which the requisite text on chemometric techniques and their applications are found together, instead of having to wander between textbooks and journals. This book boasts a separate chapter on data presentation, acquisition, and design of experiments for data generation, data preprocessing (including outlier detection, data reconciliation, and data transform), basic statistical measures, hypothesis testing, regression, and nature of the process—either stochastic or deterministic and stationary or nonstationary. One can take this chapter in the sense of an elaborate taxonomy that helps toward an easy transition to the core chapters of this book. In addition, Chapter 1, in its present form, may be treated as a prerequisite for initiating any maiden research in this area.

Presenting theoretical postulations on chemometric techniques is not the only focus of this book. It also encourages readers to apply the different algorithms provided using chemometric techniques such as principal component analysis (PCA), partial least squares (PLS), linear discriminant analysis (LDA), support vector machine (SVM), Sammon's nonlinear mapping (NLM), and different clustering techniques for diversified applications.

Process monitoring with a view to detect normal and abnormal process situations and ensure product quality are statutory obligations. The mathematical modeling of plant-wide processes along with a pilot-scale process is included in this book to carry out simulation using the MATLAB®/SIMULNK platform, which supports the development of suitable algorithms by providing a design database leading to the detection of various operating conditions in a process. Detection of normal and abnormal operating conditions are the key factors behind successful fault diagnosis, and hence the development of fault-tolerant systems. This book intends to successfully detect the process in normal operating conditions and its faults through the algorithms proposed. Fault diagnosis and development of a fault-tolerant system are not within the scope of the proposed book.

Product quality monitoring is a statutory obligation. Beverage quality grading in bulk requires automation, which may be an alternative to laboratory-based analytical procedures. One of the aims of this book is to encourage and address such a possibility. The electronic tongue signatures assigned to various beverages (called features) are utilized for their authentication/grading. An electronic tongue is an automated device engaged in characterizing liquid samples. It exploits electrochemical methods (like voltammetry) for detection of analytes present in a liquid sample using electrochemical sensors and theoretical principles of electrochemistry. Feature extraction followed by pattern matching with the reference features using chemometric algorithms can lead to an identity for a particular liquid. This is evolving as an alternative to the time-consuming, laborious, and expensive conventional laboratory-based analytical instruments. E-tongue-based methods are especially suitable for in situ and bulk quality monitoring. Recursive principal component analysis (RPCA), PLS, LDA, Sammon's NLM, and SVM models have been used in developing pattern classification and authentication algorithms for commercial tea grading and water brand classification/authentication. Guidelines regarding prototype development of a commercial tea grader are presented in this book. These principles can be extended for the design of possible low-cost and everyday appliances such as beverage quality checking devices, adulterated milk detectors, out-of-date food product detectors, and arsenic meters to quantify arsenic in contaminated water. This book possibly will benefit people engaged in engineering (product) research and development and entrepreneurs. (As the hardware design, development, and implementation needed for these appliances are beyond the scope of this book, these could not be accommodated.)

Health care of people in remote locations (especially in villages of third-world countries like India, Pakistan, Bangladesh, etc.) is a staggering issue due to high population density and lack of infrastructural facilities and experienced doctors. In such circumstances, interactive patient care monitoring tools developed using different biological/biomedical signals (like electrocardiogram [ECG], echocardiogram [ECHO], and many more) may help the local medical practitioners to a certain degree in complementing their clinical knowledge and judgments. With the networking facilities available, treatment also can be provided from any distant, convenient, and urban locations by consulting concerned experienced physicians. This book aspires to help design such interactive patient care monitoring tools. Computerized ECG signal analysis, morphological modeling of ECG, automated ECG pattern classification, and authentication are included in this book in relation to patient care monitoring efforts.

This book is applied and interdisciplinary in nature. A large readership of this book is expected to include engineers from the chemical, electrical, electronics, mechanical and manufacturing, and biomedical and the fraternity of chemistry, food technology, biotechnology, and health care industries, for various reasons.

This book poses an open invitation for an interface, which is required henceforth for practical implementation of the propositions and possibilities referred in the book. It poses a challenge to the researchers in academia toward the development of more sophisticated algorithms with broad generality at an affordable price. The book also encourages chemometric applications in diversified areas.

SIMULINK modules and MATLAB® codes (in the form of functions) for all the significant chemometric techniques used in the applications considered are provided in this book. The pseudocodes provided along with the MATLAB® functions enable the graduate students/researchers/professionals to sort out their own applications and reproduction of the results therein. The book provides data files, sources for biomedical signals and their necessary preprocessing steps, all the algorithms developed for the automated device designs, and source codes developed in MATLAB® AND SIMULINK for all the process models required to simulate normal/faulty process data. In Chapters 3 and 4, the authors have used a few plant models and necessary simulation codes already available in the public domain and avoided replicating them in this book.

In summary, any scientific enquiry must not remain indifferent to societal needs. This book is proposed with a belief that it may elicit large numbers of R&D initiatives and motivate an entrepreneurial drive in food quality assessment, process monitors, and interactive patient care monitoring tools, especially in a country like India and countries with developing economies enriched with human resources and a large underprivileged rural population.

Acknowledgments

We would like to take the opportunity to acknowledge a few people we came across in the process of developing of this book.

We are indebted to Dr. P. C. Panchariya, senior Principal Scientist, Digital Systems Group, CSIR-Central Electronics Engineering Research Institute, Rajasthan, PILANI, India, who gave Dr. P. Kundu the opportunity to work in his laboratory on mineral water and fruit juice brand classification using the electronic tongue way back in 2005. Till now, the association has remained consistently rewarding and Dr. Panchariya has been immensely supportive of our electronic-tongue-related research over these years.

Dr. M. Kundu is grateful to the Department of Science and Technology, India (under Water Technology Initiative programme, Grant No. DST/TM/WTI/2K11/328(G), dated 11-06-2012) for funding the e-tongue-based research and development, especially on arsenic quantification.

We are grateful to Dr. Rajarshi Gupta, associate professor, Department of Applied Physics, University of Calcutta, India. Without his association and timely inputs on cardiovascular signal compression and analysis, and tele-cardiology, we would not have been able to incorporate *patient care monitoring* as a part of this book. We feel privileged to have him in our professional sphere.

We wish to acknowledge Dr. Bijay Kumar Rout, associate professor, Mechanical Engineering Department, BITS Pilani, Rajasthan, India, who incited the very first impression about the applicability of multivariate statistics during our association with him way back in 2005. Dr. Rout happened to be the then colleague of Dr. M. Kundu in BITS, Pilani.

Dr. P. Kundu wishes to express his gratitude to Professor Ashis SenGupta, Applied Statistics Unit, Indian Statistical Institute, Kolkata and visiting professor, Department of Statistics, University of California, Riverside, California, for his critical and constructive suggestions regarding statistical interpretation of data and application of multivariate statistical techniques with caution in the development of several algorithms oriented toward various applications.

In Chapters 3 and 4, we have used some SIMULINK modules for *continuous crystallization, fluid catalyzed cracking* (FCC), *continuous stirred tank heater* (CSTH), and *polyethylene manufacturing* processes along with associated MATLAB® codes available in the public domain. We duly acknowledge Professor Christos Georgakis (Department of Chemical and Biological Engineering, Tufts University, Medford, Massachusetts), Professor Jeffrey D. Ward (Department of Chemical Engineering, National Taiwan University, Taipei, Taiwan), Professor Nina F. Thornhill (Centre for Process Systems Engineering, Department of Chemical Engineering, Imperial College London, London SW7 2AZ, UK), and Professor E. Ali (Chemical Engineering Department, King Saud University, P.O. Box 800, Riyadh 11421, Saudi Arabia).

We are indebted to many graduate students for their invaluable assistance directly or indirectly in performing the research work for this book. Dr. M. Kundu wishes to acknowledge Mr. Sujeevan Kumar Agir for his contribution in relation to Chapters 5 and 7 Dr. P. Kundu would like to acknowledge Jayeeta Chowdhury for her persuit of indigenous e-tongue development related to Chapter 5. We want to thank Mr. Shubhasis Adak for his contribution in preparing the manuscript. We are thankful to our colleagues, specifically Dr. Basudeb Munshi, and our critics and teachers, who have been instrumental in the creation of this book. We are thankful to all of our families and friends for their unconditional support and love.

Finally, the authors would like to thank the CRC press team, in particular, Dr. Gagandeep Singh, Editorial Manager, CRC Press, for their patience, encouragement, and invaluable suggestions, and for trusting us in the process of writing this book.

About the Authors

Madhusree Kundu started her academic pursuits with a degree in chemistry, with honors (University of Calcutta) followed by a degree and postgraduate degree in chemical engineering from the Rajabazar Science College, University of Calcutta (1990–1992). Dr. Kundu gained experience as a process engineer at Simon Carves (I) Ltd., Kolkata (1993–1998). In the next phase of her scholarly pursuit, Dr. Kundu earned her PhD from the Indian Institute of Technology, Kharagpur (1999–2004), and started her academic profession as a member of the faculty of the Chemical Engineering Group, BITS Pilani, Rajasthan (2004–2006). She joined NIT Rourkela in 2007 and has continued there as associate professor in the Department of Chemical Engineering. Apart from teaching, she has focused her research activities in chemometrics along with process modeling and simulation, fractional order system identification and control, solution thermodynamics, and fluid-phase equilibria. Dr. Kundu has authored several research articles in international refereed journals and has a few book chapters to her credit.

Palash Kundu was born in Kolkata, West Bengal, India (1959). He received a degree in electrical engineering from the Bengal Engineering College, Sibpur, West Bengal (1981), and a postgraduate degree (1984) from Jadavpur University, Kolkata. Dr. Kundu earned his PhD from Jadavpur University and is currently serving there as associate professor in the Department of Electrical Engineering. Dr. Kundu was acclaimed over his decade-long experience a an R&D and system engineer with Press Trust of India Ltd, Central Testing Laboratory (West Bengal State Electricity Board), Hamilton Research & Technology (P) Ltd., Kolkata, PBM Cybernetics (P) Ltd., Kolkata, and Satyajit Engineering Industries (P) Ltd., Kolkata. During 1997–2001, Dr. Kundu served at the Haldia Institute of Technology (affiliated with the West Bengal University of Technology [WBUT]) as teacher in-charge in the Instrumentation Engineering Department. In 2001, he joined Jadavpur University and is currently pursuing active research in the following areas: development of artificial taste sensing (electronic tongue), e-nose for the detection of explosive and hazardous gases, wireless sensor networking for industrial automation, and biomedical signal transmission and receiving. He is an avid teacher of subjects such as design and application of embedded systems, advanced instrumentation techniques, active circuits and systems, and process instrumentation and control.

Seshu Kumar Damarla was born in Andhra Pradesh (1985). He received his BTech degree in chemical engineering from Bapatla Engineering College, Bapatla, Andhra Pradesh, India (2008) and MTech (chemical engineering) from the National Institute of Technology Rourkela (NITRKL), Odisha (2011). Mr. Damarla is the recipient of the silver medal as a top achiever in his postgraduate work. He served as assistant professor in the Department of Chemical Engineering at Maulana Azad National Institute of Technology, Bhopal, for a short duration before joining the PhD programme. Currently, Mr. Damarla is pursuing his PhD on simulation, identification and control of fractional order processes at NITRKL, Odisha. He has continued his research on the applications of chemometric techniques for fault detection in large and complex industrial processes and for assessment of product quality since completing his postgraduate work. Mr. Damarla has published a couple of research articles in internationally refereed journals and also published in the proceedings of national and international conferences. Mr. Damarla has been a referee for *Acta Biotheretica*, a Springer publication and Journal of King Saud University–Science, a Elsevier publication, and is a member of the Fractional Calculus and Application Group and International Association of Engineers (IAENG).

Introduction

"Data does not equal information; information does not equal knowledge; and, most importantly of all, knowledge does not equal wisdom. We have oceans of data, rivers of information, small puddles of knowledge, and the odd drop of wisdom."

(Henry Nix, 1990)

It is clear that in this era of data explosion, we are in need of good databases, very good data, and even better algorithms which can turn our data into knowledge and eventually toward wisdom. Quality data is defined by noise-free data, free of outliers, and is ensured by data preconditioning. Rational as well as potential conclusions can be drawn from the conditioned/preprocessed data. In this backdrop, the contributions of the book are as follows.

Multivariate techniques are more effective than univariate methods such as statistical process control (SPC) charts because they deal with the mutual interactions among the responsible variables as well in their decision-making while detecting process faults and ensuring product quality. Presently adapted methods in this book for addressing the problems taken up have their roots in signal processing, machine learning, and statistical learning theory.

Process monitoring with a view to detect process faults or abnormal situations and ensuring product quality are statutory obligations. The diagnosis of abnormal plant operation is possible, if periods of similar plant performance can be located in the process historical database. Chemometric techniques are utilized for this purpose. *Chemometrics* is the application of mathematical and statistical methods for handling, interpreting, and predicting chemical data. A modified *K*-means clustering principle using principal component analysis (PCA) similarity, and moving window-based pattern matching using PCA similarity/dissimilarity, have been implemented in detecting normal and abnormal operating conditions in various processes including the *double effect evaporator, yeast fermentation bioreactor, continuous crystallizer, fluid catalyzed cracking* (FCC) process, pilot scale *continuous stirred tank heater* (CSTH) process, and *polyethylene production* process. Classification of various operating conditions in these process is carried out using simulated data derived from dynamic models of the processes considered.

This book describes use of the electronic tongue (e-tongue or ET) signature assigned to various beverages for their authentication/grading. An e-tongue is an automated device engaged in characterizing liquid samples. It exploits electrochemical methods (like voltammetry) of detection of the analytes present in a liquid sample using electrochemical sensors and theoretical principles of electrochemistry. A characteristic electrical signature generated using an e-tongue is called a feature apart from the quantitative measures of the analytes present in it. Feature extraction followed by processing using chemometric algorithms can lead to an identity for a particular liquid. Hence, any unknown liquid can be authenticated using its own features subjected to pattern matching with the reference features. This simple method is evolving as an alternative to the time-consuming, laborious, and expensive conventional laboratory-based analytical instruments. E-tongue-based methods are especially suitable for in situ and bulk quality monitoring. As part of the e-tongue device usage, recursive principal component analysis (RPCA) using Hotelling's T2 and squared prediction error (SPE) statistics as monitoring indices, and PLS-, LDA-, Sammon's nonlinear mapping (NLM)-, and support vector machine (SVM)-based algorithms have been deployed in this book in beverage grading and authentication. As a part of miscellaneous applications of chemometric monitoring, this book provides guidelines for the design of possible low-cost and everyday appliances like adulterated milk detectors, out-of-date food product detectors, and arsenic meters to quantify arsenic in contaminated water as extended applications using the machine-learning algorithms developed for beverage quality monitoring (authentication/classification). The common approach to be adapted in design is the use of different voltammetric signals and algorithms already developed so far.

Health care of people in remote locations (especially in villages) is a staggering issue due to the lack of infrastructural facilities and experienced physicians. In such circumstances, interactive patient care monitoring tools developed using different biomedical signals may help the local medical practitioners to a certain degree in complementing their clinical knowledge and judgments. With the networking facilities available, treatment and consultation with experienced physicians also can be conducted from distant, convenient, and urban locations. The design of patient care monitoring tools seems to be one of the noble applications of chemometrics and is also included in this book as a part of miscellaneous applications. The electrocardiogram (ECG) is a popular noninvasive technique for preliminary level investigation on cardiovascular assessment. It is the time-averaged representation, recorded in predefined body positions, of small electrical impulses generated on the sinoatrial (SA) node of the heart and their propagation along the specialized conduction fibers on the heart surface. There are around 18–20 clinical features in a single ECG record, which can be useful for clinical diagnosis of a cardiac patient. Due to wide variability of the clinical features in ECG with age, gender, and other demographic factors, design of a robust automatic classifier with consistent 100% accuracy is a stiff target to achieve. ECG records are corrupted by various artifacts, of which power line noise, muscle noise, and baseline wander are the dominant ones; this poses major challenges, hence making the target even tougher to achieve. The proposed monitor is a two-step procedure: preprocessing of the ECG signals followed by the design of an automated machine-learning and predictive tool to assist in medical diagnosis. Normal and anterior myocardial infarction (AMI) data are taken from the Physionet database.

Organization of the chapters

Chapter 1 is focused on data presentation, acquisition, and design of experiments for data generation, data preprocessing, basic statistical measures, hypothesis testing, and regression. Data preprocessing including outlier detection, data reconciliation, and data transformation is immensely important in chemometrics. It is essential to determine the nature of the underlying process—whether it is stochastic or deterministic, and whether the collected data are stationary or nonstationary—so that the appropriate data preprocessing and application algorithms can be developed with utmost confidence. This chapter addresses all of these issues through suitably designed problems exploiting the power of simple EXCEL and MATLAB® functions. Chapter 1, by its morphology, is very different compared to the rest of the chapters in this book. One can take this chapter in the sense of an elaborate taxonomy toward an easy transition to the core chapters of this book. More so, Chapter 1, in its present form, may be treated as a prerequisite for initiating any maiden research in this area, and that has been a motivation behind the present chapter being proposed as an integral part of the book.

Theoretical postulations of all the chemometric techniques used in this book are briefly discussed along with their pseudo/MATLAB® codes being in Chapter 2. In this book, a signal is considered "stationary" if the mean and variance over a period of interest are constant (within the expected statistical fluctuations), nonstationary otherwise. Various algorithms have been proposed considering the very nature of the signals. Chapter 2 starts with the authors' perception on *chemometrics*. Then, it travels through PCA, *similarity, dissimilarity, clustering, partial least squares* (PLS), *Sammon's mapping, moving window based* PCA, RPCA, discriminant functions like *linear discriminant analysis* (LDA), SVMs, and *decision directed acyclic graphs* (DDAGs); all are oriented toward classification/clustering/authentication/pattern matching for deployment in the design and development of various commercial appliances. Each of the techniques discussed is accompanied by the requisite MATLAB® codes/pseudocode/algorithms to implement them. The mathematical postulations along with the mention of a few seminal contributions regarding the evolution of those techniques are also included in this chapter.

Chapter 3

Chapter 3 aims at the detection of various (normal) operating conditions of chemical processes due to the following reasons:

- It is necessary to find out the best or Pareto optimum operating conditions for a process (apart from detection of an abnormal operating condition) in the face of fluctuation in product demand, raw material quality and quantity, and utility prices. Keeping in view the optimal usage of utility/energy/material leading to economic and efficient plant operation, a robust strategy to identify the various process operating conditions is necessary.
- It is important for any monitoring algorithm to be able to distinguish between change of operating conditions and disturbances of moderately fast and slow time-varying processes.
- Successful recognition of various process operating conditions and process disturbances allow the monitoring algorithms to function one step ahead and detect the process faults. A rigorous fault-detection algorithm should be complemented by a robust and efficient process operating condition detection/classifying mechanism.

Fermenting sugars with yeast is one of the important processes used for ethanol production. Yeast fermentation is a nonlinear and dynamic process that suffers heavily from multivariable interactions, changeable parameters, the effect of unknown disturbances, and so on. Realizing optimum conditions such as pH, temperature, agitation speed, dissolved oxygen concentration, and so on, in a bioreactor is a complex task. The monitoring and control of such a process is also a challenging one. The multiple effect evaporation process is widely used in agro-based industries (like tomato juice concentration), sugar manufacturing plants, the paper industry, and many others. These mathematical models are based on linear and nonlinear mass and energy balance equations and relationships pertaining to the physical properties of the solution. Crystallization from solution is a widely applied unit operation in both the batch crystallization (small scale) and continuous crystallization (large scale) modes for solid–liquid separations. Continuous crystallization is used for the bulk production of inorganic (e.g., potassium chloride, ammonium sulphate) and organic (e.g., adipic acid) materials.

Chapter 3 utilizes a modified *K-means* clustering algorithm using dissimilarity (1-combined similarity) for classifying various operating conditions in a *double effect evaporator, yeast fermentation bioreactor,* and *continuous crystallizer.* Classification of various operating conditions in the chosen process are carried out using simulated data derived from dynamic models of the processes considered. All the developed MATLAB®-based simulation codes for a *double effect evaporator* and *yeast fermentation bioreactor* are provided in this chapter.

For the *continuous crystallizer,* Ward and Yu [1] developed a Simulink block (PCSS) to model the transient evolution of a population density function for a physical system modeled by a population balance equation. The Simulink and MATLAB® files written by the authors for solving the mathematical model of the continuous crystallizer are provided in the appendix of the referred article. Their codes have been adapted here to create a historical database for the continuous crystallization process in order to develop a modified *K*-means clustering algorithm for detecting normal operating conditions.

Chapter 4

Chemical process plant design and associated control systems are becoming increasingly complex in nature and instrument-laden to meet product quality demands, and economic, energy, environmental, and safety constraints. It is important to supervise the process and detect faults, if any, while the plant is still under control so that abnormal situations can promptly be identified. In process history-based methods, a large amount of historical process data are needed. Data can be presented as *a priori* knowledge to a diagnostic system, which is called feature extraction. This extraction process can be either qualitative or quantitative in nature. Two of the major methods that extract qualitative history information are the expert systems and trend modeling methods. Methods that extract quantitative information can be broadly classified as nonstatistical and statistical methods. Model-based methods are usually based on fairly accurate deterministic process models. In contrast, data-based methods for fault detection and diagnosis depend entirely on process measurements. Chapter 4 adapts a data-based approach for finding similar operating periods (normal or abnormal) in three simulated historical databases generated using a type IV industrial FCC process, CSTH pilot plant, and industrial gas-phase polyethylene reactor process. The FCC process is one of the major conversion technologies in oil refinery and hydrocracking units. It is widely used to convert or crack the high-boiling, high-molecular weight to lighter fractions such as kerosene, gasoline, LPG, heating oil, and petrochemical feedstock. Apart from that, FCC produces an important fraction of propylene for the polymer industry. FCC is a highly nonlinear, slow, and multivariable process possessing strong interactions among the variables, subject to a number of operational constraints, and working at high temperatures and pressures. McFarlane et al. [2] presented a mechanistic dynamic simulator for the reactor/regenerator section of a Model IV FCCU, which provides a detailed model description, along with dynamic simulation and model analysis. The present authors adopted that model for generating a historical database for the FCC unit using the SIMULINK modules and associated MATLAB® codes developed by Dr. Emadadeen M. Ali (Professor, Department of Chemical Engineering, College of Engineering, King

Saud University), which are available at http://faculty.ksu.edu.sa/Emad.Ali/Pages/SimulinkModule.aspx.

A mathematical model of CSTH along with measured data captured from the original continuous stirred tank heater pilot plant are used in a hybrid simulation process to generate a simulated historical database. The full detail of the process models and simulation codes are available at http://www.ps.ic.ac.uk/~nina/CSTHSimulation/index.htm.

Polyethylene is extensively used in packaging like plastic bags, plastic films, geomembranes, containers (for example bottles), and so on. It is generally produced in three forms: low-density polyethylene (LDPE), linear low-density polyethylene (LLDPE), and high-density polyethylene (HDPE). Polyethylene can be produced by addition polymerization of ethene. An industrial gas-phase polyethylene reactor is considered here, which needs to be monitored continuously. In order to generate a historical database, the authors have adapted SIMULINK modules along with the associated MATLAB® codes, which are available at http://faculty.ksu.edu.sa/Emad.Ali/Pages/SimulinkModule.aspx.

The processes considered in this chapter have nonstationary and time-varying behavior. These kinds of processes are monitored in an adaptive fashion using sliding time windows. Two algorithms, namely a combined similarity factor-based pattern-matching algorithm and a dissimilarity factor-based pattern-matching algorithm (based on KL-expansion) using a moving window approach (which are introduced in Chapter 2) have been deployed for this purpose. When compared with the dissimilarity factor-based pattern-matching algorithm, the combined similarity factor-based pattern-matching algorithm demonstrated far better performance (100% pool accuracy and 100% pattern-matching efficacy) in distinguishing operating conditions for a variety of disturbances, faults, and process changes.

Chapter 5

Chapter 5 is dedicated toward the design of a commercial tea grader using Bureau of Indian Standards (BIS) certified tea brands available in the Indian market. Tea is the most consumed beverage (not counting water) in the world because of its health, dietetic, and therapeutic benefits. The taste of tea is one of the most important factors in its quality grading and generally has been assessed by professional tea tasters. However, the judgment conjectured with human perception may be subjective and prone to suffer from inconsistency and unpredictability due to various mundane factors. An e-tongue may be used alternatively as a tea taste recognizer and classifier. It is an automated device engaged in characterizing liquid samples. It exploits electrochemical methods (like voltammetry) for detection of the analytes present in a liquid sample using electrochemical sensors and theoretical principles of electrochemistry. An introduction of the e-tongue with a brief review on different kinds of ETs, referring to their applications, renders a proper perspective and significance to this chapter. Algorithms including moving window-based dissimilarity, moving window-based RPCA, and Fischer discriminant analysis (FDA) enhanced by the DDAG method have been used as machine-learning components of the proposed tea grader.

Guidelines regarding the prototype development of a commercial tea grader are presented in this chapter. These principles can be extended for the design of possible low-cost and everyday appliances like beverage quality checkers, adulterated milk detectors, out-of-date food product detectors, and arsenic meters to quantify arsenic in contaminated water. This chapter possibly will benefit people engaged in engineering (product) research and development and entrepreneurs. (As the hardware design, development, and implementation needed for these appliances are beyond the scope of the present book, these could not be accommodated.)

All the developed MATLAB® codes are provided in this chapter. All tea data sets for testing and some typically large programs are available on the book webpage as an e-resource.

Chapter 6

Water quality can be monitored with various motives and purposes, however a broad-based framework of monitoring improvised under variable circumstances to cater to those varying needs should be the target. This chapter delivers a design of a commercial mineral/drinking water classifier/authenticator. The BIS-certified mineral water samples (Aquafina, Bisleri, Dolphin, Kingfisher, McDowell, and Oasis) are considered unique representatives of specific types, manufactured by different manufacturers irrespective of the place of the manufacturing unit in India. The voltametric e-tongue device is used to capture signals from the said water brands to create a voltammogram, which is used for feature extraction, followed by the development of machine-learning algorithms aiming for water brand classification using PLS, NLM, and SVM. This chapter also proposes an extended framework based on electrochemical qualification and quantification of water with respect to the ingredients (and harmful ingredients as well) present in it. The framework is supposed to be a modular device and caters to the following needs:

- Natural and drinking water classification with respect to their mineral content, alkalinity, pH, conductivity, surfactant content, and so on.
- Recycled water quality monitoring with respect to the pollutants and detection of abnormal operating conditions in waste water treatment plants (WWTP).
- Detection and monitoring of heavy metals (arsenic, chromium, copper, antimony, etc.) in drinking and ground water.

The voltametric e-tongue device is supposed to be the soul of the proposed monitoring framework tailored for different purposes and adapts different kind of voltametric techniques like pulse voltammetry, cyclic voltammetry, and anodic stripping voltammetry.

All the developed MATLAB® codes are provided in this chapter. All water data sets for testing are available on the book webpage as an e-resource:

Chapter 7

Chapter 7 provides two of the finest and diversified applications of chemometrics: one in patient care monitoring and another in monitoring arsenic content in contaminated water.

Continuous measurement of patient parameters such as heart rate and rhythm, respiratory rate, blood pressure, blood-oxygen saturation, and many others have become common features in the care of critically ill patients. A patient monitor watches for and warns against life-threatening events in patients, critically ill or otherwise. Health care of people in remote locations (especially in villages of third-world countries) is a staggering issue due to high population density and the lack of infrastructural facilities and experienced doctors. In such circumstances, interactive patient care tools developed using different biomedical signals may help the local medical practitioners to a certain degree in complementing their clinical knowledge and judgments. With the networking facilities available, treatment also can be conducted from any distant, convenient, and urban location through the consultation of relevant experienced physicians (telemedicine is an integrated technology platform where a patient can be treated and monitored through a communication link by a remotely located physician). The ECG, photoplethysmography (PPG), ultrasound, computed tomography (CT) scan, and echocardiogram (ECHO) facilities installed at rural and primary health service posts can be connected to the specialty hospitals/health-care network using even simple webcams and ISDN telephone lines, hence diagnosis and consultation can be done. This book proposes synthesis and analysis of human electrocardiograms using chemometrics.

Events in ECGs and the related ECG morphology, ECG lead system, ECG recording, clinical signature encased in ECG morphology, and abnormality in ECG morphology are integral parts of ECG signal-based device development and hence of this chapter. The myocardial infarction (MI) inferior preprocessed ECG data *s0021are.txt_ld_I_full_beat* required for testing the nonstationarity programs are available on the book webpage as an e-resource.

Computerized ECG signal analysis, morphological modeling of ECG, automated ECG pattern classification, and authentication are parts of patient care monitoring efforts, and have been included in this chapter. Computerized analysis and interpretation of an ECG can contribute toward fast, consistent, and reproducible results, which can save a great deal of the time needed by physicians to manually check paper-based ECG records. ECG modeling or synthesis, a relatively new area, has been used in applications like beat classification, compression for bulk storage, and generation of synthetic ECGs for professional training. Digitized transmission and storage of ECGs can save memory space and facilitate remote monitoring of cardiac patients as well. Preprocessing of ECGs, beat segmentation to extract zones, and modeling and reconstruction of ECGs waves are parts of the ECG modeling/synthesis process that have been included here. Normal and abnormal ECG (anterior myocardial infarction or AMI) data from the Physikalisch-Technische Bundesanstalt (PTB) Diagnostic ECG database (ptbdb) have been used in ECG signal synthesis. The preprocessed AMI dataset used for the Fourier and the Gaussian model application (*s0021are.txt_ld_I_full_beat*) is available on the book webpage as an e-resource.

The datasets (beat_p240_s0468re [normal], beat_p284_s0551lre [abnormal]) needed to test the ECG beat reconstruction program are available on the book webpage as an e-resource.

Automated ECG classifiers employing a dissimilarity factor (D) based on Karhumen–Loeve (KL) expansion are deployed for classification among normal and inferior myocardial infarction (IMI) data. The required data sets [p240_s0468 (normal), p050_s0168, p050_s0177, p050_s0174, and p050_s0219} (IMI data)] are available on the book webpage as an e-resource.

This chapter aims at quantification of arsenic in water (to 5 ppb level) using a simple e-tongue instrumentation and chemometric processing tool. Anodic stripping voltammetry (ASV) is used to detect arsenic. An arsenic quantifier is proposed using the accumulated experimental data pool. Subsequently, a low-cost and indigenous possible arsenic meter scheme suitable for bulk monitoring is proposed.

References

1. J.D. Ward, C.-C. Yu (2008). Population balance modeling in Simulink: PCSS. *Computers and Chemical Engineering*, vol. 32. 2233–2242.
2. R.C. McFarlane, R.C. Reineman, J.F. Bartee, C. Georgakis (1993). Dynamic simulator for a model IV fluid catalytic cracking unit. *Computers and Chemical Engineering*, vol. 17(3), 275–300.

chapter one

Data generation, collection, analysis, and preprocessing

This chapter presents computer-based data acquisition, design of experiments for data generation with illustrations, and data preprocessing. Outlier detection, data reconciliation, and data transform are included as components of data preprocessing. Basic statistical measures and hypothesis testing and regression are also included in this chapter. This book deploys data-based techniques in designing miscellaneous applications. It is essential to determine the nature of the underlying process: whether it is stochastic and deterministic, and whether the collected data are stationary or nonstationary, so that the appropriate data preprocessing and application algorithms can be developed with utmost confidence. This chapter manifests and exploits the power of simple EXCEL and MATLAB® functions crafted through suitably designed problems. This chapter is a handy and composite guide on data. One can take this chapter in a sense of an elaborate taxonomy towards an easy transition to the core chapters of this book.

1.1 *Data: Different data types and presentation of data*

Data is a collection of particulars, such as numbers, words, measurements, observations, or descriptions of individual identities. With a cheaply available computer-based data acquisition system, a huge amount of data can be stored and processed. Data may be qualitative and quantitative. Quantitative data are of two types: discrete (e.g., rate of sampling interval is 4 samples/s) and continuous (e.g., length of wall 2.5 m). Qualitative data basically can be classified as categorical and ordinal types. Categorical data, for example demographic data, may reveal attributes such as race, gender, or the zip code. *Ordinal* data mixes numerical and categorical data. For example, rating of varieties on a scale from 0 to 5 stars (with 0 being the lowest and 5 being the highest) belongs to ordinal type data. Bar graphs, pie charts, dot plots, line graphs, scatter (x, y) plots, pictographs, histograms, frequency distribution, stem and leaf plots, and cumulative tables and graphs are used to present data perfectly.

Some of the frequently used data presentation methods in science and engineering are described in the following. Figure 1.1 presents different types of data.

Bar Graphs: A bar graph (also called a bar chart) presents data using bars of different heights. Table 1.1 presents data about students' favorite and attempted subjects in the

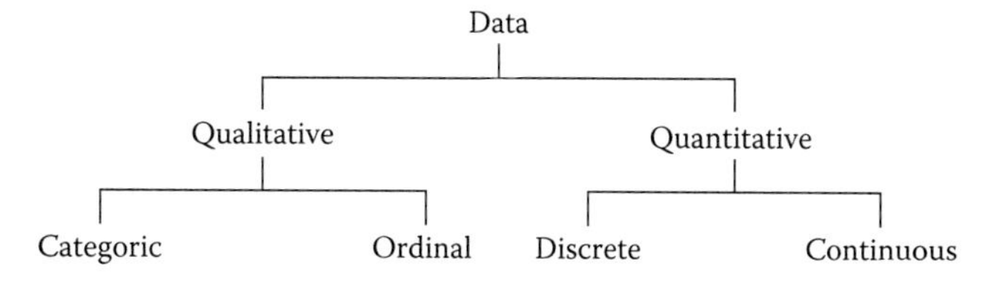

Figure 1.1 Different types of data.

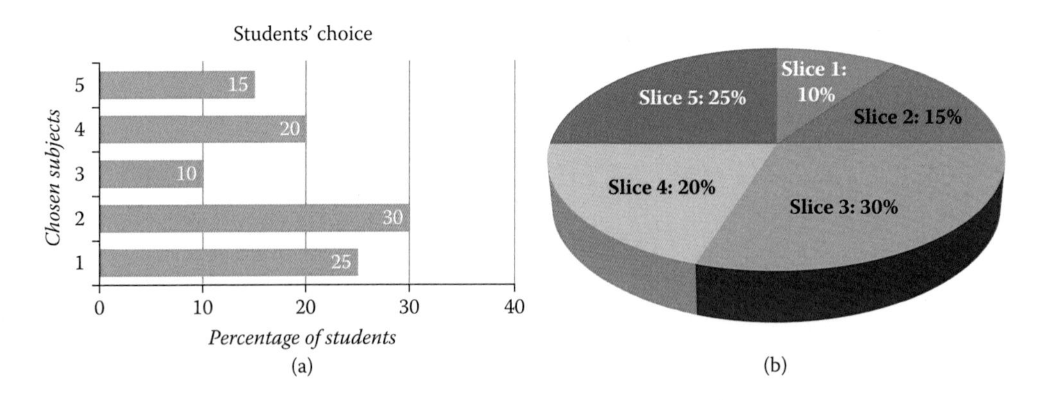

Figure 1.2 (a) Bar graph representation of students' choice. (b) Pie chart representation of students' choice.

Table 1.1 Students' exclusive choice for different subjects (in a class of 100 students)

	Favorite subject				
Subject	Mass transfer operation	Heat transfer operation	Thermodynamics	Fluid flow operations	Process control
Percentage of students	25	30	10	20	15

comprehensive examination of the Chemical Engineering curriculum, which are presented in the form of a bar graph (Figure 1.2a) and pie chart (Figure 1.2b).

Histogram: The distribution in data is presented using a histogram. When the data are in categories (such as favorite subject), one should not use a histogram; instead a bar chart/pie chart should be used. A histogram consists of rectangles or squares (in some situations), founded over discrete intervals (bins) resembling the frequency of observations. The area underneath each block is proportional to the frequency of the observations in that interval. The height of a rectangle is equal to the frequency density at the interval (the frequency divided by the width of the interval). The summation of areas under individual blocks of the histogram is equal to the number of data. The rectangles of a histogram are drawn in continuation to indicate the continuity of the original variable.

ILLUSTRATION 1.1

800 random numbers are created and 10 bins or classes are created along the x-axis between the minimum and maximum values of those random numbers. Along the y-axis class frequencies are presented as shown in Figure 1.3.

Generation of histogram using MATLAB®

Step 1. The ***randn***(n) function generates normally distributed random numbers with zero mean and standard deviation 1.0, where $n = 800$. Assign X = ***randn***(n)

Step 2. The ***rand***(n) function generates uniformly distributed random numbers with 1 unit interval, where $n = 800$. Assign X = ***rand***(n)

Step 3. ***hist***(X, 10) generates the two histograms (Figure 1.3).

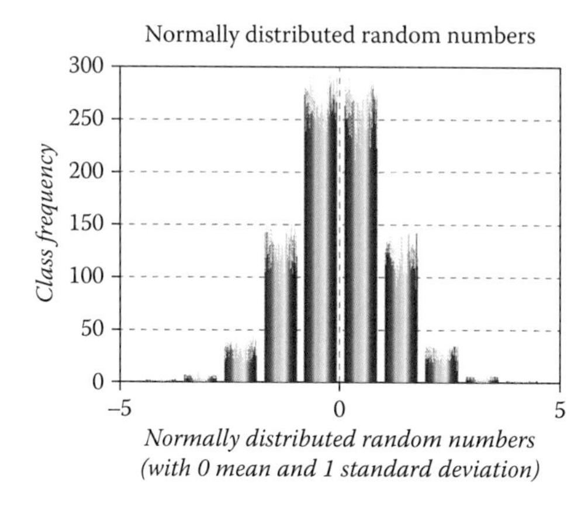

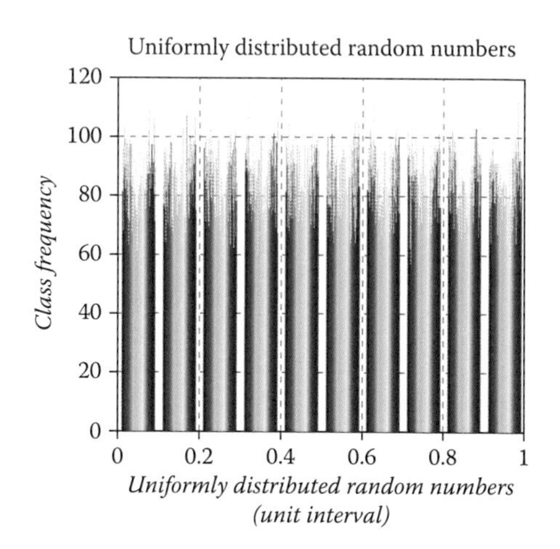

Figure 1.3 Normal and uniform distribution of random numbers using histogram.

ILLUSTRATION 1.2

Draw a histogram on students receiving grades in process control course in an interval of 10 marks.

Table 1.2 was generated in an Excel spreadsheet.

Step 1: The column "Bin" was created at an interval of 10.
Step 2: The "Frequency" column was created using the FREQUENCY (data_array, bins_array) Excel function.
Step 3: The Frequency column was used to form the histogram (Figure 1.4)

Table 1.2 Students' grades in process control

Students' marks	Bin	Frequency	Interval
12	9	0	0–9
15	19	2	10–19
37	29	0	20–29
33	39	2	30–39
42	49	2	40–49
49	59	4	50–59
55	69	5	60–69
53	79	4	70–79
57	89	3	80–89
62	99	3	90–99
66		0	100+
68			
78			
72			
77			

(*Continued*)

Table 1.2 (Continued) Students' grades in process control

Students' marks	Bin	Frequency	Interval
79			
65			
56			
82			
88			
69			
92			
93			
87			
99			

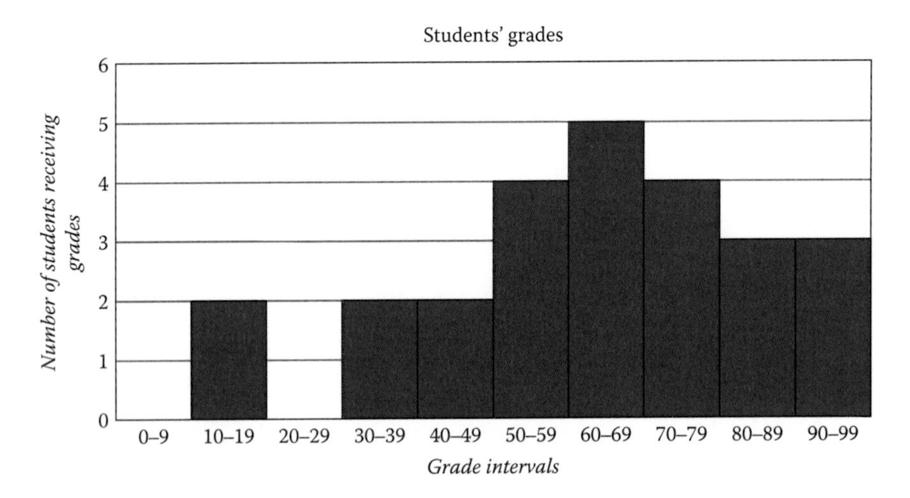

Figure 1.4 Histogram on students' grades.

1.2 Data generation: Design of experiments

Design of experiments (DOE) is a systematic and efficient method to determine the relationship between factors affecting a process output. It is used to correlate the process output and the main factors, as well as the interaction among the main factors affecting the process output. This correlation is needed to tailor the process inputs in order to optimize the output in terms of quality and quantity. There are different types of DOEs: OVAT, full factorial design, fractional factorial design (Taguchi designs), response surface design, mixture design, EVOP (evolutionary operations), and Plackett-Burman design (two-level fractional factorial design).

The objectives of experimental design are the following:

- Maximum possible inferences through minimum number of experiments
- To identify the most important inputs and the interaction among them
- Reduction of error in measurements
- Finding out an empirical output-input relationship
- Enhanced confidence in final results
- Finding out optimum inputs to maximize the output

The aforesaid objectives require data comparison, screening, regression, and optimization as parts of the DOE. The technique to be discussed in this book is "factorial design"; this illustrates the fulfillment of the previously mentioned objectives in the DOE.

1.2.1 *Factorial design and illustration*

In any process, there may be a large number of factors or inputs assumed *a priori* that may affect the output of the process. Screening of experiments is conducted to determine the inputs and interaction among them, which influence the process outputs signficantly; two-level factorial designs are of great importance in this regard. They can be augmented to form composite designs. Hence, they serve as the basic components to construct efficient data collection strategies according to the complexity of the problem under consideration. Factorial designs were used in the 19th century by John Bennet Lawes and Joseph Henry Gilbert of the Rothamsted Experimental Station [1]. Ronald Fisher argued in 1926 that factorial designs could be more efficient than studying the effect of one input at a time [2]. In general, the relationship between the inputs and outputs can be represented as

$$R = f\left(x_1, x_2, x_3 \ldots, x_p\right) + \text{err} \tag{1.1}$$

where x_i, $i = 1$ to p, are the factors, "err" is the random and systematic error, and R is the response variable. Approximating Equation 1.1 by using Taylor series expansion:

$$\begin{aligned} \text{response} &= a_0 + a_1x_1 + a_2x_2 + \cdots + a_px_p + \cdots + a_{ij}x_{ij} + \cdots + a_{jp}x_{jp} + \cdots + a_{11}x_1^2 + \cdots + \\ & a_{ii}x_i^2 + \cdots + a_{pp}x_p{}^2 + \cdots + \text{higher order terms} + \text{err} \end{aligned} \tag{1.2}$$

A polynomial response surface model is obtained where a_is' denote the parameters of the model. The first task is to determine the factors (x_i) and the interactions (x_ix_j, $x_ix_jx_k$, and still higher order interactions) that influence response R. Then, the coefficients like a_i, a_{ij}, and a_{ijk} of the significant inputs and interactions are computed. These parameters of the response surface models can be determined by least squares fitting of the model to experimental data. To perform a general factorial design, the investigator selects a fixed number of levels (two in most screening experiments) for each factor and then runs experiments with all possible combinations of levels and variables. If there are p factors, 2^p experiments must be conducted to cover all combinations. The number of experiments to be conducted grows rapidly with an increasing number of factors. While eight experiments are needed for three factors, 64 experiments are necessary for six factors. To save space, the experimental combinations in a two-level factorial experiment are often abbreviated with strings of plus and minus signs. The low and high levels of continuous variables may be coded using the – and + signs or 0 and 1, respectively. The points in a 2^2 factorial experiment can thus be represented as – +, + –, – –, and + +. It is relatively easy to estimate the main effect for a factor. To compute the effect of a factor, subtract the average response of all experimental runs for which the factor was at its low (or first) level from the average response of all experimental runs for which the factor was at its high (or second) level. Other useful investigative tools in factorial design include main effects plots, interaction plots, Pareto charts, and a normal probability plot of the estimated effects.

ILLUSTRATION 1.3

ZINC BIOLEACHING PROBLEM

An effective microbial leaching process of zinc sulfide ore (ZnS) using *Thiobacillus ferrooxidans* in aerobic conditions has been considered. Attempts have been made to optimize process parameters like initial zinc sulfide loading (pulp density), pH, temperature, and shaking speed to leach out the maximum quantity of zinc using a suspended batch culture of bacteria. An application of 2^4-level full factorial design of experiments in this regard manifested the screening of design parameters along with optimum parameter estimation. *The purpose was to identify only the important variables that affect the response (desired goal) and their interactions. It also proposed the empirical model of the yield (zinc) as a function of major design parameters and the interactions among them.* The variables and their 2^4-level full factorial designs are given in Table 1.3. The higher level of variable was designated as "+" and the lower level was designated as "–". The matrix for four variables and the interaction among them varied at two levels (+,–) and the corresponding metal leaching percentage is shown in Table 1.4. According to the design of experiments principle, six experiments were carried out at base level (initial Zn loading, 27.5 kg/m^3; pH, 3.0; temperature, 40°C; shaking speed, 75 rpm) to estimate error and standard deviation.

Table 1.3 The controlled factors of 2^4 factorial design for Zn leaching

Variable	Low level	Base level	High level
Initial Zn sulfide loading	25 kg/m^3	27.5 kg/m^3	30 kg/m^3
pH	2.5	3	3.5
Temperature	35°C	40°C	45°C
Shaking speed	60 rpm	75 rpm	90 rpm

Time: 192 h; Particle size: 135 µm; Inoculum size: 20% (v/v).

The standardized effects of individual factors (Zn sulphide loading, pH, temperature, and shaking speed) and interactions are presented in Table 1.5. Specimen calculations are presented here.

1.2.1.1 The effect of Zn loading

The main effect factor is $\bar{y}_{i+} - \bar{y}_{i-}$ where $\bar{y}_{i+}$ is the average response or yield of metal zinc for all high levels (+) of the ith variable (Zn sulfide loading) and $\bar{y}_{i-}$ is the average response or yield of metal zinc for all low levels (–) of the ith variable (Zn sulfide loading).

Hence, the Zn loading effect (x_1) is

$$= \begin{Bmatrix} (y_2 + y_4 + y_6 + y_8 + y_{10} + y_{12} + y_{14} + y_{16}) / 8 \\ -(y_1 + y_3 + y_5 + y_7 + y_9 + y_{11} + y_{13} + y_{15}) / 8 \end{Bmatrix}$$

$$= \begin{Bmatrix} (68 + 53.3 + 59.84 + 47.784 + 64.328 + 51.386 + 56.168 + 44.825) / 8 \\ -(72.6 + 58.4 + 63.888 + 51.392 + 68.679 + 55.246 + 59.967 + 48.238) / 8 \end{Bmatrix}$$

$$= 54.95 - 60.55 = -5.59$$

Table 1.4 2^4 full factorial design for zinc leaching with extraction percentage of zinc for each combination of inputs

Observation	Coded factors										Extraction (%)
	x_1	x_2	x_3	x_4	x_1x_2	x_1x_3	x_1x_4	x_2x_3	x_2x_4	x_3x_4	
1	−	−	−	−	+	+	+	+	+	+	72.6
2	+	−	−	−	−	−	−	+	+	+	68.0
3	−	+	−	−	−	+	+	−	−	+	58.4
4	+	+	−	−	+	−	−	−	−	+	54.3
5	−	−	+	−	+	−	+	−	+	−	63.8
6	+	−	+	−	−	+	−	−	+	−	59.84
7	−	+	+	−	−	−	+	+	−	−	51.39
8	+	+	+	−	+	+	−	+	−	−	47.78
9	−	−	−	+	+	+	−	+	−	−	68.68
10	+	−	−	+	−	−	+	+	−	−	64.33
11	−	+	−	+	−	+	−	−	+	−	55.25
12	+	+	−	+	+	−	+	−	+	−	51.39
13	−	−	+	+	+	−	−	−	−	+	59.97
14	+	−	+	+	−	+	+	−	−	+	56.17
15	−	+	+	+	−	−	−	+	+	+	48.24
16	+	+	+	+	+	+	+	+	+	+	44.83
17	0	0	0	0	0	0	0	0	0	0	63.8
18	0	0	0	0	0	0	0	0	0	0	63.4
19	0	0	0	0	0	0	0	0	0	0	63.1
20	0	0	0	0	0	0	0	0	0	0	63.21
21	0	0	0	0	0	0	0	0	0	0	63.10
22	0	0	0	0	0	0	0	0	0	0	63.84

"−": Low level, "+": High level, x_1: initial Zn sulfide loading, x_2: pH, x_3: temperature, x_4: shaking speed.

Table 1.5 Model coefficients (Equation 1.2) and standardized effects of variables

Inputs and interaction	Standardized effect
x_1	−5.59
x_2	−12.73
x_3	−7.61
x_4	−3.42
x_1x_2	0.455
x_1x_3	0.488
x_1x_4	0.233
x_2x_3	1.11
x_2x_4	0.376
x_3x_4	−0.003
$x_1x_2x_3$	−0.04

x_1: initial Zn sulphide loading, x_2: pH, x_3: temperature, x_4: shaking speed.

Similarly, the pH effect $(x_2) = -12.73$, temperature effect $(x_3) = -7.61$, and shaking speed effect $(x_4) = -3.42$.

1.2.1.2 *Two-factor interaction effects*

One factor may be influencing the levels of the other factor in different measures. If the factors do not behave additively, they interact. The interaction between the two factors is two-factor interaction. A measure of the two-factor interaction between the two factors x_1 and x_2 is provided by the average difference between the average effect of one factor at one level of the second factor and its average effect at the other level of the second factor. Two-factor interaction between x_1 and x_2 is denoted by x_1x_2.

From Table 1.4, one can calculate x_1x_2 in the following way:

$$x_1x_2 = \frac{1}{2}\left[\left\{\frac{(y_4 - y_3)}{2} + \frac{(y_8 - y_7)}{2} + \frac{(y_{12} - y_{11})}{2} + \frac{(y_{16} - y_{15})}{2}\right\} - \left\{\frac{(y_2 - y_1)}{2} + \frac{(y_6 - y_5)}{2} + \frac{(y_{10} - y_9)}{2} + \frac{(y_{14} - y_{13})}{2}\right\}\right]$$

$$= \frac{1}{4}\left[(y_1 + y_5 + y_9 + y_{13} + y_4 + y_8 + y_{12} + y_{16}) - (y_3 + y_7 + y_{11} + y_{15} + y_2 + y_6 + y_{10} + y_{14})\right]$$

$$= 0.455$$

Similarly, for other two-factor interactions like x_1x_3, x_1x_4, x_2x_3, x_2x_4, and x_3x_4, the results are tabulated in Table 1.5.

1.2.1.3 *Three-factor interaction effects*

Most of the three- and four-factor interactions are insignificant in comparison to the main factor and two-factor interaction effects. Three-factor interactions are calculated using the sign of two-factor interactions as provided in Table 1.4. The signs of the main effects are generated using the signs indicating the factor levels. The signs of the interactions are generated by multiplying the signs of the corresponding experiment levels (main effect signs). A measure of the three-factor interaction between the three factors x_1, x_2, and x_3 is provided by the difference between the average effect of the x_1x_2 factor at one level of the x_3 factor and their (x_1x_2) average effect at the other level of the x_3 factor. One can calculate $x_1x_2\ x_3$ in the following way:

$$x_1x_2\ x_3 = \frac{1}{2}\left[\left\{\frac{(y_8 - y_7)}{2} - \frac{(y_6 - y_5)}{2} + \frac{(y_{16} - y_{15})}{2} - \frac{(y_{14} - y_{13})}{2}\right\} - \left\{\frac{(y_4 - y_3)}{2} - \frac{(y_2 - y_1)}{2} + \frac{(y_{12} - y_{11})}{2} - \frac{(y_{10} - y_9)}{2}\right\}\right]$$

All the main factors are statistically significant factors since they have nonzero means (Table 1.5). All the interactions are statistically insignificant because they do not differ much from normal distribution (zero means). Therefore, there seems to be no significant interaction between individual factors. The main effects of all four individual factors used for the study shown in Table 1.5 are negative. This shows that higher leaching of zinc sulfide ore is obtained at a low pulp density of 25 kg/m^3. The reduction in the rate of bacterial leaching at a higher pulp density can be attributed to ineffective homogeneous mixing of solids and liquids leading to gas transfer limitations because the liquid becomes too thick (high viscosity) for efficient gas transfer to the cells. Table 1.5 also reveals higher percentage leaching at low pH. The tolerance of acidophiles to most metals in low pH media probably results from effective competition by H^+ ions for negatively charged sites at the cell surface. In bioleaching of Zn (II) from concentrates, pH sharply decreases with time, which results in suppression of bacterial activity. In order to avoid this negative effect, lime and/or calcite is added to the leach suspension. Another important observation from Table 1.5 is that percentage of leaching decreases at higher temperatures. The intensity of bacterial leaching for zinc sulfide ore depends on the rate of supply and dissolution of O_2 and CO_2. The dissolution of O_2 and CO_2 decreases with the increased temperature. Table 1.5 also shows that the yield is low at higher shaking speed levels. Normally, the enzymes responsible for bioleaching of zinc sulfide ore will be secreted more at low shaking speeds. From Table 1.5, the lack of interaction between the factors except that between pH and temperature is evident.

The six observed recoveries at the center (base level) were 63.80%, 63.4%, 63.1%, 63.214%, 63.103% and 63.842% (Table 1.4). The average recovery of these six center points is 63.40983%. The average of the 16 runs for the base design (Table 1.4) is 57.6716%. Since these two averages are not very similar (difference of 5.8%), it is suspected that there is a curvature present in the response surface. However, it is not clear which factor(s) contributes to the curvature except its existence. For the purpose of this study (screening of factors), it can be assumed that the linearity assumption holds here approximately. The overall average yield percentage is 60.5. Table 1.5 can be utilized to determine the Pareto chart (Figure 1.5). A Pareto chart is a special type of bar chart where the values being plotted are arranged in descending order. It provides the descending order of absolute effects exerted by different variables and interactions among them. It helps to set the priority while designing the experimental process.

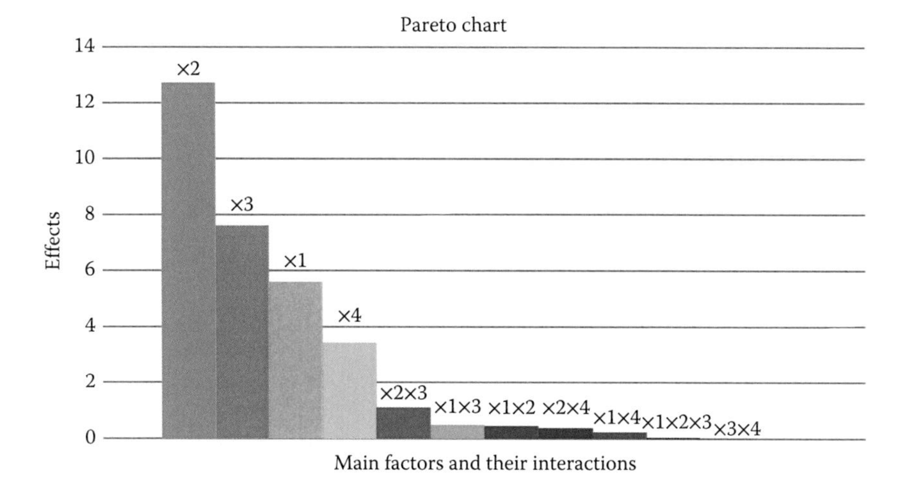

Figure 1.5 Pareto chart indicating the effect of factors in the yield of leached out zinc.

Regression of Equation 1.2 after substituting 16 Y values presented in Table 1.4 (as *yield* matrix) and the values of initial Zn loading, pH, temperature, and shaking speed according to the stratagem presented for the dimensionless coded factors (as matrix *coeff*), the model coefficients are determined; hence, the empirical model relating percentage zinc extraction as a function of Zn loading, pH, temperature, and shaking speed of the bioreactor evolves as Equation 1.3 with an average prediction error of 5.64% between the experimental and predicted zinc leaching.

$$Y = 60.5 + 0.315x_1 - 2.916x_2 - 0.04x_3 - 0.017x_4 \tag{1.3}$$

Equation 1.3 is the resulting polynomial response surface model and can be utilized for finding out optimum inputs to maximize the output. MATLAB® code and an Optim tool screenshot is presented below.

MATLAB® code for Illustration 1.3

```
function f=leaching(X)
E=load('coeff.txt','-ascii');
F=load('yield.txt','-ascii');
a1=E(:,1);
a2=E(:,2);
a3=E(:,3);
a4=E(:,4);
a=[a1 a2 a3 a4];
y=F(:,1);
ypred=60.5+a*[X(1) X(2) X(3) X(4)]'
err=y-ypred
f=sum(abs(err))/16
end
```

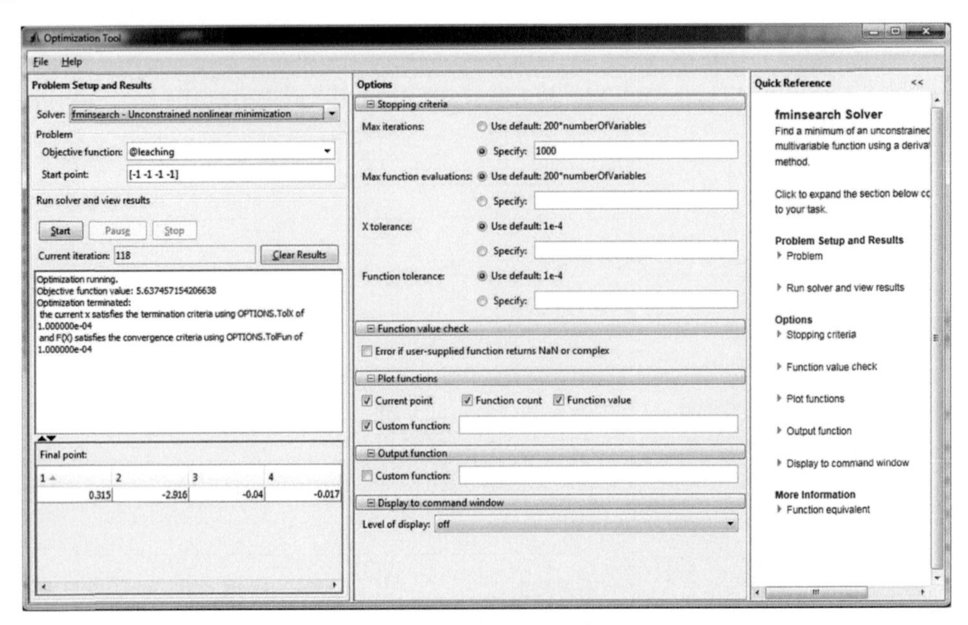

OVAT (one-variable-at-a-time) experimentation is popular. It involves variation in the level of an input (with levels of all other inputs being fixed) to find the level that imparts an optimal output. This procedure is then repeated for each of the remaining inputs. The inputs that were varied in previous sets of experiments are kept at levels that gave optimal responses. For the present case to achieve the same accuracy for finding out a main effect using OVAT, one has to conduct eight experiments for the lower level of that variable and another eight experiments for the higher level of that variable, keeping other input factors or variables constant. Hence, for four factors, a total of sixty-four experiments have to be conducted (a fourfold increase). The OVAT approach necessitates more experiments than the factorial design plans. Experiments must be duplicated and the results must be averaged to reduce the effects of measurement errors. The OVAT approach does not provide information on the impact of the interaction among inputs on the response.

1.3 Computer-based data acquisition

Data collection has become an established technology over the years and its progress is being reflected in different spheres, whether it be laboratory experimentation, the virtual laboratory, or plant supervision and control. Laboratory experiments are an integral part of engineering education, and offer a thorough understanding of the theoretical aspects as well as an opportunity to develop a hands-on skill. Most modern laboratory experiments are designed in such a way that measurement instruments are interfaced with small computers as online data acquisition components using a data acquisition (DAQ) card at the interface. The optimized use of information and communication technology (ICT) in e-learning is gaining popularity by extending the experience of conventional classroom teaching and laboratory-based experimentation to students located remotely. To impart smart engineering education, web-enabled simulated and remote control virtual laboratory experiments are interfaced to PCs using suitable DAQ hardware and web-enabled application software. Before the DAQ card came into the picture, a physical variable used to be measured (in any manufacturing plant) by online/offline sensors followed by conversion to an electrical signal (a voltage or current signal) by a suitable transducer. Then, it used to be transmitted through an analog to digital (A/D) converter and the digitized signal would pass to the computer. The computer would record the data (while used simply as data logger), monitor process status (used in supervisory control), and generate control commands (while used in direct digital control mode; DDC). In DDC, the control commands are converted to analog signals by the digital to analog (D/A) converter and sent to the final control elements/actuators.

A modern measurement system, in general comprises of sensors/transducers (along with a signal conditioning circuit), DAQ card, and a display/PC (with programmable software). Using low-cost microcontrollers and data acquisition cards, it is easier than ever to acquire data quickly and in digitized form. A DAQ card consists of an A/D converter, a D/A converter, and signal-conditioning hardware (like a low/high-pass filter, noise attenuator, etc.). DAQ has certain advantages like storage, retrieval of data, and real-time processing of data. The sensor/transducer plays a pivotal role in modern PC-based data acquisition apart from the PC and A/D and D/A converters. Separate subsections are dedicated to offer a brief introduction to all of the components.

1.3.1 Sensor/transducer

A sensor detects and responds to some changes occurring in its surroundings due to physical, chemical, and biological causes. To make the sensor output meaningful, a conditioning of that signal and its conversion to electrical (current/voltage) form is necessary, which is implemented using a transducer. A signal conditioning circuit makes the signal noise-free, linear, and attenuated/amplified as required. Sensors are of two categories: online sensors, which are preferred since they provide process information quickly without any intermission or sampling delays, are devoid of fewer human errors, and allow for arbitrary frequencies of measurements, and offline sensors, which are used because of the difficulty and expense of constructing a sampling system for some process variables. Cost may be of critical consideration for the selection of a sensor and the way measurements should be taken (online or offline). There is a trade-off between a high-cost, frequent online measurement and a relatively low-cost, infrequent offline measurement. Sensors must have several characteristics like accuracy, precision/repeatability, range, durability, reliability, and response time.

1.3.2 Analog-to-digital (A/D) converter

An A/D converter, also called an encoder, is a device that alters an analog signal into a digital signal, usually a numerically coded signal. It is needed as an interface between an analog component, such as a sensor or transducer, and a digital component such as a computer. The conversion of an analog signal into a digital signal (a binary number) is an approximation since an analog signal can take on an infinite number of values, whereas the numbers that can be formed by a finite set of digits are limited. This approximation process is called *quantization.*

1.3.3 Digital-to-analog (D/A) converter

A D/A converter, also called a *decoder,* is a device that changes a digital signal into an analog signal. This converter is needed as an interface between a digital component and an analog component (a physical device) to operate the physical device, such as a control valve or a pump.

1.4 Basic statistical measures and regression

To derive meaningful information from data, one needs to examine the statistical soundness of the data generated. In order to interpret available data, the statistical attributes of measurement must be characterized. Data statistics may be sample based or population based. Population is the largest collection of items of interest, and the sample represents a subsection of a population. In other words, the sample should represent the population with fewer but sufficient numbers of components. One population can have several samples with different sizes. Population is often characterized by basic parameters like μ (population *mean*) and σ (population *standard deviation*). The frequency distribution of data of an experimental outcome, the confidence in the new data in the face of uncertainty in measurement, and the decisions to be made about the acceptability of the new outcome are all matters of concern and should be addressed before using the data from any source. To find a causal relation between the observation and the factors influencing it, regression is the means.

1.4.1 Mean, median, mode

There are many averages in statistics. The *mean* is the average, where all the entry values are added up and then divided by the number of entries. The *median* is the middle value in the list of entries. To find the median, the entries have to be relisted in numerical order. The *mode* is the value that occurs most often. If no entry is repeated, then there is no mode for the list. *Range* is the difference between the largest and smallest entry values.

ILLUSTRATION 1.4

Find the mean, median, mode, and range for the following list of values: **13, 18, 13, 14, 13, 16, 14, 21, 13**

The mean is the usual average:

(13 + 18 + 13 + 14 + 13 + 16 + 14 + 21 + 13) ÷ 9 = **15**

The median is the middle value: hence, one has to rewrite the list in order:

13, 13, 13, 13, 14, 14, 16, 18, 21

There are nine numbers in the list, the middle one will be the (9 + 1) ÷ 2 = 10 ÷ 2 = fifth number. Hence, the median is **14**.

The mode is the number that is repeated more often than any other: **13** is the mode.

The largest value in the list is 21, and the smallest is 13; the range is 21–13 = 8.

1.4.2 *Variance and standard deviation*

For n data values, the average of the squared differences of each entry from the mean value of the n entries is *variance*. For n data values, which are the following:

- Population based: divide by n when calculating variance
- Sample based: divide by n-1 when calculating variance

The *standard deviation* is the spread of numbers around the mean. Its symbol is σ (sigma). This is expressed as follows:

$$\text{The population-based standard deviation} = \sigma = \sqrt{\frac{1}{n}\sum_{i=1}^{n}(x_i-\mu)^2} \tag{1.4}$$

$$\text{The sample-based standard deviation} = \sigma = \sqrt{\frac{1}{n-1}\sum_{i=1}^{n}(x_i-\bar{x})^2} \tag{1.5}$$

The MATLAB® function *datastats* determines all of the data statistics.

1.4.3 *Covariance and correlation coefficient*

The linear relationship between two entries (x and y) in a data sample, if any is expressed by *covariance*:

$$S_{xy} = \frac{1}{n-1}\sum_{i=1}^{n}(x_i-\bar{x})(y_i-\bar{y}) \tag{1.6}$$

ILLUSTRATION 1.5

Data set = [13 13 13 13 14 14 16 18 21]; find the statistics of the data set.

MATLAB CODE

```
xdata=[13 13 13 13 14 14 16 18 21]'
xds = datastats (xdata)
```

Answer:

Field	Description	
num	The number of data values	9
max	The maximum data value	21
min	The minimum data value	13
mean	The mean value of the data	15
median	The median value of the data	14
range	The range of the data	8
std	The standard deviation of the data	2.83

Population covariance is defined in the following way:

$$\sigma_{xy} = \frac{1}{n}\sum_{i=1}^{n}(x_i - \mu_x)(y_i - \mu_y) \tag{1.7}$$

Physically, covariance is the extent to which a dependent variable and associated independent variable move in conjugation.

The *correlation coefficient* of two entries (x and y) is a normalized linear relationship between the dependent and its associated independent variable or scaled covariance between them.

$$\text{Sample-based correlation coefficient: } r_{xy} = \frac{S_{xy}}{S_x S_y} \tag{1.8}$$

$$\text{Population-based correlation coefficient: } \rho_{xy} = \frac{\sigma_{xy}}{\sigma_x \sigma_y} \tag{1.9}$$

The linear correlation coefficient is referred to as the *Pearson product moment correlation coefficient*. The value of r is such that $-1 \leq r \leq +1$. It is used to express a measure of fitness of an adopted regression equation to correlate dependent variables and the associated independent variable.

1.4.4 Frequency

For handling large amount of data, the data range is divided into a number of *classes*. The number of times a particular incidence ocurrs in a class is called the *class frequency*. Frequency data can be plotted as *column charts*, *pie charts*, and *histograms*. Such data can be used to determine the number of times a particular incidence occurred in the past and that information can be used to describe the future probability of similar event. If frequency is denoted by $f(g)$, then probability = $f(g)/n$, where n is the total number of occurrences. Frequency distribution can be related to probability distribution.

1.4.5 Distribution

Any measured values have a range or spread/distribution. Normal, binomial, and Poisson distributions are the three important distributions. The most important attributes to distribution are mean, variance, and standard deviation. Different types of mechanisms produce experimental data with different distributions. The mathematical name of the bell-shaped distribution is the *normal distribution* or *Gaussian distribution,* which is defined by the *probability density* equation ($pd\{x\}$), a function of variable x with mean μ and standard

ILLUSTRATION 1.6

Frequency and probability distribution using relative frequency of students securing different grades in process control.

The frequency and relative frequency of grades received in process control are as follows:

Marks (M) range	Category	Class width Δx	Class frequency $f(g)$	Frequency density $fd(g) = f(g)/\Delta x$	Relative frequency = Probability** $P(x)$
$0 \le M < 10$	Fail	10	$f(g) = 0$	0	
$10 \le M < 20$	Fail	10	$f(g) = 0$	0	
$20 \le M < 30$	Fail	10	$f(g) = 0$	0	
$30 \le M < 40$	Fail	10	$f(g) = 2$	2/10 = 0.2	2/100 = 0.02
$40 \le M < 50$	P grade	10	$f(g) = 6$	6/10 = 0.6	6/100 = 0.06
$50 \le M < 60$	D grade	10	$f(g) = 22$	22/10 = 2.2	22/100 = 0.22
$60 \le M < 70$	C grade	10	$f(g) = 30$	30/10 = 3.0	30/100 = 0.3
$70 \le M < 80$	B grade	10	$f(g) = 25$	25/10 = 2.5	25/100 = 0.25
$80 \le M < 90$	A grade	10	$f(g) = 10$	10/20 = 0. 5	10/100 = 0.1
$90 \le M < 100$	Ex	10	$f(g) = 5$	5/20 = 0. 25	5/100 = 0.05
			Total of all frequencies = 100		Total = 1.0

The probability, $P(x)$ of a future outcome x is a prediction of relative frequency of that outcome on the basis of past experience of relative frequency of that incident. The area under frequency distribution is equal to the sum of individual frequencies. The area under probability distribution is equal to 1, the sum of all individual probabilities.

deviation σ. It is expressed in shorthand as ($N\{\mu, \sigma^2\}$). The probability density $pd(x)$ of a particular value x is largest when the value is close to μ. Then the probability falls off as a bell-shaped curve on either side of μ. Most of the random experimental errors may be considered to follow a *normal distribution*. A normal distribution is defined by the following probability density equation:

$$pd(x) = \frac{1}{\sqrt{2\pi\sigma^2}} \exp\left\{-\frac{(x-\mu)^2}{2\sigma^2}\right\} \tag{1.10}$$

ILLUSTRATION 1.7

Calculate the probability of a single experimental result falling within a value between 77 and 79 for a normal distribution of data with mean 80 and standard deviation 2.

Solution:

Using Equation 1.10, first calculate the probability when $x = 77$, with $\mu = 80$ and $\sigma = 2$:
Then calculate the probability when $x = 79$, with $\mu = 80$ and $\sigma = 2$:

$$pd(x) = 0.06485 \text{ (by Equation 1.10)}$$

$$pd(x) = 0.1761$$

Hence, the total probability = (0.06485 + 0.1761) = 0.2409.
The probability of the experimental result to stay within 77–79 is 24.09%.

ILLUSTRATION 1.8

Calculate the probability of a single experimental result to fall within a value between 78 and 82 for a normal distribution of data with mean 80 and standard deviation 2.

Solution:

First calculate the probability of the experimental result lying in the range 78–79. Set $x = 78$ with $\mu = 80$ and $\sigma = 2$ using Equation 1.10:

$$pd(x) = 0.12106$$

Then calculate the probability when $x = 79$, with $\mu = 80$ and $\sigma = 2$:

$$pd(x) = 0.1761$$

The total probability that the experimental result falls between 78 and 80 is = 0.2971 (0.12106 + 0.1761).

Hence, the probability of the experimental results falling between 78 and 82 is = (0.2971 × 2) = 0.594 or 59.4%.

There are some common probabilities associated with all normal distributions and they are as follows:

- 68.3% of all data points fall within ± 1.0 standard deviation from the mean value ≅ 68.3% of all data points fall from $(\mu - 1.0 \times \sigma)$ to $(\mu + 1.0 \times \sigma)$
- 90.0% of all data points fall within ± 1.64 standard deviation from the mean value ≅ 90.0% of all data points fall from $(\mu - 1.64 \times \sigma)$ to $(\mu + 1.64 \times \sigma)$
- 95.4% of all data points fall within ± 2.0 standard deviation from the mean value ≅ 95.4% of all data points fall from $(\mu - 2.0 \times \sigma)$ to $(\mu + 2.0 \times \sigma)$
- 99.7% of all data points fall within ± 3.0 standard deviation from the mean value ≅ 99.7% of all data points fall from $(\mu - 3.0 \times \sigma)$ to $(\mu + 3.0 \times \sigma)$

Normal distribution with a 0.0 mean and 1 as standard deviation is called *standard normal distribution*. Any random variable X from a normal distribution with mean μ and standard deviation σ can be transformed into a z score via the following equation:

$$Z = \frac{(X - \mu)}{\sigma} \tag{1.11}$$

The abscissa of a standard normal distribution curve is usually written as z; the probability density is expressed as $pd(z)$. The probabilities of a value falling between specific z values, $p(z_1 < z < z_2)$, can be calculated using a standard Z table, which is actually an area under the standard normal distribution between the values z_1 and z_2. Frequently, real-world phenomena follow a normal (or near-normal) distribution. This provides the experimenters an opportunity to use the normal distribution as a model for assessing probabilities associated with their experimental outcome. The analysis involves two steps:

1. Transform raw data into z-scores.
2. Find the probability using standard normal distribution tables.

Binomial distribution applies to a physical situation where there are only two possible results for each event. *Poisson's distribution* approximates the binomial distribution when the probability of a particular event is low.

ILLUSTRATION 1.9

The distribution of marks for an exam is described by a normal distribution with mean 50 and standard deviation 11.6. For a group of 200 students, estimate the number of students falling into each of the ranges of marks using Excel.

Solution:

- Use the class boundaries 9.5, 19.5, 29.5, etc.
- Use the function NORMDIST to calculate the cumulative probability up to each range boundary. *NORMDIST (19.5, 50, 11.6, TRUE)* for example calculates the cumulative probability up to 19.5.
- Calculate the probability of each range by taking the difference between the successive cumulative probabilities.
- Multiply each probability by 200.

The distribution of students' marks is as follows:

MARKS RANGE (M)	CLASS BOUNDARY	CUMULATIVE PROBABILITY	PROBABILITY	NUMBER OF STUDENTS
$0 \le M < 19$	19.5	0.00399	0.00399	0.6~0
$20 \le M < 29$	29.5	0.0373	0.0331	6.62~7
$30 \le M < 39$	39.5	0.1806	0.1433	28.66~29
$40 \le M < 49$	49.5	0.483	0.3024	60.48~60
$50 \le M < 59$	59.5	0.7956	0.3126	62.52~63
$60 \le M < 69$	69.5	0.955	0.1584	31.68~32
$70 \le M < 79$	79.5	0.995	0.04	8
$80 \le M < 89$	89.5	0.9997	0.0047	0.94~1
$M > 90$	100	1.0	0.0003	0.06~0

1.4.6 *Uncertainty*

Any scientific/engineering measurements have *uncertainty*. Statistics helps infer the best estimates of a measurement. The *true value* of a variable is a measurement with a perfect measuring process. In the absence of any perfect measuring process, the observed values found are the best estimates of the true value. The uncertainty in the measurement is, therefore, the best estimate of the possible error. The error may be a *systematic* error or a *random* error. Systematic error is due to the use of inappropriate technique. On repetition, systematic error will push the measured value in the same direction around the mean value of it. Random error results in the spread of the mean value of the measurement due to lack of precision in measurement even in a carefully designed experimentation. *Precision* and *accuracy* are the two terms that need to be clarified in the context of uncertainty. The precision of a measurement is the best estimate of random error in a measurement. Precision dictates how close the measured values are to each other. Uncertainty is expressed as a range where the estimated value is likely to exist within a confidence limit independent of any true value. Suppose the measured density of alkanolamine reported is 1.00794 ± 0.00007, where ± 0.00007 is the measure of uncertainty. Often, the uncertainty of a measurement is found by repetitive measurements to get a good estimate of the standard deviation of the values.

Accuracy of measurement is the closeness of the agreement between the result of a measurement and a true value of the measured. It is assumed that a true value can be defined, known, and realized perfectly. Accuracy is a qualitative concept—it can be *high* or *low*, and it should not be used quantitatively. Accuracy is how close a measured value is to the actual (true) value.

The uncertainty in a single experimental measurement is given by standard deviation σ of the experimental variation. With increasing number of replicate runs, the mean value of replicates approaches the μ (true mean) as a consequence of the central limit theorem. The distribution of sample mean values containing n measurements is also normally distributed with a mean value of μ and standard error $\sigma / \sqrt{n}$ (as a consequence of the central limit theorem). If μ and σ both are known then the following statistics are applicable:

- 90.0% of $\bar{x}$ values would fall in the range: $\left(\mu-\left\{1.64\times{}^{\sigma}\!/_{n}\right\}\right)$ to $\left(\mu+\left\{1.64\times{}^{\sigma}\!/_{n}\right\}\right)$
- 95.0% of $\bar{x}$ values would fall in the range: $\left(\mu-\left\{1.96\times{}^{\sigma}\!/_{n}\right\}\right)$ to $\left(\mu+\left\{1.96\times{}^{\sigma}\!/_{n}\right\}\right)$
- 99.0% of $\bar{x}$ values would fall in the range: $\left(\mu-\left\{2.58\times{}^{\sigma}\!/_{n}\right\}\right)$ to $\left(\mu+\left\{2.58\times{}^{\sigma}\!/_{n}\right\}\right)$

In most cases, the aim of the experimentation is to find the best estimate of an unknown true value (μ) by using sample mean $\bar{x}$; we can claim with

- 90.0% confidence that μ will fall within a range: $\left(\bar{x}-\left\{1.64\times{}^{\sigma}\!/_{n}\right\}\right)$ to $\left(\bar{x}+\left\{1.64\times{}^{\sigma}\!/_{n}\right\}\right)$
- 95.0% confidence that μ will fall within a range: $\left(\bar{x}-\left\{1.96\times{}^{\sigma}\!/_{n}\right\}\right)$ to $\left(\bar{x}+\left\{1.96\times{}^{\sigma}\!/_{n}\right\}\right)$
- 99.0% confidence that μ will fall within a range: $\left(\bar{x}-\left\{2.58\times{}^{\sigma}\!/_{n}\right\}\right)$ to $\left(\bar{x}+\left\{2.58\times{}^{\sigma}\!/_{n}\right\}\right)$

1.4.7 *Confidence interval*

When there is an uncertainty in the experimental method itself, the true value is the mean value (μ) of a sample containing n measurements. *Confidence interval* is the range with which we would be X% confident of finding a true value (μ) provided the standard deviation is known. If the population standard deviation is not known, sample standard deviation (s) can be assumed as population standard deviation. In this case, finding a true value with the same X% will require a *t-value,* which depends on the sample size. The incorporation of "*t-distribution*" results in a wide range of uncertainty in estimating true value (μ). The lower the *t*-value, the higher the uncertainty in estimating the true value. The use of *t*-distribution for sample data (having sample mean ($\bar{x}$) and standard deviation (s)) is analogous to the use of normal distribution for population data (having mean (μ) and standard deviation (σ)).

The *critical t-value* ($t_{(T,\alpha\, df)}$) depends on three factors:

- Whether there is a two-tailed or a one-tailed distribution (T). When the distribution is symmetrical on both sides of the mean value then $T = 2$
- Significance level (α) or confidence level X%. (Significance level is the maximum probability of being wrong.)
- Number of degrees of freedom ($df = n - 1$); n is the size of the sample.

The critical *t*-value can be obtained from the *t-table*. The Excel function *TINV* can also be used to find out the critical *t*-value ($t_{(T,\alpha\, df)}$) with any level of confidence and degree of freedom.

Suppose $T = 2$, $\alpha = 0.5$, and $n = 6$; *calculate the t-value for a two-tailed distribution (answer = 0.0653).*

Solution:

- Open Excel Sheet
- Go to Formula
- Use *TINV*

$$\text{Confidence interval } CI(\mu,\ X\%) = \bar{x} \pm \left\{t_{2,\,\alpha,\,df} \times {}^{s}\!/\!\sqrt{n}\right\} \tag{1.12}$$

ILLUSTRATION 1.10

In a particular experiment to measure the arsenic contamination in drinking water, the experimental procedure has a standard deviation of 1.5 parts per billion (ppb). A number of analysts have repeated the same experiment. The true value of the arsenic concentration is 20 ppb. Each of the analysts has made four measurements.

I. Estimate the range of values within which 95% of the analysts' measurement for the true value are likely to appear.
II. Estimate the range of values within which 95% of the experimental means $\bar{x}$ are likely to fall.

ANS I.

95% of the measurements will appear in the range:

$$(\mu - 1.96 \times \sigma) \text{ to } (\mu + 1.96 \times \sigma)$$

$$(20 - 1.96 \times 1.5) \text{ to } (20 + 1.96 \times 1.5) = 17.06 \text{ to } 22.94$$

ANS II.

95.0% of $\bar{x}$ values would fall in the range: $\left(\mu - \left\{1.96 \times \sigma/_{n}\right\}\right) \text{ to } \left(\mu + \left\{1.96 \times \sigma/_{n}\right\}\right)$

$$= \left(20 - \left\{1.96 \times {}^{1.5}/_{4}\right\}\right) \text{ to } \left(20 + \left\{1.96 \times {}^{1.5}/_{4}\right\}\right)$$

$$= \underline{19.265 \text{ to } 20.375}$$

ILLUSTRATION 1.11

A set of six replicate measurements of arsenic content in water have been made, revealing the contamination level to be 25, 27, 23, 20 26, and 24 ppb. Assume the standard deviation of the measurement procedure is known *a priori*, and is σ = 3.0. Calculate, with 95% confidence, the range of values within which the unknown true value μ lies.

Solution:

The mean of six measurements = 24.17.

We can claim with 95% confidence that the unknown true value lies within the following range:

$$= \left(\bar{x} - \left\{1.96 \times \sigma/_{n}\right\}\right) \text{ to } \left(\bar{x} + \left\{1.96 \times \sigma/_{n}\right\}\right)$$

$$= \left(24.17 - \left\{1.96 \times {}^{3}/_{6}\right\}\right) \text{ to } \left(24.17 + \left\{1.96 \times {}^{3}/_{6}\right\}\right)$$

$$= (24.17 - 2.4) \text{ to } (24.17 + 2.4) = \underline{21.8 \text{ to } 26.6}$$

1.4.8 Hypothesis Testing

The scientific method of experimentation is the basis of modem science and technology. A system to be investigated may be characterized by inputs/factors, outputs, and mechanisms. Each factor has two or more levels and a change of level will cause a particular outcome. A decision has to be made about the acceptability of the new outcome of an experimental investigation. A *statistical hypothesis* is an assumption or a guess about the population expressed as a statement about the probability distributions of parameters of the populations. A *hypothesis* is a combination of two mutually exclusive statements covering all outcomes, which is to be stated before the experiment and is being tested in order to accept or reject the experimental outcome. A *null* hypothesis dictates no factor or input effect and an *alternative* hypothesis dictates that the factor effects the outcome. There are two possible ways to reach the wrong conclusions about the experimental investigation via hypothesis testing:

1. Type I error: Claimed factor effect when no such situation actually exists
2. Type II error: Failure in detecting a factor effect when such a situation really exists

The probability of getting a type I error is the *p-value* or calculated probability. The significance level (α) for the test is the maximum acceptable value of type I error. *If* (α) *then the alternative hypothesis is true.*

For example, as part of a monitoring procedure, a researcher wishes to test whether the arsenic contamination of water exceeds 50 ppb. If the true arsenic content is μ, the appropriate hypothesis for the test would be:

ILLUSTRATION 1.12

If the uncertainty in the measurement procedure, σ, is not known in Illustration 1.11, calculate with 95% confidence the range of values within which the unknown true value lies.

Solution:

- Use the MATLAB® datastats function

The mean of six measurements ($\bar{x}$) = 24.17.

The standard deviation of theses six replicates (s) = 2.48

- The range of values within which the unknown true value lies (with 95% confidence):

$$CI(\mu, X\%) = \bar{x} \pm \left\{ t_{2,\,\alpha,\,\mathrm{df}} \times s/\sqrt{n} \right\} = 24.17 \pm \left\{ t_{2,\,0.05,\,\mathrm{df}=(6-1)} \times 2.48/\sqrt{6} \right\}$$

$$= 24.17 \pm \left\{ 2.57 \times 2.48/\sqrt{6} \right\} = 24.17 \pm 2.602$$

$$= 21.568 \text{ to } 26.77$$

The uncertainty in this case is greater than the previous case (Illustration 1.11) (because σ was known in Illustration 1.11).

Null hypothesis H_0: Arsenic content does not exceed 50 ppb; $\mu \leq 50$ *ppb*
Alternative hypothesis H_A: Arsenic content does exceed 50 ppb, $\mu > 50$ *ppb*

Two kinds of errors may occur while testing the hypothesis:

Type I error: $P\{\text{reject } H_0 | H_0 \text{ is true}\}$
Type II error: $P\{\text{fail to reject } H_0 | H_0 \text{ is false}\}$

The probability of type I error is determined by p-value for a given set of experimental results. First α is selected to compute the confidence limit for testing the hypothesis. α is the maximum probability of type I error that would be acceptable.

If $p \leq \alpha$, then accept the alternative hypothesis.

If $p > \alpha$, then there is insufficient evidence to accept the alternative hypothesis.

Tails: In defining a hypothesis system with respect to a possible factor effect, there remain two possibilities:

Two-tailed: The factor or input may operate in either direction towards a possible outcome.
Null hypothesis H_0: Arsenic content equals 50 ppb; $\mu = 50$ *ppb*
Alternative hypothesis H_A: Arsenic content does not equal 50 ppb, $\mu \neq 50$ *ppb*

One-tailed: To achieve an outcome, the factor is only imposed in one possible direction.
Null hypothesis H_0: Arsenic content does not exceed 50 ppb; $\mu \leq 50$ *ppb*
Alternative hypothesis H_A: Arsenic content exceed does exceed 50 ppb, $\mu > 50$ *ppb*

There are different parametric tests or hypothesis tests. They are called parametric tests because actual experimental values are used in the calculations of the following test procedures.

- *F*-test (to test the differences in variance between data samples (sets))
- *t*-test (to test the differences in mean value between data samples (sets))
- Correlation testing (to test for a linear relationship between data variables)

For each test, it is possible to calculate test statistics; for example, a *t*-statistic for the *t*-test. Each of the statistics is calculated from the experimental results. A large value for the test statistic implies a large factor effect. For each test, there are tables of critical values. The choice of critical value for a given experiment will depend on the value of the significance level (α), and the required degree of freedom available in the experimental data. The alternative hypothesis will be accepted (with a significance level of α) if the value of the test statistic is greater than the critical value. In contrast, the alternative hypothesis will be accepted if $p \leq \alpha$.

F-test: An *F-test is performed to test for significant difference in variance between two samples drawn from normal distribution. F-statistics* is the ratio of the two sample variances. In order to conform to any of the *hypotheses*, the *F-value* calculated is compared with the critical *F-value* corresponding to a one-tailed or two-tailed distribution having *significance level* of 0.05. If the *F-value* is greater than *F-critical* the alterative hypothesis is true.

$F_{stat} = \frac{S_i^2}{S_j^2}$ where S_i^2 and S_j^2 are the measured sample variances of groups *i* and *j*, and they are the best estimates of population variances σ_i^2 and σ_j^2 representing the group samples *i* and *j*. In practice, the experimenter has two sets of data, and he can determine S_i^2 as well as S_j^2. However, the experimenter is unaware of whether or not the two sample groups belong to the same population. There are two possible hypothesis descriptions:

Two-tailed hypothesis
Null hypothesis H_0: The source populations are the same ($\sigma_i^2 = \sigma_j^2$)
Alternative hypothesis, H_A: The source populations are different ($\sigma_i^2 \neq \sigma_j^2$)

One-tailed hypothesis
Null hypothesis, H_0: Variance of group *i* is not greater than variance of group *j* ($\sigma_i^2 \leq \sigma_j^2$)
Alternative hypothesis, H_A: Variance of group *i* is greater than variance of group *j* ($\sigma_i^2 > \sigma_j^2$)

$$\text{In Excel, FSTAT} = \left\{ \text{VAR}(\text{array } i) \middle/ \text{VAR}(\text{array } j) \right\}$$

***F* critical value:** $F_{critical} = F_{(T,\,\alpha,\,dfi,\,dfj)}$, where T is the number of tails: one-tailed or two-tailed. α is the significance level required, which is 0.05 by default. *dfi* and *dfj* are the degrees of freedom of the variances for the numerator and denominator. The degree of freedom is $df = n - 1$, in general, where n is the size of sample. In principle, a difference between σ_i^2 and σ_j^2 can be achieved in two ways: either ($\sigma_i^2 > \sigma_j^2$) or ($\sigma_i^2 < \sigma_j^2$). Hence, the values of F are either greater than 1 or less than 1. Therefore, two $F_{critical}$ values exist: $F_{critical} > 1$, when $\sigma_i^2 > \sigma_j^2$ and $F_{critical} < 1$, when $\sigma_i^2 < \sigma_j^2$.

In Excel, *FINV* (α, dfi, dfj) provides the upper one-tailed critical value. Hence, for a two-tailed $F_{critical}$ the following equation applies:

$$F_{(2,\,\alpha,\,dfi,\,dfj)} = F_{(1,\,\alpha/2,\,dfi,\,dfj)} \tag{1.13}$$

ILLUSTRATION 1.13

In order to evaluate the performance of two experimenters (A and B), it was decided to compare the variances in their generated data (sample groups) consisting of 21 similar measurements for each of them (all the experimental conditions remain the same). The recorded variances are as follows: $S_A^2 = 85.0$ and $S_B^2 = 195.5$.

APPLY

I. A two-tailed test, with an alternative hypothesis, which states that there is a difference between the variances of the generated experimental sample groups by the two analysts.
II. A one-tailed test, with an alternative hypothesis, which states that the variance of the sample group generated by analyst A is less than that generated by analyst B.

Solution:

I. For a two-tailed test:

$$F_{\text{stat}} = \frac{S_B^2}{S_A^2} = \frac{195.5}{85.0} = 2.3$$

Calculate F_{critical}:

Use Excel FINV (α, df$A$, df$B$), which provides the upper one-tailed critical value.

For a two-tailed F_{critical} $\left(\text{apply the equation } \left(F_{(2,\,\alpha,\,\text{df}A,\,\text{df}B)} = F_{(1,\,\alpha/2,\,\text{df}A,\,\text{df}B)}\right)\right)$

$F_{(t,\alpha,\,\text{df}A,\,\text{df}B)} = F_{(2,\,0.025,\,20,\,20)}$, $\text{df}_A = \text{df}_B = n-1 = 20$, and $\alpha = 0.025$

$F_{\text{critical}} = 2.46$. Since $F_{\text{stat}} < F_{\text{critical}}$, we do not reject the null hypothesis; there is not enough evidence to discriminate between the two analysts.

p-value (two-tailed) $= 2 \times \text{FDIST}(F_{\text{stat}}, \text{df}_A, \text{df}_B) = 0.068$, and the p calculated is greater than α, so we accept the null hypothesis.

II. For a one-tailed test:

$F_{\text{stat}} = 2.3$, one-tailed critical value with $F_{(t,\alpha,\,\text{df}A,\,\text{df}B)} = F_{(1,\,0.05,\,20,\,20)}$, $\text{df}_A = \text{df}_B = n-1 = 20$, and $\alpha = 0.05$.

$F_{\text{critical}} = 2.12$. In this case $F_{\text{stat}} > F_{\text{critical}}$, so we accept the alternative hypothesis. There is enough evidence that variance of the group B sample is greater than the group A sample.

p-value (one-tailed) $= \text{FDIST}(F_{\text{stat}}, \text{df}_A, \text{df}_B) = 0.034$, and the p calculated is less than α, so we accept the alternative hypothesis.

***t*-Test**: William Gossett developed the theory of the t-test and t-value to analyze the statistics of small data samples and published under the name of student, hence the name student's t-test. *A t-test is performed to locate the mean value of a data sample, or differences in mean values of different data samples, and follows the procedure of the hypothesis test.* A t-test is a hypothesis test to decide the following:

- Whether an unknown true value (μ) of a single sample set is likely to differ from a specific value (μ_0).
- Whether unknown true values ($\mu_{A,}$, $\mu_{B,}$) of two samples sets are likely to differ from each other.

Before carrying out the test, it is necessary to state the null and alternative hypothesis and decide upon the significance level (α) for the test. There are two ways to carry out a t-test: the t-statistics method and the p-value. The t-value is a measure of deviation from the null hypothesis. *A larger t-value in comparison to the t-critical value suggests the alternative hypothesis to be true.*

One-sample t-test

In a one-sample test, the mean $\bar{x}$ of the sample group (data set) is the best estimate for an unknown true mean value μ. The aim of the one-sample t-test is to decide whether μ is likely to differ from a specific value μ_0.

Hypothesis for two-sample two-tailed t-test:

Null hypothesis, $H_0 : \mu = \mu_0$

Alternative hypothesis, $H_A : \mu \neq \mu_0$
Accept alternative hypothesis if $t_{stat} > t_{critical}$

Hypothesis for one-sample one-tailed t-test:
Null hypothesis, $H_0 : \mu \leq \mu_0$
Alternative hypothesis, $H_A : \mu > \mu_0$
Accept alternative hypothesis if $t_{stat} > t_{critical}$

$t_{stat} = \dfrac{(\bar{x} - \mu_0)}{s/\sqrt{n}}$ where s is the sample standard deviation and n is the number of data values in the sample group.

Two-sample t-test

In a two-sample test, the means $\overline{x_A}$ and $\overline{x_B}$ of the two sample groups (data sets) are the best estimate for their unknown true mean values μ_A and μ_B. The aim of the two-sample t-test is to decide whether μ_A and μ_B are likely to differ from each other.

Hypothesis for two-sample two-tailed t-test:
Null hypothesis, $H_0 : \mu_A = \mu_B$
Alternative hypothesis, H_A: $\mu_A \neq \mu_B$
Accept alternative hypothesis if $t_{stat} > t_{critical}$

Hypothesis for two-sample one-tailed t-test:
Null hypothesis, $H_0 : \mu_A \leq \mu_B$
Alternative hypothesis, H_A: $\mu_A > \mu_B$
Accept alternative hypothesis if $t_{stat} > t_{critical}$

$t_{stat} = \dfrac{\left(\overline{x_A} - \overline{x_B}\right)}{s' \times \sqrt{1/n_A + 1/n_B}}$, where n_A and n_B are the number of data values in sample group A and B. s' is the pooled standard deviation.

$s' = \sqrt{\dfrac{\{(n_A - 1)S_A^2 + (-1)S_B^2\}}{(n_A + n_B - 2)}}$. In this case, the assumption has been made that both samples have been drawn from populations that have similar variances. The pooled standard deviation is the best estimate of the common standard deviation. S_A and S_B are sample standard deviations.

For calculating $t_{critical:}$ degrees of freedom, $\text{df} = (n_A + n_B - 2)$.

ILLUSTRATION 1.14

Following are two sets of refractive index values measured from the samples collected from a fermentation broth. As a part of monitoring the fermentation process, the biomass concentration of the fermentation broth is determined in terms of the refractive index values of liquor. Determine with 99% confidence whether the two sets of data represent the same or different batches.

Solution:

Type I sample	184	135	129	95	132	67	152
Type II sample	348	255	244	179	250	126	233.9

- First, the *F*-test is performed to test significant differences in variance between data samples (sets).

For two-tailed hypothesis:

Null hypothesis, H_0: There is no significant deviation between the variances ($S_A^2 = S_B^2$).

Alternative hypothesis, H_A: There is significant deviation between the variances ($S_A^2 \neq S_B^2$).

Mean values: $\overline{x_A} - 127.7$, $\overline{x_B} = 233.7$

Sample standard deviations: $S_A = 37.8$, $S_B = 75.5$

$$F_{\text{stat}} = \frac{S_B^2}{S_A^2} = \frac{75.5^2}{37.8^2} = 3.98$$

$F_{(t,\alpha,\,\text{df}A,\,\text{df}B)} = 6.977$, $df_A = 6$, $df_B = 6-1 = 5$, and $\alpha = 0.025$, $t = 2$

$F_{\text{critical}} > F_{\text{stat}}$, hence, the null hypothesis is true.

p-value (two-tailed) $= 2 \times \text{FDIST}(F_{\text{stat}}, df_A, df_B) = 0.4648$, and the *p* calculated is greater than α, so we accept the null hypothesis. Hence, the null hypothesis is true.

Secondly, the *t*-test is performed to locate the differences in mean values of different data samples.

Null hypothesis, H_0: Means of the source populations are equal $\mu_A = \mu_B$.

Alternative hypothesis, H_A: Means of the source populations are not equal $\mu_A \neq \mu_B$.

There is no significant deviation between the variances. We can use pooled standard deviation.

$$s' = \sqrt{\frac{\left\{(n_A - 1)S_A^2 + (-1)S_B^2\right\}}{(n_A + n_B - 2)}} = 58.05 \tag{1.14}$$

$$t_{stat} = \frac{\left(\overline{\overline{x_A}} - \overline{x_B}\right)}{s' \times \sqrt{1/n_A + 1/n_B}} = -3.28 \tag{1.15}$$

Number of tails '*t*' = 2, degrees of freedom = $df = (n_A + n_B - 2) = 11$, $\alpha = 0.01$ (99% confidence)

Critical t-value $t_{(T,\alpha,df)} = 3.11$ (using Excel function TINV)

$t_{stat} < -t_{critical}$, so we can accept the alternative hypothesis.

We can claim with 99% confidence that the two sets of refractive index values conform to two different batches.

1.4.9 Correlation

Correlation is a measure of the extent of the relation being shared by two variables. The *linear correlation coefficient* ($0 \leq r \leq 1$) between the two variables reveals the extent to which the data follows a linear relationship. The square of the correlation coefficient is also a measure of correlation and is called the *coefficient of determination*. For a perfect linear relation, $r^2 = 1$ and $r = \pm 1$, perfect positive and negative correlation. The hypothesis test for a linear correlation examines whether a straight line with a non-zero slope will give a better fit using linear regression than a line with zero slope.

Null hypothesis: No linear relationship between the correlating variables.
Alternative hypothesis: Linear relationship between the correlating variables.

The *test statistics* is the r value and can be calculated using sample size, sample mean, and standard deviation. The Excel functions CORREL and PEARSON can be used to calculate correlation coefficient r. The r_{critical} value can be obtained from the table of Pearson's correlation coefficient, which depends on

- Whether there is a two-tailed or a one-tailed distribution (T). When the distribution is symmetrical on both sides of the mean value then $T = 2$.
- Significance level (α).
- Number of degrees of freedom ($df = n - 2$); n is the number of data pairs.

If the calculated r value $\geq +r_{\text{critical}}$ or r value $\leq -r_{\text{critical}}$, the alternative hypothesis is accepted.

1.4.10 Regression

Regression was introduced by Karl Pearson, who adapted a related idea presented by Francis Galton in the 1880s. Unlike correlation, *regression analysis* is a statistical process for estimating the causal relationships among variables. A regression analysis involves independent, dependent variables and parameters. In regression, the target is to derive a functionality relating the dependent variable as a function of the independent variables called the regression function. Regression is broadly classified under two categories: *parametric* and *nonparametric*. *Linear regression* and ordinary linear least squares are parametric regression, where the regression function is defined by the minimum number of parameters estimated using the data of interest. The earliest form of regression was the method of least squares, which was published by Legendre in 1805 and by Gauss in 1809. Nonparametric regression refers to the techniques where the structure of the regression function is not predetermined, but instead falls under a set of functions decided by the data. The sample size required is considerably larger than the parametric regession. This book utilizes regression in the following ways: for data smoothing (local regression smoothing using the MATLAB® Curve Fitting Toolbox) and for classification using multivariate linear regression model partial least squares (PLS) (discussed in Chapter 2).

In the general multiple regression model, if there are p independent variables,

$$y_i = \beta_1 x_{i1} + \beta_2 x_{i2} + \cdots + \beta_p x_{ip} + \epsilon_i \tag{1.16}$$

ILLUSTRATION 1.15

The following table shows the marks obtained by 10 postgraduate students in the process plant simulation (PPS) and advanced process control (APC). The evaluation was done out of 50 marks. *It seems the hypothesis that "a student efficient in PPS is equally efficient in APC" holds.* Investigate with a one-tailed correlation between the performance in the two subjects using the data given.

Subject	1	2	3	4	5	6	7	8	9	10
PPS	31	35	29	32	30	36	37	35	33	37
APC	28	34	32	35	38	40	42	46	45	47

Solution:

Use the Excel function CORREL and calculate the correlation coefficient $r_s = 0.7806$.
The one-tailed critical value for $\propto = 0.05$, $r_{\text{critical}} = 0.549$.
Since $r_s > r_{\text{critical}}$, we accept the alternative hypothesis (linear relationship between the performance in the two subjects).

where x_{ij} is the ith observation on the jth independent variable, and where the first independent variable takes the value 1 for all i (so β_1 is the regression intercept). The least squares parameter estimates are obtained from p normal equations. The normal equations are written as

$$\sum_{i=1}^{N}\sum_{k=1}^{p} X_{ij}X_{ik}\beta_k = \sum_{i=1}^{N} X_{ij}y_i, \quad j = 1,\ldots\ldots p \tag{1.17}$$

The residual can be written as

$$\epsilon_i = y_i - \beta_1 x_{i1} + \beta_2 x_{i2} + \cdots + \beta_p x_{ip} \tag{1.18}$$

In matrix notation, the normal equations are written as

$$\left(X^T X\right)^{-1} \beta = X^T Y \tag{1.19}$$

The fit parameter vector is

$$\beta = \left(X^T X\right)^{-1} X^T Y \tag{1.20}$$

It is important to check the statistical significance of the estimated parameters and degree of fitness of the regression function. Statistical significance can be determined by an F-test of the overall fit, followed by t-tests of individual parameters.

1.4.11 Chi-squared test

The family of chi-squared, χ^2, tests is based on the frequency of occurrence of an event. The tests assess whether the observed distribution of occurrence is a random phenomenon or caused by a factor. $\chi^2 = \sum_i \frac{(O_i - E_i)^2}{E_i}$, where O_i is the observed frequency in category i and E_i is the expected frequency in i. The relevant hypotheses for χ^2 tests are as follows:

Null hypothesis: The difference between the expected and observed frequencies of occurrence of an event is by sheer chance.
Alternative hypothesis: The difference between the expected and observed frequencies of occurrence of an event is not by chance but due to a factor.

Critical χ^2 values depending on degrees of freedom and levels of confidence are obtained from an χ^2 table. The χ^2 test is a one-tailed test. $\chi^2_{\text{critical}} = \chi^2_{(1,\,\alpha,\,df)}$. If $\chi^2 \geq \chi^2_{\text{critical}}$, the alternative hypothesis is acceptable, and if $\chi^2 < \chi^2_{\text{critical}}$, the null hypothesis should not be rejected. The Excel function *CHISQ.INV* provides a one-tailed χ^2_{critical} value = CHISQ.INV $((1-\alpha)\ \text{df})$, where df is the degree of freedom = number of observation - 1.

1.5 *Stochastic and stationary processes*

A deterministic process is governed by certain physical laws, and if the condition of its state is specified at initial conditions, one can precisely specify/predict the states of it for a later period. Nondeterministic processes are of two types: stochastic or probabilistic processes, and chaotic processes. Most of the natural processes are nondeterministic in nature. A stochastic process has such governing principals that its state at a later period of time can't be forecasted even if the initial conditions are known; instead one can predict

ILLUSTRATION 1.16

A set of six measurements of arsenic content in water have been made, revealing the contamination level to be 15, 22, 20, 14, 18, and 31 ppb. The expected contamination level is 20 ppb. Is it possible to say with 95% confidence that the difference between the observed and expected contamination level occurred just by chance?
Solution:

The results are tabulated below.

Step I: calculation of χ^2
Step II: calculation of χ^2_{critical} using *CHISQ.INV* $((1-\alpha)\ \text{df})$

Test Sample	1	2	3	4	5	6	Sum
Observed (O_i)	15	22	20	14	18	31	120
Expected (E_i)	20	20	20	20	20	20	120
$\frac{(O_i - E_i)}{E_i}$	1.25	0.2	0	1.8	0.2	6.05	9.5

$\chi^2 = 9.5$

$\chi^2_{\text{critical}} = 11.07$, where $\alpha = 0.05$, $\text{df} = 6 - 1 = 5$

Since $\chi^2 < \chi^2_{\text{critical}}$, the null hypothesis should not be rejected, which means the differences between the observed and expected measurements occurred by sheer chance and are not due to any factor.

the probability of future states of the process. It is very difficult to distinguish between nondeterministic and deterministic processes with erroneous initial conditions because in both cases, the future state is unpredictable. For practical purposes, we may consider unpredictable systems (whether deterministic or nondeterministic) as stochastic systems, since we deploy the same statistical properties in studying such systems regardless of their governing physical laws. A stochastic process can be treated as a probabilistic counterpart of a deterministic process.

A *time series* can be defined as statistical data that are collected and recorded over successive increments of time. Time series such as electrical signals that are continuous in time have been treated in this book as discrete time series because only digitized values at discrete time intervals are considered. The field of time series infuses many areas of science and engineering, particularly statistics and signal processing. This book utilizes various signals (multisensor data) to design appliances like food quality monitors, patient care monitors, and many others. The stationarity/nonstationarity of the time series data (signals) influences the nature and extent of rigor involved in developing various machine learning algorithms as an integral part of beverage quality monitor design and patient care monitor design. This section will characterize time series with respect to their stationary/nonstationary nature.

The building blocks in the analysis of time series data are random or stochastic processes. A purely random process is a stochastic process, $\{\epsilon_t\}_{t=-\infty}^{\infty}$, where each element is (statistically) independent of the other elements and each element has an identical distribution. As a simple example, suppose ϵ_t has a Gaussian distribution with mean μ and variance σ^2 at instant t, and the same is true for all integers t (or equivalently $\epsilon_t \sim N(\mu, \sigma^2)$). Mean and variance are the basic first- and second-order statistics of a time series. The order relates to the highest power in the integral defining the statistic: for expectation it is $\int xf(x)dx$ and for variance it is $\int x^2 f(x)dx$. Mathematically, the autocovariance between X_t and X_s for some processes is defined as cov $(X_t, X_s) = \mathbb{E}\left|\{X_t - \mathbb{E}(X_t)\}\{X_s - \mathbb{E}(X_s)\}\right|$. Statistics of higher than second order are mostly ignored in time series analysis.

A stationary process means that the mean and the variance of the process are independent of t (that is they are time invariant) and the autocovariance between X_t and $X_{t+\tau}$ can depend only on the lag τ (τ is an integer and needs to be finite). Hence, for stationary processes $\{X_t\}$, the definition of autocovariance is $\gamma(\tau) = \text{cov}\ (X_t, X_{t+\tau})$. Autocorrelation ($\rho$) of a process is merely a normalized version of the autocovariance (it has values between –1 and 1). $\rho(\tau) = \gamma(\tau)/\gamma(0)$ for integer τ and $\gamma(0) = \text{cov}\ (X_t, X_t) = \text{Var}(X_t)$. Hence, $\gamma(\tau)$ and $\rho(\tau)$ are always zero for $\tau \neq 0$. Therefore, they do not depend on τ. Thus, the purely random process must be at least second-order stationary.

Probability models like moving average models, autoregressive models, and autoregressive moving average (ARMA) models are extensively used for modeling time series. The other way of looking at time series is to determine how much energy is contained within a time series as a function of frequency (the spectrum). In this book, probability models have been used to determine the stationarity/nonstationarity of the time series process using the stochastic process theoretic. A time series can be assumed to be made up of two parts: an overall level, mean μ, and a random error component, ϵ. The random error part is called white noise. Let us take the representative moving average model of order q, $MA(q)$, to model the residual of a time series.

A simple model is constructed by linear combinations of lagged elements of a purely random process $\{\in_t\}_{t=-\infty}^{\infty}$ with $\mathbb{E}(\epsilon_t)=0$. A moving average process $\{X_t\}$ of order q is defined as

$$X_t = \beta_0 \in_t + \beta_1 \in_{t-1} + \ldots\ldots\ldots + \beta_q \in_{t-q} = \sum_{i=0}^{q} \beta_i \in_{t-i} \tag{1.21}$$

For the $MA(q)$ process the mean is

$$\mathbb{E}(X_t) = \mathbb{E}\left(\sum_{i=0}^{q} \beta_i \in_{t-i}\right) = \sum_{i=0}^{q} \beta_i \mathbb{E}(\in_{t-i}) = 0 \tag{1.22}$$

because $\mathbb{E}(\epsilon_t)=0$ for any value of t.

The variance and autocovariance are as follows:

$$\text{var}(X_t) = \text{var}\left(\sum_{i=0}^{q} \beta_i \in_{t-i}\right) = \sum_{i=0}^{q} \beta_i^2 \text{var}(\in_{t-i}) = \sigma^2 \sum_{i=0}^{q} \beta_i^2 \tag{1.23}$$

since $var(\in_t)=\sigma^2$ for all t.

$$\begin{aligned}\gamma(\tau) = \text{cov}\ (X_t,\ X_{t-\tau}) &= \text{cov}\left(\sum_{i=0}^{q} \beta_i \in_{t-i}, \sum_{j=0}^{q} \beta_j \in_{t-\tau-j}\right) \\ &= \sum_{i=0}^{q}\sum_{j=0}^{q} \beta_i \beta_j \text{cov}\left(\in_{t-i}, \in_{t-\tau-j}\right) \\ &= \sigma^2 \sum_{i=0}^{q}\sum_{j=0}^{q} \beta_i \beta_j \delta_{j,i+\tau}\end{aligned} \tag{1.24}$$

where $\delta_{j,i+\tau}$ is the Kronecker delta which is 1 for $j=i+\tau$ and 0 otherwise. Hence, continuing the summation provides

$$\gamma(\tau) = \sum_{i=0}^{q-\tau} \beta_i \beta_{i+\tau} \tag{1.25}$$

The autocovariance of a $MA(q)$ process is a convolution of $\{\beta_i\}$ with itself, and it is 0 for $\tau > q$. When one computes the sample autocovariance, it cuts off at a certain lag q (i.e., it is effectively zero for lags of q + 1 or higher), then one can postulate the MA(q) model in Equation 1.21 as the underlying probability model. *Hence, stationarity can be checked by calculating autocovariance which will be effectively zero after the order (=lag) of the time series process.* The autocorrelation function can be calculated from *autocovariance*. For a time series that is white noise or series of random numbers, its autocorrelation functions (ACF, ρ_k) should theoretically be zero, which proves the absence of *heteroscedasticity* (time-varying variance) in that specific time series. The value of ρ_k always lies between –1 and +1. The sample ACFs are not exactly zero but follow a sampling distribution that can be approximated by a normal curve with mean zero and standard error of $1/\sqrt{n}$, n being the number of observations in the series. For white noise, 95% of the ACFs should be within $\pm 1.96/\sqrt{n}$.

The standard error of autocorrelation (S_{pk}) is the basis of the decision about whether sample autocorrelation is zero or significant. A sample autocorrelation is considered to be significantly different from zero with 95% confidence, if it is larger in magnitude than twice its standard error. The ± 2 limit of the standard error values are often called the confidence limit for sample autocorrelation.

One can quantify the preceding qualitative checks for correlation using formal hypothesis tests, such as the Ljung-Box-Pierce Q-test and Engle's ARCH test (MATLAB-GARCH toolbox). Both tests are most often used as post-estimation lack-of-fit tests applied to the fitted residuals. Under the null hypothesis of no serial correlation, the Q-test statistic follows a chi-square distribution (asymptotically). Engle's test inquires for the presence of ARCH effect; the null hypothesis concludes a time series to be a random sequence of Gaussian disturbances (i.e., no ARCH effects exist). This test statistic also follows a chi-square distribution. This book uses the GARCH toolbox in association with MATLAB to determine the stationarity/nonstationarity of electrical signals in Chapters 5 and 6.

1.6 Data preprocessing

Data pretreatment is a major concern in data-based application/algorithm development, and it includes outlier detection, data reconciliation, data smoothing, and application of transforms on data (if required for a specific application). Data corruption may be caused by failures in sensors or transmission lines, process equipment malfunctions, erroneous recording of measurement and analysis of results, or external disturbances. These faults would cause the data to have spikes, jumps, or excessive oscillations. The general strategy is to detect data that are not likely in conformity with other measurements/information *(outlier detection)* and to substitute these data with estimated values that are in agreement with other data/process-related information *(data reconciliation).* Sometimes data are transformed, which allows the data to provide more information that is not available in their original form (*transform*). Transformation and transform are often used interchangeably, though incorrectly. Data are often transformed for better interpretability. The logarithmic, square root, and multiplicative transformations are widely used transformations, where the data dimension and domain remain unchanged, unlike data realizing various transforms.

1.6.1 Outlier detection

An *outlier* is an observation in a data set which appears to be inconsistent with the remainder of that set of data [3]. Data in a particular instance has two attributes: contextual and behavioral. For example, in time-series data, time is a contextual attribute. The frequency of rainfall for a particular location is a behavioral attribute. The anomalous behavior is defined using the values for the behavioral attributes within a specific context. There are various types of outliers such as point outliers, contextual outliers, and collective outliers. A data instance can be considered a point anomaly with respect to the rest of the data, which is very common in experimental data generation. If a data instance is anomalous in a specific context, but not otherwise, then it is termed a contextual anomaly. When a collection of related data instances is irregular with respect to the entire data set, it is termed a collective outlier, as in the case of data pertaining to faulty operating conditions in a process/manufacturing plant. Often the data contains noise that tends to be similar to the actual anomalies; hence, it is difficult to distinguish and remove it from the anomaly. The anomalies need to be unusual in a fascinating way, and the supervision process redefines fascinating or unusual incidents. It is more difficult to get anomalous leveled data than to get data with a normal level/tag. Substantial effort and expertise is required to generate training data for normal, as well as anomalous, categories in model-based outlier detection. Abnormal data may appear without any precedence.

There are three fundamental approaches for anomaly detection: a clustering approach, a classification approach, and a novelty approach (aims at detecting previously unobserved patterns in the data). Anomaly detection techniques may also be classified under methods applicable for univariate and multivariate data. Another fundamental distinction in anomaly detection is parametric and nonparametric methods. Statistical parametric methods either assume a known underlying distribution of the observations (which itself is a shortcoming for such techniques; it does not hold good for high-dimensional real data) or, at least, they are based on statistical estimates of unidentified distribution. Even when the assumptions seem reasonably justified, several hypothesis test statistics are applied to detect anomalies; choosing the best statistic is often not a straightforward and trivial task. The parametric methods identify those observations as outliers, which deviate from the model assumptions. For example, it is assumed in process quality control (SPC) [4] that the data instances that more than a 3 σ distance away from the distribution mean μ (where σ is the standard deviation for the distribution) are declared as outliers. The $(\mu \pm 3\sigma)$ region contains 99.7% of the data instances. Several statistical anomaly detection techniques assume that the data follows a Gaussian distribution. Grubb's test has been used to detect anomalies in a univariate data set under the assumption that the data is generated by a Gaussian distribution [5]. A z-value is created for each instance of the data set and the z-value created is compared with a threshold (generated using a t-distribution with significance level of $\left(\alpha/_{2N}\right)$; N is the size of the data). Several variants of Grubb's test suitable for multivariate data have been proposed.

Anomaly detection using a regression model has been used extensively for time series data. The ARIMA (autoregressive integrated moving average) family of models is used for the estimation and filtering of process autocorrelation. The residuals of the ARIMA model are conditionally independent and approximately normally distributed, to which application of traditional SPC (statistical process control) schemes finds a wide variety of industrial application [6].

Traditional SPC is closely related to univariate and online outlier detection methods. The main idea of SPC is to detect deviations in a process from its predefined target after the incidence of deviation. The most important tools of SPC are control charts (like an $\bar{x}$ control chart, s chart, and R chart). A control chart consists of the control statistic and the control limits. The data are serially examined over the time length. If at any point in time, the control statistic lies within the control limits, the process is still in control and the procedure continues at the next point of time. If the control statistic exceeds the control limit, the algorithm stops and the process is considered out of control. In univariate SPC, the underlying assumptions are twofold: the normal distribution of the continuous quality characteristic and identically and independently distributed data prevailing for a fairly long period.

Traditional control charts have become obsolete for autocorrelated processes [7]. For such cases, three approaches have been adapted. The first approach concentrates on adjusting parameters and control limits to accommodate the autocorrelation. The second approach adapts a moving window of observations from the univariate process and converts it as a multivariate case to be monitored using multivariate control charts. The third, and the most popular, approach, is a residual-based method, which applies traditional control charts (Shewhart chart, EWMA chart, CUSUM chart) to the residuals after fitting the original data to a time series model (like ARIMA); this method has been extended for multivariate data [8]. Modeling the process's raw data using vector autoregression (VAR) models followed by the application of T^2 control charts to residues has proven to be an effective method for the statistical control of autocorrelated multivariate processes [9]. The traditional MSPC (multivariable SPC) can only be implemented if the joint distribution of all p variables is a multivariate normal distribution. It is also assumed that the variables are independent and identically distributed according to normal distributions $N_P\left(\mu, \sum\right)$ with mean vector ($\mu_1, \mu_2, \mu_3 \ldots \mu_P$) and covariance matrix $\sum$. The Hotelling T^2 chart is the analog of the Shewhart chart in MSPC. Cumulative sum (cusum) and EWMA control charts were also developed in MSPC [10]. Another type of chart used in MSPC, showing variable contributions to SPE (standard prediction error or Q statistic) and T^2 (D statistics) statistics, is contribution plots.

The simplest nonparametric statistical technique uses histograms to detect anomalous univariate data, which is a two-step procedure. The first step involves building of a histogram based on the normal training data. In the second step, there is a check for whether a test datum falls out of the bins of the histogram and is detected as an outlier. The effectiveness of outlier detection depends on the chosen size of the bin in histogram building. For multivariate data, an attribute-wise histogram is constructed. During testing, the anomaly contribution for each attribute value of the test instance is calculated as the height of the bin that contains the attribute value. The feature-wise anomaly contributions are combined to obtain an overall anomaly score for the test instance. *The interaction among the attributes of an outlier may give rise to an exceptional outlier that cannot be detected using histogram-based techniques for multivariate data.* The histogram-based technique for univariate and multivariate data has been utilized by the intrusion detection community [11]. A probability density approximation using a nonparametric method is the Parzen windows estimation using the kernel function. The key steps of anomaly detection here involve the determination of the probability distribution function (pdf) of normal instances and the test instances that fall in the low probability area of the pdf. Ben-Gal [6] has reviewed different nonparametric outlier detection methods, including the distance-based (supervised) and cluster-based techniques (unsupervised). Generally, unsupervised methods can be used either for noise removal or anomaly detection, and supervised methods of anomaly detection are utilized in specific applications.

For continuous processes, the desired values of the important process variables are mostly time-invariant stationary signals and principal component analysis (PCA) provides a useful framework for outlier detection. The latent structure method or dimensionality reduction method using PCA can be used to detect outliers in multivariate data, where characteristic variables are closely correlated. A covariance or correlation matrix is subjected to PCA decomposition. A PCA-based reference model for a process is developed with the data collected in normal operating conditions. Any abnormal operating conditions with respect to the process will be reflected in the new scores when compared with the reference model. The application of PCA in outlier detection is discussed in Chapter 3. Multiway PCA (MPCA) can be used for outlier detection as well as for data reconstruction, especially for batch processes. For time-varying industrial processes with their changing mean, variance, and correlation structure, recursive PCA (RPCA) can be effective for outlier detection. Applications of LDA (linear discriminant analysis) and support vector machines (SVMs) in outlier detection can be posed/exploited as binary or multiclass classification problems and are discussed in Chapters 5 and 6.

Outlier detection techniques have been exploited in numerous applications. Chandola et al. [11] undertook a comprehensive review on outlier detection, in which they conducted a bibliographic survey on various applications of anomaly detection techniques considered by the statistics, computer science, database, data-mining, and machine-learning communities in domains like fraud detection, industrial damage detection/system, public health management, image processing, lexicography, sensor networking, speech recognition, and robotics.

Outlier detection techniques find potential application in process fault detection and quality monitoring. Statistical process control (SPC) is closely related to univariate outlier detection methods. Hodge and Austin [12] suggested that motoring faults on a factory production line adapting outlier detection techniques can ensure product quality control. Chemical and manufacturing processes (e.g., refinery fluid catalytic cracking processes/iron and steel making, water treatment plants, Tennessee Eastman process) deal with multivariate high-dimensional data, and there is a growing demand for better performance and reliability. Fault-tolerant system design is getting increasing attention; therefore this book showcases the developments in classification, clustering, and proximity-based anomaly/attribute/pattern detection algorithms. Rule-based techniques, state estimation, diagraph, modeling and simulation, and neural networks have been deployed for fault detection/quality monitoring in process plants [13]. Most chemical/manufacturing processes are very large and complex. As the number of variables become larger, it may be difficult to collect consistent information regarding faults. In such cases maintenance of a knowledge base (historical database containing normal and abnormal data) becomes increasingly difficult.

There are various issues regarding outlier detection. Sometimes the boundary between normal and abnormal data becomes blurry, making the detection of anomalous behavior difficult. Very often, the decision regarding the degree of deviation with respect to a normal data set is subjective and is specific to the domain of application. Certain processes like catalyst deactivation, equipment aging, sensor and process drifting, and preventive maintenance and cleaning processes showing slow time-varying dynamics; normal behavior evolves with time and the notion of normal data pertaining to the current process is not sufficiently representative for future use. In bioprocesses, microbes sometimes adapt to the severity of anomalous conditions. Here, the notion of abnormal data pertaining to the current situation is not representative enough for the future use of anomaly detection. All such issues remain the motivation behind the development of better techniques and still better algorithms.

1.6.2 Data reconciliation

Data reconciliation is an integral component of monitoring a process that owes its formal roots to an inspiring article by Kuehn and Davidson [14]. Efficient and safe plant operation can only be achieved if the operators are able to monitor all key process variables. Measured process data are not always accurate enough; they are affected by random and gross errors. Random errors follow a Gaussian distribution. Gross error can be generated through (i) instrument systematic bias which is generated due to instrument miscalibration and sensor drift that consistently provides erroneous measurement, (ii) measurement device failure, and (iii) nonrandom events affecting the process, such as process leaks. It is important to estimate the process states with the information provided by the raw measurements using conservation equations and physical and chemical laws (process models) while taking into account the process constraints. *Data reconciliation (DR) is a model-based technique that checks the consistency between the measurements and conservation equations.* The reconciled data is used in achieving optimal process monitoring, control, and optimization; instrument maintenance; equipment performance analysis; and management decision-making. In general, the optimal estimates of process states (corresponding to raw measurements) by DR can be considered a solution to a constrained optimization problem (least squares or maximum likelihood problem), where the measurement error is being treated as an objective function and conservation equations as constraints.

Process variable classification, steady state and dynamic data reconciliation, parameter estimation, and gross error detection are all included as topics in classical data reconciliation literature. Process variables are basically of two types, measured and unmeasured. Measured variables are also of two types, redundant (that can be estimated) and nonredundant (that cannot be estimated otherwise). It is the redundancy of the measurements that make the process of reconciliation possible [15–16]. Jiang et al. [17] presented principles of DR along with illustrations.

Various process-related information is available due to the immense improvement in data collection, compression, storage, and interpretation technologies over the years. Extraction and estimation of meaningful information from the process historical database leading to a data-based model (knowledge based) seems to be a natural and logical choice instead of using first principal models, which may be unreliable for complex nonlinear processes and suffer from uncertainty. Currently, high-dimensional data from sensor arrays have various applications. It is not possible to render physical attributes to this high-dimensional data. Multivariate statistical modeling can be used for finding the conditioned/reconciled data, which may find application in process monitoring for environmentally benign, safe, and economic operation, and production specifications. The most commonly used model in this regard is principal component analysis (PCA). There is synergy between PCA and DR [18] that uses a PCA-based energy monitoring system in the Tennessee Eastman process with a combination of PCA tools, data reconciliation, and flow-sheet simulation. PCA is an orthogonal transform, which when applied to high-dimensional and correlated process data collected under normal operating conditions or steady state gets abridged to uncorrelated process data in fewer dimensions without losing meaningful information.

This book emphasizes the multivariate statistical models for reconciliation/conditioning of data instead of classical DR methodologies. Let us find the synergy between PCA and classical DR. A DR model is defined by material and energy balance equations, usually expressed by a system of linear equations, $A\bar{\bar{X}} = b$ (A is the incidence matrix, the X vector contains, variables, and b is the source vector). The reconciled data $\bar{\bar{X}}$ is formulated by the following equation using classical data reconciliation:

$$\bar{\bar{X}} = \left(I - VA^T\left(AVA^T\right)^{-1}A\right)X + VA^T\left(AVA^T\right)^{-1}b \tag{1.26}$$

where X represents the measured variables, I is the identity matrix, and V is the variance matrix of error. The PCA model is derived from a covariance matrix of normalized data pairs. The covariance matrix (C) is decomposed using eigenvector decomposition ($C = U^T \wedge V$), where the columns of U are eigenvectors of the covariance matrix, the diagonal elements of $\wedge$ are the eigenvalues, and the columns of V represent the right singular vectors. According to the number of significant principal components (p), the first p columns of eigenvectors are selected, and the reconciled measurement is formulated as $\bar{X}_{PCA} = U_P^T X$. For nonstationary processes and slow time-varying processes, where the covariant structure and correlation among the variables changes, the projection matrix (loading vector) does change. (This is unlike PCA-based projection, which is applied mainly for stationary processes.) It may be categorized under adaptive data reconciliation using moving window-based PCA and recursive PCA, where the covariance/correlation matrix is constantly updated. Following is an illustration of linear steady-state data reconciliation, where all the flow rates are measured in a network. The data are re-estimated using first principal and PCA-based models.

1.6.3 *Data smoothing and filtering*

1.6.3.1 *Smoothing signal*

There is a need to correlate the experimentally obtained data with a suitable framework. The proposed correlation can be interpolated to estimate missing data; it can detect outliers

ILLUSTRATION 1.17

Consider the following water supply network (Figure 1.6). Seven flow rates are measured seven times (data, Z.mat). The measured flow rates and reconciled flow rates as per classical DR and PCA models, along with their second-order statistics (standard deviation), are tabulated below.

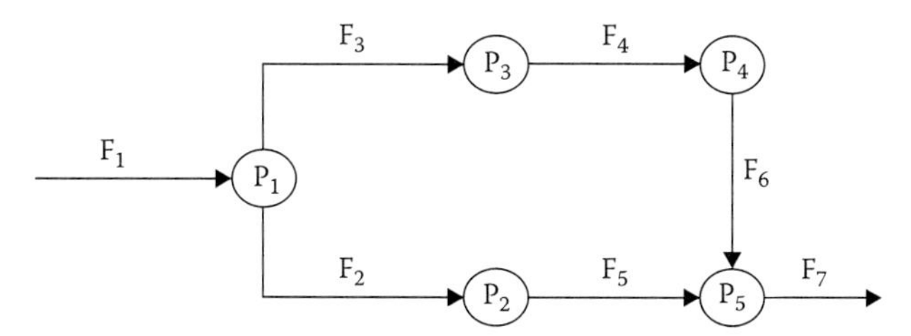

Figure 1.6 Water supply network.

Stream identification no.	Measured flow rate, F_i	Standard deviation in raw F_i	Reconciled F_i using classical DR	Standard deviation in DR recon. F_i	Reconciled F_i using PCA	Standard deviation in PCA recon. F_i
1	120	0.327	119.0	0.270	144.4	0.202
2	80	0.767	76.3	0.338	94.1	0.587
3	45	0.665	42.7	0.311	54.6	0.697
4	45	0.630	42.7	0.311	53.3	0.469
5	76	0.612	76.2	0.339	89.7	0.553
6	40	0.641	42.7	0.311	48.5	0.676
7	110	0.784	119.0	0.270	131.1	0.735

In classical DR, we assume the variables are uncorrelated and the variance matrix V has variances (standard deviation square) along the principal diagonal when all other off-diagonal elements are zero. The vector of raw flow-rate measurements is $X = [F_1 \;\; F_2 \;\; F_3 \; F_4 \; F_5 \; F_6 \; F_7]'$. The mass balance equation around each node (P) is as follows:

For P_1: $F_1 - F_2 - F_3 = 0$
For P_2: $F_2 - F_5 = 0$
For P_3: $F_3 - F_4 = 0$
For P_4: $F_4 - F_6 = 0$
For P_5: $F_5 - F_6 - F_7 = 0$

Process model constraints can be written as $A\bar{\bar{X}} = 0$, where A is the incidence matrix of dimension 5×7. In A, each row represents a node, and each column represents a flow stream. Each element in A is either +1, −1 or 0, depending on whether the corresponding flow is an input stream, an output stream, or not associated with this node. The reconciliation of flow rate in water network problem becomes

$$\text{Minimizing } J(\bar{X}) = (X - \bar{X})^T V^{-1} (X - \bar{X}) \text{ subjected to } A\bar{\bar{X}} = 0 \qquad (1.27)$$

Equation (1.27) can be solved using Lagrange's multiplier. The unconstrained objective function becomes

$$\text{Minimizing } J(\bar{X}) = (X - \bar{X})^T V^{-1} (X - \bar{X}) - 2\lambda^T A\bar{X} \qquad (1.28)$$

where $\lambda^T = [\lambda_1, \lambda_2, \lambda_3, \lambda_4, \lambda_5]$.

The necessary conditions to obtain the minima of (1.28) are

$$\partial J / \partial \bar{X} = -2V^{-1}(X - \bar{X}) - 2A^T\lambda = 0 \qquad (1.29)$$

$$\partial J / \partial \lambda = A\bar{X} = 0 \qquad (1.30)$$

Premultiplying both terms in Equation (1.29) by a covariance matrix results in

$$(X - \bar{X}) + VA^T\lambda = 0 \qquad (1.31)$$

Premultiplying each term of Equation (1.31) by the incidence matrix A and applying $A\bar{\bar{X}} = 0$:

$$AX + AVA^T\lambda = 0 \qquad (1.32)$$

$$\lambda = -(AVA^T)^{-1} AX \qquad (1.33)$$

Putting λ in Equation 1.31 provides $\bar{X} = X - VA^T (AVA^T)^{-1} AX$ as the reconciled value. The covariance matrix of the reconciled flow rates is determined to check their statistical fitness, and it is found to be a symmetric matrix, having variances along the diagonal and off-diagonal elements as correlations. The reconciled flow rates have less standard deviation in comparison to raw measurement; hence, they are more precise.

In a PCA-based model, a raw data matrix (Z) consisting of seven flow rate measurements for seven times (7×7 matrix; Z.mat) is used. The MATLAB® function princomp is used to find the eigenvector matrix (only the first principal component was chosen) followed by the determination of the reconciled value of flow rates as $\left(U_P^T X.\right)$ The PCA reconciled values show standard deviation in between the classical DR reconciled flow rates and raw measurements. The MATLAB® code (Dr11.m) is as follows:

```
Dr11.m
%DR1 Problem on cooling water network,
X = [120.0; 80.0; 45.0; 45.0; 76.0; 40.6; 110.0]; % measured
flowrate in the network
% variance in raw measurement
Vv = [0.1069 0 0 0 0 0 0; 0 0.5886 0 0 0 0 0; 0 0 0.442 0 0 0 0; 0 0
0 0.397 0 0 0; 0 0 0 0 0.374 0 0; 0 0 0 0 0 0.411 0; 0 0 0 0 0 0 0.615];
A=[1 -1 -1 0 0 0 0; 0 1 0 0 -1 0 0; 0 0 1 -1 0 0 0; 0 0 0 1 0 -1 0;
0 0 0 0 1 1 -1]; % Incidence matrix
b=inv(A*Vv*A');
Xhat=X-Vv*A'*b*A*X % estimated X (Flowrate)
I=eye(7);
W=(I-Vv*A'*b*A) % optimal projection of process data into a
multivariate model
C=W*Vv*W' % statistical fitness of the estimated flow rates in terms
of covariannce. Square-roots of the diagonal elements of the C
matrix provides the standard deviation of the reconciled flow rates
>>>>>>>>>>>>>>>>>>>>>>>>>>>>>>>>>>>>>>>>>>>>>>>>>>>>>>>>>>
% DR2 problem using PCA
% network Flow rate data in file Z.mat (A data set of water flow
rate measurements by 7 times at each of the 7 nodes, hence resulting
in a 7×7 matrix. One can simulate the Z.mat by incorporating the
standard deviation of measurement at each of the nodes reported in
the experimental data, as have been done by the authors]
load Z
[COEFF, SCORE, latent]=princomp(Z);
ZZhat=COEFF(:,1)'*Z % reconciled flowrate
```

and remove noise infused in the raw signal/data. This can be accomplished by fitting the data with a suitable parametric model. Alternatively, nonparametric models can be used for data analysis. The term nonparametric refers to the flexible form of the regression curve. There are other notions of nonparametric statistics, which refer mostly to distribution-free methods. In the present context, however, neither the error distribution nor the function form of the mean function is prespecified. The flexibility of a method is extremely helpful for preliminary and exploratory data analysis [19]. Filtering (averaging), as well as local regression, smooths noisy data. The smoothing process attempts to estimate the average of the distribution of raw signal. The estimation is based on a specified number of adjacent response values. It is expected that the noise components in each measurement cancel out in the averaging process.

First consider a linear model with one predictor:

$$y = f(x) + \epsilon_i \tag{1.34}$$

We want to estimate f, the trend or smoother, assuming the data are ordered so that $x_1 < x_2 < x_3 ..... < x_n$. In the case of multiple observations at a given x_i, a weight w_i needs to be introduced.

There are different methods of smoothing: moving average smoother, running median smoother, kernel smoothers, and local regression-based smoothing (use the MATLAB® Curve Fitting Toolbox). The smoothing process is considered local because each smoothed value is determined by neighboring data points defined within the span. Savitzky-Golay, a least squares smoothing regressor, can be thought of as a generalized moving average. The smoother coefficients are determined by performing an unweighted linear least squares fit using a polynomial of a given degree. The Lowess and Loess methods use locally weighted linear regression to smooth data.

1.6.3.2 *Filtering signal*

When a random signal passes through a deterministic system, its statistical properties get modified. A filter can be thought of as a deterministic system, where the input is a random signal, and the output is also a random signal with desired statistical properties. A linear time-invariant filter finds application in data-based monitoring device development, process control and random measurement noise. Depending on the requirements of low-frequency content and high-frequency content of signals, low-pass or high-pass filters

ILLUSTRATION 1.18

APPLICATION OF LOCAL REGRESSION-BASED SMOOTHING FILTERS

ECG signals (wave11.txt and wave12.txt, adapted from the Physionet database), each with the dimensions 230 × 2, are smoothed. The raw and smoothed signals are presented in Figure 1.7. The MATLAB® code Smooth_wave.m is as follows:

```
clc
clear all
fname=input('Enter raw data ecg file:','s')
fname1=strcat(fname,'.txt')
pause
X=load(fname1,'-ascii')
Y=[X(:,1) X(:,2)];
N=size(Y)
m=input('Enter options for method:1 for moving average , 2 for
lowess , 3 for loess , 4 for sgolay , 5 for rlowess')
if (m==1)
  method= 'moving'
end
if (m==2)
  method='lowess'
end
if (m==3)
  method='loess'
end
if (m==4)
```

```
    method='sgolay'
  end
  if (m==5)
    method='rlowess'
  end
  % Smoothing of raw ECG DATA
  figure
  subplot (2, 1, 1);
  YY = smooth(Y(:,2),7, method);
  YY=[Y(:,1),YY(:)]
  plot(Y(:,1),Y(:,2),'r', YY(:,1),YY(:,2),'b')
  grid on
  fname2=strcat(fname,'_filtered.txt')
  save(fname2,'YY','-ascii')
  pause
  %% only smoothed data
  subplot (2, 1, 2);
  plot(YY(:,1) , YY(:,2))
  grid on
  %%% Data files required: wave11.text, wave12.txt
```

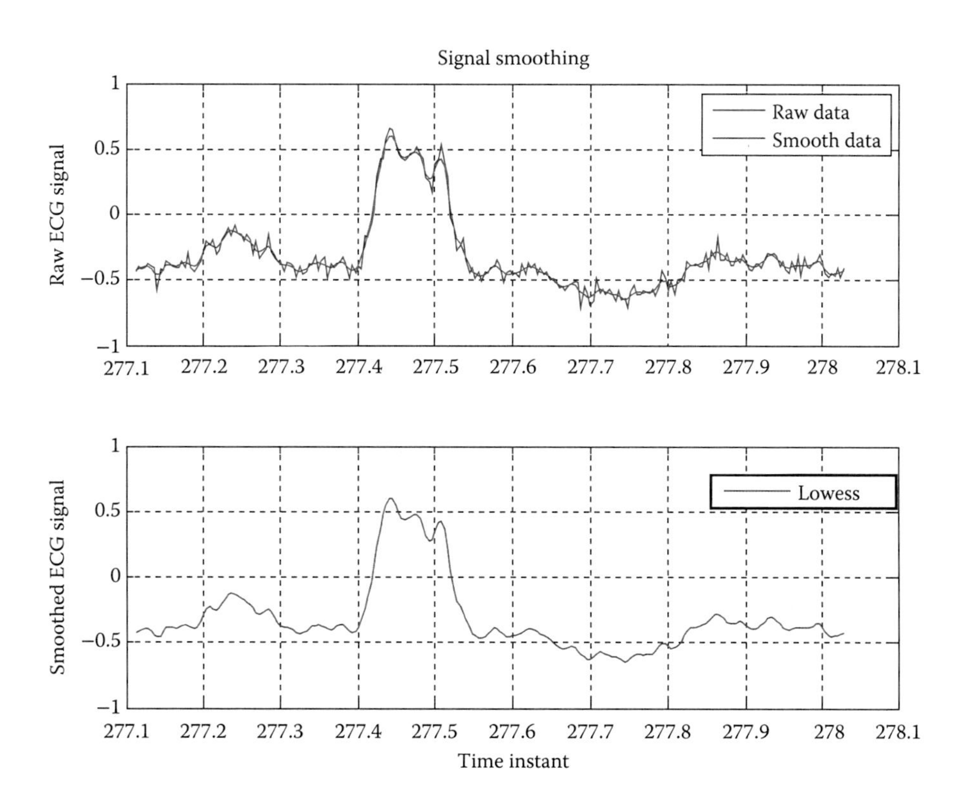

Figure 1.7 Smoothing ECG signals.

are designed. Simple noise-filtering tools have been developed by using time series polynomial model representations like MA, AR, ARMA, etc. A filtered signal $y(n)$ at sampling interval n is related to a noisy signal $x(n)$ in the following way:

$$y(n) = b_1x(n) + b_2x(n-1) + \ldots + b_{nb+1}x(n-nb) - a_2y(n-1)\ldots - a_{na}(n-na) \tag{1.35}$$

where na and nb are the lengths of past sampling time spans of y and x signals, respectively. Assuming zero initial conditions, Equation 1.35 is the simple time domain representation of a digital filter. If $nb = na = 2$, Equation 1.35 takes the following form for an arbitrary $y(5)$:

$$y(5) = b_1x(5) + b_2x(4) + b_3x(3) - a_2y(4) - a_3y(3) \tag{1.36}$$

To compute the estimated value $y(5)$ for sensor reading $x(5)$, the weighted sum of the three most recent sensor readings and two most recent estimated (filtered) values are used. When the filters are generated using the sensor readings only (all $a_i = 0$) and the use of previous estimates only (all $b_i = 0$), they are called the MA and AR filters, respectively. When both AR and MA terms are included, the resulting filter is ARMA or autoregressive moving average filter. The least squares estimation of filter coefficients estimates smooth data. The MATLAB system identification toolbox can be used for identifying (characterizing) the unknown data and smoothing it. Based on the prediction error method (which minimizes the error between the actual and predicted output data) it is possible to construct this type of linear dynamic model for any structure. Because of their limited range of applicability, one has to adapt process model parameters to the altered process condition. Details of the time-series identification are not included here because they are beyond the scope of the book.

1.6.4 *Transform and transformation*

The application of transforms on a signal extracts additional information from the signal that is not readily available in its pristine form. A raw signal after application of transforms becomes a processed signal. A time versus amplitude representation of the signal is not always the most desired representation. In many cases, the most illustrative information is hidden in the frequency content of the signal as expressed by the use of the fast Fourier

ILLUSTRATION 1.19

APPLICATION OF FILTERS

A SISO LTI model is simulated with arbitrary input (using a pseudorandom, binary signal). The output data (equivalent to raw plant data) is filtered through armax/arx/oe. The raw plant data is compared with filtered data (Figure 1.8). The MATLAB® code arxTest_1_demo.m is as follows:

```
arxTest_1_demo.m, Developed December 2010
%gp33 is a SISO LTI model
%lsim Simulate LTI model responses to arbitrary inputs,lsim(sys,u,t)
% simulated data equivalent to plant data, which is modelled with arx,
% armax, oe
```

```
u3=idinput(1000,'prbs',[0 1],[-2,2]); % u =
idinput(N,type,band,levels), Generate input signals
%N= no of inputs, band=[low high],levels = [minu, maxu]: signal u
will always be between minu and maxu
%type = 'rgs': Gives a random, Gaussian signal.
%type = 'rbs': Gives a random, binary signal. This is the default.
%type = 'prbs': Gives a pseudorandom, binary signal.
%type = 'sine': Gives a signal that is a sum of sinusoids.
t = 0:1:(length(u3)-1);
s=tf('s');
gp33=0.87*(11.61*s+1)*exp(-s)/((3.89*s+1)*(18.8*s+1));
Input1 = [u3];
Output1=lsim(gp33,Input1,t);
%lsim Simulate LTI model responses to arbitrary inputs,lsim(sys,u,t),
%Linear time invariant models (LTI)
data = iddata(Output1,Input1);%Class for storing time-domain and
frequency-domain data
%data = iddata(y,u,Ts)
na = 2*ones(1,1);
nb = 2*ones(1,1);
nc = 2*ones(1,1);
nk = ones(1,1)
% Three types of model; ARMAX, ARX, oe
% ArM = ARX(data,[na nb nk]);
% ArM1=oe(data,[na nb nk])
ArM2=armax(data,[na nb nc nk])
ypred=idsim(Input1,ArM2);
% hold on
plot(ypred,'-b')
pause
plot(Output1,'r')
pause
% plot(Output1,ypred,'o')
plot(t,Output1,'or',t,ypred,'-b')
```

transform, Z-transform, and wavelet transform. These transforms allow further insight into the data. The words transformation and transform are often used interchangeably. However, the semantic meanings of the two words are different. Synonyms of the word transformation are alteration, evolution, change, and reconfiguration. Transformation is the process that alters the data, leaving them unreduced in the same domain (like the Box-Cox transformation, 1964) [20], and the processes that profoundly change the nature, structure, domain, and dimension of data should be called transforms (e.g., Wigner-Ville, Choi-Williams, short-time Fourier transform).

It is not an exaggeration to say that statistics is based on various transformed data. Basic statistical précises such as the sample mean, variance, z-scores, histograms, etc., are all transformed data. Some more advanced summaries such as principal components, linear discriminants, and support vectors are also examples of transforms. This book exploits several statistical transforms that provide an interface between available data and various monitoring appliance design. In many cases, the N sample values of a given signal (N components of the signal vector) are subjected to an orthogonal transform like PCA to get

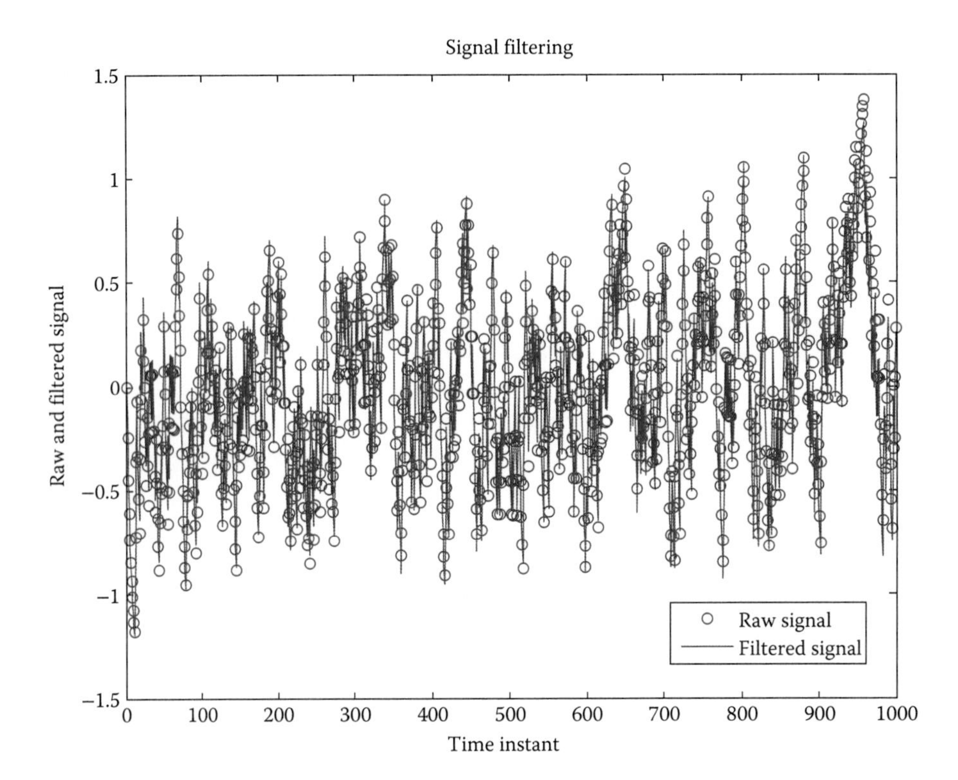

Figure 1.8 Filtering signal of the SISO LTI process.

another set of N components that can carry the same information as that of original signal. After the orthogonal transform, the signal components are completely decorrelated, and the energy/information contained in the signal is maximally concentrated along a small number of components. Transforms in statistics, in general, are utilized for several reasons, since transformed data (i) are easy to store (compressed), report, and analyze, (ii) comply better with a specific modeling framework, and (iii) allow for additional insight to the physics of the underlying processes not available in the domain of nontransformed data. There are various types of transforms possible in mathematics:

Integral transforms: Hilbert transform, Hankel transform, Abel transform, Laplace transform, Wavelet transform
Discrete transforms: Discrete Fourier transform (DFT), Fast Fourier transform
Discrete-time transforms: Discrete-time Fourier transform (DTFT), Z-transform
Data-dependent transforms: Karhunen–Loève transform

Box and Cox [20] introduced a family of transformations, indexed by real parameter λ, applicable to positive data $X_1, \ldots, X_n$:

$$Y_i = \begin{cases} \dfrac{X_i^{\lambda} - 1}{\lambda} & \text{where } \lambda \neq 0 \\ \log X_i & \text{where } \lambda = 0 \end{cases} \tag{1.37}$$

This transformation is mostly applied to responses in linear models exhibiting nonnormality and/or heteroscedasticity. For properly selected λ, data $Y_1, \ldots, Y_n$ after transformation appear to be more normal and amenable to standard modeling techniques. The parameter λ is selected by maximizing the log-likelihood,

$$(\lambda - 1)\sum_{i=1}^{n} \log X_i - \frac{n}{2} log\left[\frac{1}{n}\sum_{i=1}^{n}\left(Y_i - \overline{Y_i}\right)^2\right] \tag{1.38}$$

where Y_i is given in Equation 1.37 and $\overline{Y_i} = \frac{1}{n}\sum_{i=1}^{n} Y_i$.

Most often the dynamic phenomena to be studied are modeled by the differential equations (either a lumped parameter or discretized system), and they are either initial value problems or the solutions are constrained by boundary conditions. To find their explicit solutions in the realm of mathematics, make use of the Laplace transform (LT). The LT, when applied to linearized differential equations (in a continuous time domain), converts them into a set of algebraic equations in the s domain (s is the variable in the complex s-plane: $(s = a + jb)$, and hence, facilitates the formation of a transfer function or the input-output form of the process model. The inverse Laplace transform ensures that the model solution is reverted to the time domain. The discrete counterpart of the Laplace transform is the z-transformation, which is suitable for computerized implementation of a control system and discrete simulation.

Time frequency analysis (TFA) is concerned with simultaneous analysis of a signal in the time and frequency domains. TFA tools are commonly used in signal processing for the analysis of multiscale systems (e.g., seismic data analysis, speech processing, multiscale filtering, etc.). The widely accepted methods in time frequency analysis are short-time Fourier transform (STFT), Winger-Ville distributions, and wavelet transforms. In Chapter 7, electrocardiogram (ECG) modeling, or synthesis, is presented using the Fourier model for which raw ECG data has been preprocessed using a wavelet transform. Therefore, we have decided to provide a brief introduction on FT and wavelet transforms here.

Fourier analysis was originally concerned with representation and analysis of periodic phenomena, via Fourier *series*, and later it was extended to nonperiodic phenomena via the Fourier *transform*. In fact, the journey from Fourier series to the Fourier transform is accomplished merely by considering nonperiodic phenomena as a limiting case of periodic phenomena while the period tends to infinity. (The Fourier transform is an extension of the Fourier series that results when the period of the represented function is extended and allowed to approach infinity.) A discrete set of frequencies in the periodic case becomes a continuum of frequencies in the nonperiodic case. Thus, it gives rise to the concept of a *spectrum*, and with it evolves the most important principle of the subject: every signal has a spectrum. One can fully analyze the signal either in the time (or spatial) or frequency domain.

The Fourier transform is called the *frequency domain representation* of the original signal and is used as a tool to analyze the frequency components of the signal [21]. The Fourier transform of a function of time itself is a complex-valued function of frequency, whose absolute value represents the amount of the frequency present in the original function, and whose complex argument is the phase offset of the basic sinusoid in that frequency. Motivation for the Fourier transform comes from the study of Fourier series. In the study of

Fourier series, complicated but periodic functions are written as the sum of simple waves mathematically represented by sines and cosines (discussed later in Chapter 7). Due to the properties of sine and cosine, it is possible to recover the amplitude of each wave in a Fourier series using an integral. In many cases, it is desirable to use Euler's formula (which states that $e^{2\pi i\theta} = \cos(2\pi\theta) + i\sin(2\pi\theta)$) to write Fourier series in terms of the basic waves $e^{2\pi i\theta}$. The discrete time Fourier transform (DTFT) is the member of the Fourier transform family that operates on periodic, discrete signals. The DTFT itself is a continuous function of frequency, but discrete samples of it can be readily calculated via the discrete Fourier transform (DFT).

Given a sequence of N samples $f(x)$ indexed by $x = 0, 1, \ldots N-1$, the discrete Fourier transform (DFT) is defined as $F(u)$ (where $u = 0, 1, \ldots N-1$):

$$F(u) = \frac{1}{N}\sum_{x=0}^{N-1} f(x)e^{\frac{-j2\pi ux}{N}}, \; u = 0, 1, \ldots N-1 \tag{1.39}$$

$F(u)$ are often called Fourier coefficients or harmonics, the independent variable x represents time, and u represents the frequency. Under suitable conditions $f(x)$ is obtained from F via the inverse transform (IDFT):

$$f(x) = \sum_{u=0}^{N-1} F(u)e^{\frac{j2\pi ux}{N}}, \; x = 0, 1, \ldots N-1 \tag{1.40}$$

Conventionally, the two elements of the sequence $f(x)$ and $F(u)$ are referred to as time domain and frequency domain data, respectively. In general, both $f(x)$ and $F(u)$ are complex.

Let us consider the following function:

$$f_1(t) = \cos(2\pi.5.t) + \cos(2\pi.25.t) + \cos(2\pi.50.t) \tag{1.41}$$

The function $f_1(t)$ and the corresponding $F_1(u)$ are shown below (Figure 1.9), which provides excellent localization in the frequency domain but poor localization in the time domain.

There are limitations of FT:

- Cannot provide simultaneous time and frequency localization
- Not very useful for analyzing time-variant, non-stationary signals
- Not appropriate for representing discontinuities

STFT is a well-known technique in signal processing to analyze nonstationary signals. The short-time Fourier transform (STFT) uses a sliding window to find a spectrogram, which provides information on both time and frequency (Figure 1.10). STFT segments the signal into narrow time intervals (narrow enough to be considered stationary) and takes the Fourier transform of each segment. The window/FFT size depends on the FFT's resolution. Each FT provides the spectral information of a separate time slice of the signal, providing simultaneous time and frequency information.

The definition of STFT is as follows:

$$\text{STFT}_f^u(t', u) = \int \left[f(t).W(t-t') \right].e^{-j2\pi ut}dt = f(t').e^{-jut'} \tag{1.42}$$

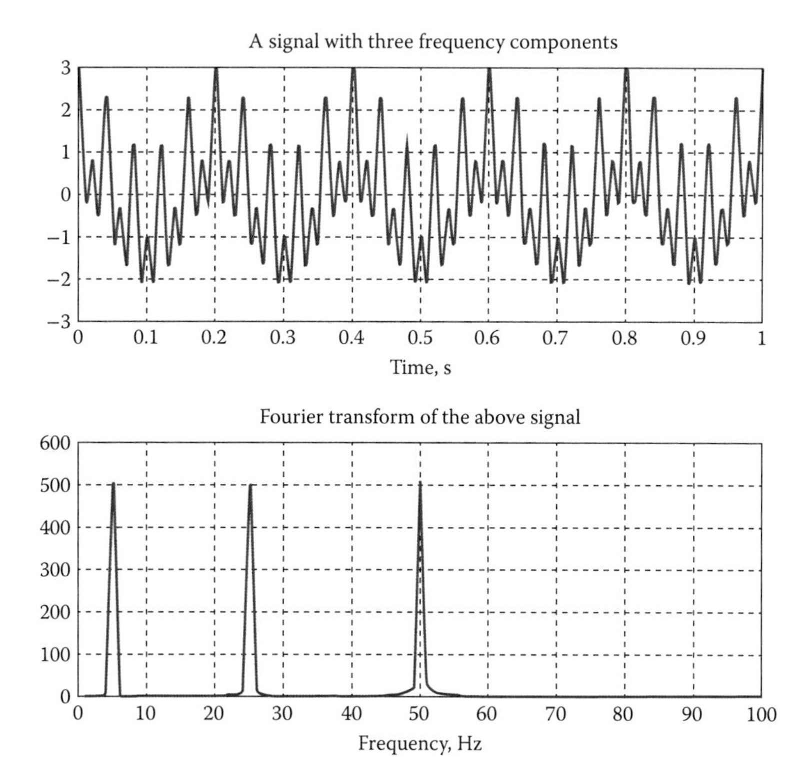

Figure 1.9 Signal f_1 in time and frequency domain.

where $f(t)$ = signal to be analyzed, t' = time parameter, u=frequency parameter, $\mathrm{STFT}_f^u(t', u)$ = STFT of $f(t)$ computed for each window centered at $t = t'$, and $W(t - t')$ = window function centered at $t = t'$

$W(t) \to 1$: STFT tends to FT, providing excellent frequency localization, but no time localization.

$W(t) \to \delta(t)$: Results in a time signal with a phase factor, providing excellent time localization without frequency localization.

However, another problem still persists: the length of the window limits the resolution in frequency. A wavelet transform with varying windows seems to be a solution to the problem. The wavelet transform is a tool for carving up functions, operators, or data into components of different frequency, allowing one to study each component separately

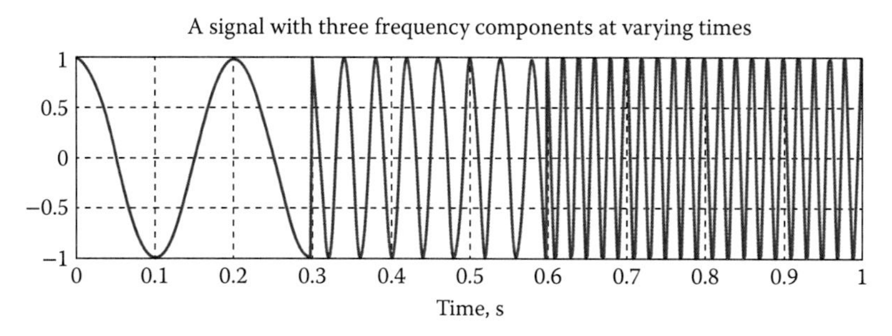

Figure 1.10 Signal with varying frequency in varying time frame.

[22]. The basic idea of the wavelet transform is to represent any arbitrary function $f(t)$ as a superposition of a set of such wavelets or basis functions. These basis functions, or baby wavelets, are obtained from a single prototype wavelet called the mother wavelet, by dilations or contractions (scaling) and translations (shifts). Wavelets are widely used to remove the noise from signals by extracting the low-frequency content and removing the high-frequency content above a threshold value. The denoised and compressed signal is obtained by reconstructing the signal by applying an inverse wavelet transform to the scaling and threshold wavelet coefficients.

Wavelets were developed as an alternative to STFT for characterizing nonstationary signals. Wavelets provide an opportunity to localize events in both time and frequency by using windows of different lengths while in STFT the window length is fixed.

The integral wavelet transform can be represented as

$$W\left(a,b=\int x(t)\psi_{(a,b)}(t)dt\right) \tag{1.43}$$

$$\psi_{(a,b)}(t)=\frac{1}{\sqrt{a}}\psi\left(\frac{t-b}{a}\right) \tag{1.44}$$

where ψ represents a mother wavelet, $x(t)$ is the original signal, and a and b are the scaled and translation parameters, respectively. $\frac{1}{\sqrt{a}}$ is used to ensure that the energy of the scaled and translated signals are the same as the mother wavelet. A scaled parameter specifies the location in frequency domain and a translation parameter determines the location in the time domain.

The discrete wavelet transform is used to reduce the computational burden without losing significant information. To obtain the discretized wavelet transform, scale and translation parameters are discretized as $a=2^{-j}$, which is called the binary dilation or dyadic dilation, and $b=k2^{-j}$ is the binary or dyadic position. The discretized wavelet function becomes

$$\psi_{jk}(t)=2^{j/2}\psi\left(2^{j}t-k\right) \tag{1.45}$$

Any square-integrable signal (a square-integrable function, also called a quadratically-integrable function, is a real or complex-valued measurable function for which the integral of the square of the absolute value is finite: https://en.wikipedia.org/wiki/Absolute_value).

Thus, if $\int_{-\infty}^{\infty}\left|f(x)\right|^{2}dx\neq\infty$ can be represented by successively projecting it on scaling and wavelet functions. The scaling function is shown as

$$\phi_{jk}(t)=2^{j/2}\phi\left(2^{j}t-k\right) \tag{1.46}$$

The scaling coefficients α_{jk} (Equation 1.47), which are the low-frequency content of the signal, are obtained by the inner product of the signal with the scaling function. The wavelet coefficients d_{jk} (Equation 1.48), which are the high-frequency content of the signal, are obtained by the inner product of the original signal with the wavelet function ψ.

$$\alpha_{jk} = \int x(t)\phi_{jk}(t)dt \tag{1.47}$$

$$d_{jk} = \int x(t)\psi_{jk}(t)dt \tag{1.48}$$

The discrete wavelet transform (DWT) analyzes the signal at different frequency bands with different resolutions by decomposing the signal into a coarse approximation and detailed information. The input signal $x(n)$ is filtered by a low-pass filter $L(n)$ and a high-pass filter $H(n)$ in parallel, obtaining the projection of the original signal onto a wavelet function and a scaling function. Dyadic downsampling is applied to the filtered signal by taking every other coefficient of the filtered output. Therefore, the signal can be subsampled by two simply by discarding every other sample. This decomposition halves the time resolution since only half the number of samples now characterizes the entire signal. However, this operation doubles the frequency resolution, since the frequency band of the signal now spans only half the previous frequency band, effectively reducing the uncertainty in the frequency by half. The same procedure is repeated for the next scale to the downsampled output of $L(n)$, since the low-pass output includes most of the original signal content. α_j and d_j at any scale can be obtained as follows:

$$\alpha_j = L\alpha_{j-1} \text{ and } d_j = Hd_{j-1} \tag{1.49}$$

The original signal can be computed recursively by adding the wavelet coefficients at each scale and the scaling coefficients at the last scale. Plenty of books and research articles are available on wavelet (analysis) transforms [22,23].

References

1. F. Yates, K. Mather (1963). Ronald Aylmer Fisher. *Biographical Memoirs of Fellows of the Royal Society*, vol. 9, 91–120.
2. R. Fisher (1926). The arrangement of field experiments. *Journal of the Ministry of Agriculture of Great Britain*, vol. 33, 503–513.
3. R. Johnson (1992). *Applied Multivariate Statistical Analysis*. Upper Saddle River, NJ: Pearson Prentice Hall.
4. W.A. Shewhart (1931/1980). *Economic Control of Quality of Manufactured Product*. Milwaukee, WI: ASQC Press.
5. F.E. Grubbs (1969). Procedures for detecting outlying observations in samples. *Technometrics*, vol. 11, 1–21.
6. I. Ben-Gal (2005). Outlier detection. In *Data Mining and Knowledge Discovery Handbook: A Complete Guide for Practitioners and Researchers*, O. Maimon, L. Rockach (Eds.). USA: Springer.
7. L.C. Alwan, H.V. Roberts (1988). Time-series modelling for statistical process control. *Journal of Business & Economic Statistics*, vol. 6(1), 87–95.
8. X.H. Amherst (2010). Topics in multivariate time series analysis: Statistical control, dimension reduction visualization and their business applications. Ph.D Dissertation, University of Massachusetts.

9. X. Pan, J. Jarrett (2014). The multivariate EWMA model and health care monitoring. *International Journal of Economics and Management Sciences*, vol. 3(2), 176–183.
10. D.C. Montgomery (2008). *Introduction to Statistical Quality Control*. New York, NY: John Wiley & Sons.
11. V. Chandola, A. Banerjee, V. Kumar (2009). Anomaly detection: A survey. *ACM Computing Surveys*, vol. 41(3), 1–72.
12. V. Hodge, J. Austin (2004). A survey of outlier detection methodologies. *Artificial Intelligence Review*, vol. 22(2), 85–126.
13. V. Venkatasubramanian, R. Rengaswamy, S.N. Kavuri, K. Yin (2003). A review of process fault detection and diagnosis: Part III: Process history based methods. *Computers and Chemical Engineering*, vol. 27, 327–346.
14. D.R. Kuehn, H. Davidson (1961). Computer control II. Mathematics of control. *Chemeical Engineering Progress*, vol. 57, 44–47.
15. M.C. Romagnoli, M. Sanchez (1999). *Data Processing and Reconciliation for Chemical Process Operations*. Cambridge, MA: Academic Press.
16. M. Bagajewicz (2001). Data processing and reconciliation in chemical process operations (book review). *AIChE. Journal*, vol. 47(4), 962–963.
17. X. Jiang, P. Liu, Z. Li (2012). A data reconciliation based approach to accuracy enhancement of operational data in power plants. *Chemical Engineering Transactions*, vol. 35, 1213–1218.
18. B. Farsang, S. Nemeth, J. Abonyi (2014). Synergy between data reconciliation and principal component analysis in energy monitoring. *Chemical Engineering Transactions*, vol. 39, 721–726.
19. W. Hardle (1991). *Smoothing Techniques: With Implementation in S*. New York, NY: Springer-Verlag.
20. G.E.P Box, D.R. Cox (1964). An analysis of transformations. *Journal of the Royal Statistical Society, Series B (Methodological)*, vol. 26(2), 211–252.
21. C. Eleanor (2008). *Discrete and Continuous Fourier Transforms: Analysis, Applications and Fast Algorithms*. Boca Raton, FL: CRC Press.
22. M. Stephane (2008). *A Wavelet Tour of Signal Processing: The Sparse Way*. Cambridge, MA: Academic Press.
23. I. Daubechies (1992). *Ten Lectures on Wavelets*. Philadelphia, PA: SIAM.

chapter two

Chemometric techniques: Theoretical postulations

This chapter is devoted to the theoretical postulations of chemometric techniques. Out of various types of chemometric techniques available, we concentrate only on selected multivariate statistical techniques to be utilized for the process operating condition (normal) classification, detection of abnormal process operating conditions (fault), beverage quality monitoring, and patient care monitoring proposed in subsequent chapters of the book. It may be noted that presenting theoretical postulations on multivariate statistics is not the focus of this chapter. Multivariate statistical/chemometric techniques have been adapted in requisite and appropriate algorithm development. Each of the techniques is accompanied by the requisite MATLAB® code/pseudocode/algorithms to implement it. The mathematical postulations, along with the mention of a few seminal contributions regarding the evolution of those techniques, are also included in this chapter.

2.1 Chemometrics

In his seminal article of 1995, Wold (1995) described chemometrics as an art of extracting chemically relevant information from data produced in chemical experiments using the tools of mathematics and statistics [1]. And most importantly, he viewed that chemometrics should continue to be motivated by the unraveling of chemical problems, rather than by using "method" development to unravel them. After more than 20 years, his ideas on chemometrics still stand, with updates and inclusions regarding its definition, formalism, treatment, and application in accordance with present needs and the experience of the foregoing years.

The term *chemometrics* was introduced by Wold (1995) [1]. The foundation of the International Chemometrics Society in 1974 led to the first description of the discipline. In the following years, several conference series were organized (including Computer Application in Analytics (COMPANA), Computer-Based Analytical Chemistry (COBAC) and Chemometrics in Analytical Chemistry (CAC)) on chemometrics. Some novel chemometric journals such as the *Journal of Chemometrics* (Wiley) and *Chemometrics and Intelligent Laboratory Systems* (Elsevier) are still dedicated towards the publication of research related to chemometrics. The earlier definition of chemometrics is as follows:

> The chemical discipline that uses mathematical and statistical methods (a) to design or select optimal measurement procedures and experiments and (b) to provide maximum chemical information by analyzing chemical data [1].

The discipline of chemometrics was originated in chemistry and found its application in NIR (near infrared spectroscopy) calibration, UV-vis calibration, chromatographic pattern analysis, QSAR (quantitative structure–activity relationship) models, and molecular mechanics. The giant pool of data generated by modern analytical instrumentation is one reason that analytical chemists, in particular, were drawn to chemometric methods. The availability of personal computers by the beginning of 1980s, and the phenomenal improvements in data acquisition, storage, processing, and interpretation, have catalyzed the growth of chemometrics as a discipline.

With time, chemometrics has expanded beyond unraveling chemical problems; it is also used to analyze and interpret various sensory/multi-sensory data related to biology, astronomy, forensic sciences, clinical data, and many more fields. The method of unraveling those data, started to become somewhat generic in nature. There have been specific ways of familiarizing chemometrics. To date, chemometrics is evolving (as a discipline) out of a combined effort to cater to the needs of analytical chemists; multivariate statistics applications; biomedical signal processing, health informatics, fault detection, and diagnosis. This has also lead to fault-tolerant system design, machine-learning-related research, and a variety of new interfaces being developed (like the interface between bio-informatics and chemometrics, courtesy of analytical data provided by spectroscopy, chromatography, and gene expression profiles or protein sequences). These various efforts are moving towards being consolidated under the single platform of chemometrics. Participation of researchers/scientists from chemical science and engineering, computer science, statistics, and mathematics have strengthened this confluence, one of the finest present-day tributaries of which is *analytics*. Analytics is the discipline that uses mathematical and multivariate statistical methods for systematic analysis of data.

Data collection (design of experiments), presentation, preprocessing (signal processing, outlier detection, and smoothing, filtering, and reconciliation of data), feature extraction (either by visual checking or by mathematical processors/using transforms), and pattern recognition (using multivariate statistical software/machine learning algorithms) are the frameworks of chemometrics. This subject is oriented towards the better understanding of a process/system and development of applications like fault detection/diagnosis in plants, environment/health/food and beverage quality monitoring, clinical trials, and many more. Sensing technology and instrumentation should be another pivotal component of chemometrics, since a fair amount of knowledge in these areas is the source of the sustenance of chemometrics as a discipline. The mathematical/statistical/computational methods needed are to be oriented/customized according to the purpose of chemometrics as a discipline and they may be better stated as chemometric techniques. In fact, it is to be noted that chemometrics is an interdisciplinary forum with expanding limits; it cannot be nourished and proliferated without interdisciplinary participation and expertise.

2.2 *Principal component analysis (PCA)*

Principal component analysis (PCA) is a potential chemometric tool to develop a data-based process model so that meaningful information can be derived from it. PCA finds successful application in diverse fields like image processing, computer graphics, process monitoring and control, social science, and so many others. Sudjianto and Wasserman [2] and Trouve and Yu [3] have applied principal component analysis to extract features from large data sets. Statistical process monitoring (SPM) is a widely used technique for fault diagnosis of chemical processes to improve process quality and productivity [4]. PCA is a popular method for SPM [5,6]. The basic strategy of PCA is to discard the noise and collinearity between process variables, while preserving the most important information of the original data set. To use PCA for process monitoring, a PCA model is first established based on process data collected under normal operating conditions. Then, the control limits of monitoring statistics (e.g., T^2, Q, which are not related to the newly projected data; they are related to the fitness of the PCA model) are calculated, and thus, the process can be monitored online by utilizing these statistic [7]. Kano et al. [8] implemented various PCA-based statistical process monitoring methods using simulated data obtained from the Tennessee Eastman plant. PCA has been utilized in its various forms like recursive PCA, multiway PCA, moving window PCA, probabilistic PCA, consensus, and hierarchical PCA for process monitoring/fault detection/food and beverage quality monitoring, and many more.

2.2.1 PCA decomposition of data

Following are the salient features of PCA:

- PCA is a multivariate statistical technique that deals with high-dimensional and highly correlated data (like industrial process data reflecting relatively few underlying mechanisms that drive the process), especially where the number of feature variables is much greater than the number of samples.
- The PCA-based model seems to be suitable for complex processes, such as bioprocesses, where the development of a physical model based on first principles is not always possible due to the lack of model parameters and clarity regarding the physical phenomena happening in the process.
- PCA is used for data compression without compromising meaningful possession of it.
- PCA is a kind of subspace modeling technique that projects the original data along a few selected directions (principal component directions) orthogonal to each other, along which the data's variability is maximal.
- The number of principal components is often much smaller than the number of original variables.
- The original data can be expressed as a linear combination of those few uncorrelated and orthogonally projected data.
- The detection of principal component directions is an unsupervised technique governed by eigenvector decomposition.

The basic steps of PCA decomposition of a data set are as follows:

Our goal is to map the raw data vector $X = [x_1 x_2 x_3 x_4 \ldots\ldots x_m]^T$ having m features onto $Z = [z_1 z_2 z_3 z_4 \ldots\ldots z_k]^T$, where $k < m$. The vector can be represented as a linear combination of a set of orthonormal vectors v_i, such that $X = \sum_{i=1}^{m} z_i v_i$, where the coefficients z_i can be found from the following equation:

$$z_i = \mathrm{v}_i^t X \tag{2.1}$$

This corresponds to a rotation of the coordinate system from the original X to a new set of coordinates given by Z. To reduce the dimensions of the data only a subset ($k < m$) of the basis v_i is kept. The basic vectors v_i are called principal components, which are equal to the eigenvectors of the covariance matrix of the data set. However, the reduction of dimensionality from m to k causes an approximation error. The sum of squares of the errors over the whole data set is minimized if we select the vectors v_i that correspond to the largest eigenvalues of the covariance matrix, and the method is called the cumulative percentage variance (CPV) method.

Let X be the data matrix which is undergoing eigenvector decomposition:

$$[V, \Lambda] = \text{eig}\,(X) \tag{2.2}$$

where $\Lambda = [\lambda_1, \lambda_2, \lambda_3 \ldots \lambda_m]$, $\lambda_1, \lambda_2, \lambda_3 \ldots \lambda_m$ are in decreasing magnitude and V is the eigenvector. $P = [v_1,\ v_2 \ldots\ldots\ldots]^t$ is the loading matrix with $P^T P = 1$.

$$t = XP \tag{2.3}$$

$$e = X - TP \tag{2.4}$$

where t is the score matrix and e is the residual error.

$$\text{CPV} = \frac{\sum_{j=1}^{k} \lambda_j}{\sum_{j=1}^{m} \lambda_j} \times 100 \tag{2.5}$$

If n and m are the number of rows and the number of columns of the given feature matrix, the computational complexity of evaluating the covariance matrices is nm^2. If we use singular value decomposition (SVD) to decompose the matrices, we can directly use the feature vectors, and avoid the multiplications, in evaluating covariance matrices. The computational complexity of eigenvector decomposition is $nm \min(n, m)$.

2.2.2 *Principle of nearest neighborhood*

After PCA decomposition, a new data matrix D of dimension $n \times k$ is obtained where n is the number of observations:

$$D = \begin{bmatrix} z_{11} & \cdots & z_{1k} \\ \vdots & \ddots & \vdots \\ z_{n1} & \cdots & z_{nk} \end{bmatrix} \tag{2.6}$$

With the matrix D defined, the next step is directed towards classification of substances. The PCA-score data sets are grouped into a number of classes following the rule of the nearest-neighbor clustering algorithm. Let $(C)_{i=1 \ldots l}$ denote l pattern classes in n number of measurements, represented by the single prototype vector $(y)_{i=1 \ldots l}$. The maximum value of l can reach up to n. The mean or class centroids of $(y)_{i=1 \ldots l}$ vectors have m numberof latent features, each of which represents a unique feature in reduced dimension space. The distance between an incoming pattern x and the prototype vectors are $D_j = \|x - y_i\|,\ 1 \le i \le l$. The minimum distance will classify x at C_j for which D_j is minimum:

$$D_j = \min \|x - y_i\|, \ \text{for } 1 \le i \le l \tag{2.7}$$

For an online system, it may be inferred that the incoming pattern represented by unknown type has similarity with one of the *lth* class of known types. The MATLAB princomp and pcacov programs are used for PCA decomposition.

2.2.3 Hotelling T^2 and Q statistics

The statistical metrics T^2 and Q are the two important limiting indices apart from the score of a PCA that can be followed to know the exact process status [9,10]. T^2 is a measure of a multivariate distance data point to the origin measured in the model plane, whereas Q is a statistic/square prediction error (SPE); SPE finds the variation that is not being captured by the PCA model. SPE is a measure of the distance from the model plane to an observation lying outside the model plane. The T^2 statistic is a scaled two-norm of a score vector, and the Q statistic is a two-norm of a residual vector in the residual subspace. Consequently, a disturbance that involves a change in the relations between the variables will increase the SPE since the model does not cover the direction of the disturbance. A disturbance in the model plane, i.e., of the same nature as the identification data, will turn up as an increase in T^2. SPE and T^2 statistics assume that the data are normally distributed, which may not be true for all cases. In most cases though, the distributions of the model residuals are approximately normal due to the central limit theorem. The SPE and T^2 measured as such do not depend on a certain distribution, but only on their confidence limits. Figure 2.1 provides a geometrical interpretation of SPE and T^2 [11]. SPE and T^2 [11] provide complementary information, and they can be computed in the following way:

$$T^2 = t^T \Lambda^{-1} t = X^T P \Lambda^{-1} P^T X \tag{2.8}$$

$$t_{\text{new}} = x_{\text{new}} P(P'P)^{-1} \quad Q = e^T e = X^T (1 - PP^T) X \tag{2.9}$$

The approximated control limits of T^2 and Q statistics, with a confidence level α, can be determined from the normal operating data in several ways by applying the probability distribution assumptions [9,12,13]. The threshold limit of SPE with α level of significance can be computed according to the following equations [9]:

$$Q_\alpha = \theta_1 \left[\frac{h_0 c_\alpha \sqrt{2\theta_2}}{\theta_1} + 1 + \frac{\theta_2 h_0 (h_0 - 1)}{\theta_1^2} \right]^{1/h_0} \tag{2.10}$$

$$h_0 = 1 - \left(\frac{2\theta_1 \theta_3}{3\theta_2^2} \right) \tag{2.11}$$

$$\theta_i = \sum_{j=m_1+1}^{m} \lambda_j^i \tag{2.12}$$

$$c_\alpha = 1.645 \tag{2.13}$$

where m_1 is the number of largest principal components chosen by the CPV measure, m is the number of features, and n is the number of data sample. c_α is the normal deviate corresponding to the $(1-\alpha)$ percentile. The threshold limit of T^2 with α level of significance can be computed according to the following equation using the F distribution [14]:

$$T_\alpha^2 = \frac{(n^2-1)m_1}{n(n-m_1)} F_\alpha(m_1, n-m_1) \tag{2.14}$$

where $F_\alpha(m_1, n-m_1)$ is the upper limit of the α percentile of the F distribution with degree of freedoms m_1 and $n-m_1$.

The MATLAB code for calculating T^2, the threshold limit T_{upper}^2, Q, and the threshold limit Q_{upper} is provided in Chapter 5.

2.3 Similarity

Similarity of newly collected plant data (via an online measuring system) to the data sets pertaining to various operating conditions can be expressed in terms of *Euclidean* and *Mahalanobis* distances. The Euclidean distance or Euclidean metric is the "ordinary" distance (i.e., straight line) between two points in Euclidean space. With this distance, Euclidean space becomes metric space. The associated norm is called the Euclidean norm. Older literature refers to the metric as the Pythagorean metric. The Mahalanobis distance is a measure of the distance between a point *P* and a distribution *D*, introduced by P.C. Mahalanobis in 1936 [15]. It dictates the number of standard deviations away the point *P* is from the mean of *D*. The distance is zero if *P* is at the mean of *D*, and grows with *P* moving away from the mean. Apart from them, some new criterions have been proposed to deter-

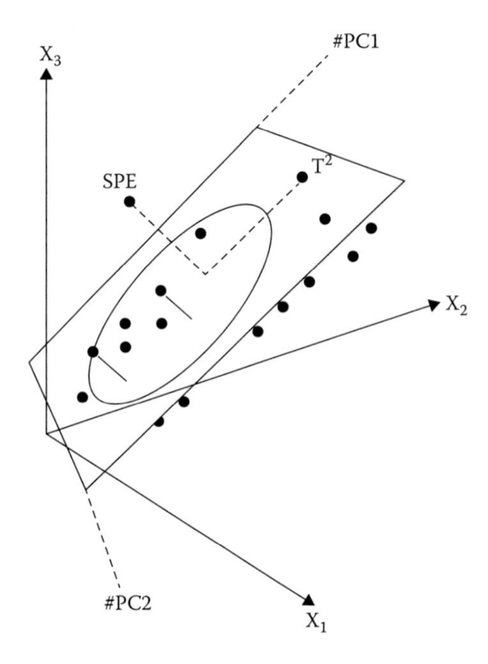

Figure 2.1 The geometrical interpretation of the SPE and T^2 measures. x_1, x_2 and x_3 are the original variables and #PC1 and #PC2 are the principal components. The points represent samples or observations. (From W.J. Krzanowski, *J. Am. Statis. Assoc.*, 74, 703–707, 1979.)

mine the similarity or dissimilarity among the incoming process data and process historical database, aiming towards the detection of process normal/faulty conditions.

2.3.1 PCA similarity

The PCA similarity factor among data sets, which differ (are altered) due to the entry of new data samples, is developed by choosing the largest k principal components of each multivariate time series data set that describe at least 95% of variance in each data set [11]. These principal components are the eigenvectors of the covariance matrix pertaining to the data sets. The PCA similarity factor between these two data sets is defined by Equation 2.15:

$$S_{PCA} = \frac{1}{k}\sum_{i=1}^{k}\sum_{j=1}^{k}\cos^2\theta_{ij} \tag{2.15}$$

where k is the number of selected principal components in both data sets, and θ_{ij} is the angle between the ith principal component of x_1 and j^{th} principal component of x_2. When the first two principal components explain 95% of variance in the data sets, S_{PCA} may not capture the degree of similarity between two data sets because it weights all PCs equally. Obviously S_{PCA} has to be modified to weight each PC by its explained variance. The modified S_{PCA}^{λ} is defined as

$$S_{PCA}^{\lambda} = \frac{\sum_{i=1}^{k}\sum_{j=1}^{k}\lambda_i^1\lambda_j^2\cos^2\theta_{ij}}{\sum_{i=1}^{k}\sum_{j=1}^{k}\lambda_i^1\lambda_j^2} \tag{2.16}$$

where λ_i^1, λ_j^2 are the eigenvalues of the first and second data sets respectively. Program 2.1 is the MATLAB code for finding PCA similarity.

Program 2.1 MATLAB code for PCA similarity factor

```
% The function 'PCA _ Similarity' compares two data sets, X1 and X2,
and returns 1 if they are similar (or exactly equal) and 0 if they
are dissimilar. So the range of PCA similarity factor is [0, 1]. This
function determines ordinary and weighted similarity between two
data sets as X1 and X2. The program also returns PCs of X1 and X2.
TPCA and TPCA1 correspond to ordinary and weighted similarity.

function [TPCA, TPCA1, PC1, PC2]= PCA_Similaity(X1,X2)

TPCA=[];
TPCA1=[];
 SPCA=[];
SPCA1=[];
size(X1);

%Find PCs of X1 and X1

[pc1,~,latent1,~]  = pca(X1);
[pc2,~,latent2,~]  = pca(X2);

PC1=pc1(:,1:2);
PC2=pc2(:,1:2);

% Find eigenvalues of X1 and X2
```

```
e1=latent1;
e2=latent2;
E1=sum(e1);
E2=sum(e2);
e11=(e1/E1)*100;
e12=(e2/E2)*100;
kn=size(e1);

S=0;

for i=1:kn(1,1)
   S=S+e11(i);
   if S>95
       break
   end
end

k=i;

if k<2
    k=2;
else
    k=i;
end

% Consider 1st two eigenvalues corresponding to 1st two principle eigen-
vectors

e1=e1(1:k);
e2=e2(1:k);

n1=size(PC1);
n2=size(PC2);

T=[];

for i=1:n1(1,2)

    th=[];

    for j=1:n2(1,2)

        a=PC1(:,i);
        b=PC2(:,j);
        b=b';
        d=0;

        for k=1:n1(1,1)
            d=d+a(k)*b(k);
        end

        d1=sum(a.^2);
        d2=sum(b.^2);
        th=[th;(d/(d1*d2))];

    end

    T=[T th];

end
% Find ordinary similarity between X1 and X2

Spca=0;
```

```
for i=1:n1(1,2)
    for j=1:n2(1,2)
        Spca=Spca+(T(i,j) )^2;
    end
end
Spca=Spca/n1(1,2);
% Find weighted similarity between X1 and X2
Spca1=0;
for i=1:n1(1,2)
    for j=1:n2(1,2)
        Spca1=Spca1+e1(i)*e2(j)*(T(i,j))^2 ;
    end
end
l=0;
for k=1:n1(1,2)
    l=l+e1(k)*e2(k);
end
Spca1=Spca1/l;
SPCA=[SPCA Spca ];
SPCA1=[SPCA1 Spca1];
TPCA=[TPCA;SPCA];
TPCA1=[TPCA1;SPCA1];
end
```

2.3.2 *Distance-based similarity*

In addition to the above similarity measure, the distance similarity factor can be used to cluster multivariate time series data. The distance similarity factor compares two data sets that may have similar spatial orientation [16]. The distance similarity's worth is tested when the data pertaining to different operating conditions may have similar principal components. The distance similarity factor is defined as

$$S_{dist} = 2\times\frac{1}{2\pi}\int_{\varphi}^{\infty} e^{-\frac{z^2}{2}}\,dz = 2\times\left|1-\frac{1}{2\pi}\int_{-\infty}^{\varphi} e^{-\frac{z^2}{2}}\,dz\right| \tag{2.17}$$

where $\varphi=\sqrt{\left(\overline{x_2}-\overline{x_1}\right)\sum_1^{*-1}\left(\overline{x_2}-\overline{x_1}\right)^T}$, $\overline{x_2}, \overline{x_1}$ are samples meaning row vector, Σ_1 is the covariance matrix for data set X_1, and Σ_1^{*-1} is a pseudoinverse of the covariance matrix of X_1. Data set X_1 is assumed to be a reference data set. In Equation 2.17, a one-side Gaussian distribution is used because $\varphi \geq 0$. The error function can be calculated by using ANT software or standard error function tables. The integration in Equation 2.17 normalizes S_{dist} between 0 and 1. Program 2.2 presents the MATLAB code for distance similarity factor.

Program 2.2 MATLAB code for distance similarity factor

```
% 'Distance_Similarity' compares two data sets (X1 and X2) that have the
same spatial orientation but are located far apart.
function [sdist]= Distance_Similarity(X1,X2)

x1=mean(X1);
x2=mean(X2);

R2=(X2'*X2)/(length(X2)-1);
[U2,S2]=eigs(full(R2));
D2=zeros(1,length(S2));

for i=1:1:length(S2)
    D2(i)=S2(i,i);
end

CumS2=zeros(1,length(D2));

for j=1:1:length(D2)

    CumS2(j)=D2(j)/trace(diag(D2));

end

sum5=0;

for k1=1:1:length(CumS2)

    sum5=sum5+CumS2(k1);
    if sum5>=0.99
        break;
    end

end

S=diag(D2(1:k1));
psinv=inv(S);
pii=(((x2(1:k1)-x1(1:k1))*psinv*(x2(1:k1)-x1(1:k1))')^(1/2));
sdist=erfc(pii);

end
```

2.3.3 Combined similarity factor

The combined similarity factor (SF) combines S_{PCA}^{λ} and S_{dist} using the weighted average of the two quantities and is used for clustering of multivariate time series data. The combined similarity is defined as

$$SF = \alpha_1 S_{PCA}^{\lambda} + \alpha_2 S_{dist} \tag{2.18}$$

Selection of α_1 and α_2 is up to the user, but ensure that their sum is equal to one. Program 2.3 presents the MATLAB code for calculating the combined similarity factor.

Program 2.3 MATLAB code for combined similarity factor

```
% This program takes on two data sets X1 and X2 to find the combined
similarity between them.

function [SPCA SDIST SF]=Combined_SF(X1,X2, alpha1, alpha2)
% PCA similarity
```

```
[Similarity_PCA Similarity_w_PCA1 PC1 PC2]=similarity_pca(X1,X2);
TPCA=[Similarity_PCA];
TPCA1=[Similarity_w_PCA1];

% Distance similarity

[sdist]=simil_dist(X1,X2);
Sf=alpha1*TPCA1+alpha2*sdist;
SF=Sf;
SDIST=sdist;
SPCA=TPCA1;

end
```

2.3.4 *Dissimilarity and Karhunen-Loeve (KL) expansion*

Karhunen-Loeve (KL) expansion is a familiar technique applied for feature extraction and data dimensionality reduction [17]. In this book, the KL expansion has been deployed as a dissimilarity factor for finding patterns in time series data sets. Two multivariate data sets are considered, which are as follows:

$$X_1 = \begin{bmatrix} x_{11} & x_{12} & \cdots & x_{1n} \\ x_{21} & x_{22} & \cdots & x_{2n} \\ \vdots & \cdots & \ddots & \vdots \\ x_{m1} & x_{m2} & \cdots & x_{mn} \end{bmatrix}_{m1 \times n1} \quad X_2 = \begin{bmatrix} x_{11} & x_{12} & \cdots & x_{1n} \\ x_{21} & x_{22} & \cdots & x_{2n} \\ \vdots & \vdots & \ddots & \vdots \\ x_{m1} & x_{m2} & \cdots & x_{mn} \end{bmatrix}_{m_2 \times n_2} \tag{2.19}$$

The dimension of the X_2 matrix is=$m_2 \times n_2$
The correlation matrix of the data sets X_i is

$$R_i = \frac{1}{N_i - 1} X_i^T X_i, \text{where } i = 1,2 \tag{2.20}$$

The combined data matrix is $X = [X_1; X_2]$.
The correlation matrix of the combined data matrix $X_{m_3 \times n}$, $[m_3 = m_1 + m_2]$ is

$$R = \frac{1}{m_3 - 1} XX^T = \frac{m_1 - 1}{m_3 - 1} R_1 + \frac{m_2 - 1}{m_3 - 1} R_2 \tag{2.21}$$

From eigenvalue decomposition, the following relation can be deduced as

$$RP_0 = P_0 \Lambda \tag{2.22}$$

where P_0 is an orthogonal matrix and Λ is a square matrix containing eigenvalues on its diagonal. Using orthogonal matrix P_0 and diagonal matrix Λ, the transformation matrix is derived as

$$P = P_0 \Lambda^{-0.5} \text{ and } P^T RP = I \tag{2.23}$$

Multiplication of original matrix X_i with the transformation matrix P yields the transformed matrices as

$$Y_i = \sqrt{\frac{m_i - 1}{m_3 - 1}} X_i P;\ i = 1,2 \tag{2.24}$$

The covariance matrix of Y_i is

$$S_i = \frac{1}{m_i - 1} Y_i^T Y_i = \sqrt{\frac{m_i - 1}{m_3 - 1}} P^T R_i P, \text{where}\ i = 1,2 \tag{2.25}$$

It can be proved that the addition of covariance matrices S_1 and S_2 results in an identity matrix I. Application of eigenvalue decomposition on covariance matrices produces the following relations:

$$S_i w_j^i = \lambda_j^i w_j^i \tag{2.26}$$

$$S_2 w_j^1 = (1 - \lambda_j^1) w_j^1 \tag{2.27}$$

where w is the eigenvector. From the above two equations, we can get

$$\lambda_{j,2} = (1 - \lambda_{j,1}) \tag{2.28}$$

As a result, (since eigenvectors of a correlation matrix represent directions of principal components and eigenvalues are equivalent to the variances of principal components) both transformed data sets have the same set of principal components, and the corresponding eigenvalues of the correlation matrices are oppositely ordered. Thus, the dominant eigenvector of Y_1 is the weakest eigenvector of Y_2, and vice versa. When data sets are quite similar to each other, the eigenvalues λ_j^i must be near 0.5. On the other hand, when data sets are quite different from each other, the largest and the smallest eigenvalues should be near one and zero, respectively. Hence, the dissimilarity factor may be defined as follows:

$$D = \frac{4}{P} \sum_{j=1}^{P} (\lambda_j - 0.5)^2 \tag{2.29}$$

The dissimilarity factor ranges from zero to one. The value of the dissimilarity factor is near zero when the given two data matrices are alike and near one if they are different.

The KL expansion is applied for classification in Chapter 5. Program 2.4 presents the MATLAB code for calculating the dissimilarity factor based on KL expansion.

Program 2.4 MATLAB code for dissimilarity factor based on KL expansion

```
% This program takes on two data sets X1 and X2 and calculates the dis-
similarity between them.

function [D]=DISSIM(x1,x2)

Ni=length(x1);
R1=(x1'*x1)/(Ni-1);
R2=(x2'*x2)/(Ni-1);

x11=[x1;x2]';
x21=[x1;x2];

N=length(x1);

R11=(x11*x21)/(N-1);
```

```
R=((Ni-1)/(Ni-1))*R1+((Ni-1)/(Ni-1))*R2;

[V,S]=eig(R11);

D=zeros(1,length(S));

for i=1:1:length(S)
    D(i)=S(i,i);
End

CumS=zeros(1,length(D));

for j=1:1:length(D)
    CumS(j)=D(j)/trace(diag(D));
End

sum=0;

for k=1:1:length(CumS)
    sum=sum+CumS(k);
    if sum>=0.95
        break;
    end
end

P=V(:,1:end);
Yi=(sqrt((Ni-1)/(N-1))*x1*P)/sqrt(S);
Si=(Yi'*Yi)/(Ni-1);
[V1,D1]=eig(Si);
sum1=0;
k=length(D1);

for i=1:1:k
    sum1=sum1+(D1(i,i)-0.5)^2;
end

D=(4/k)*sum1;

end
```

2.3.5 *Moving window-based pattern matching using similarity/dissimilarity factors*

In this approach, the snapshot or test data with an unknown start and end time of the operating condition (in sample wise/batch manner) moves through historical data or a template (reference) and the similarity/dissimilarity between them is quantified by distance. A PCA-based combined similarity and dissimilarity factor is calculated using KL expansion [18–21]. Johannesmeyer et al. (2002) have thoroughly reviewed pattern matching in a historical database in the context of data mining (aerospace, seismic data interpretation, stock market analysis, relationships between chemical structures, and biological functions in molecular biology and medicine) and pattern recognition in multivariate time series data (image processing, speech and character recognition, and process monitoring applications) [19 and references therein]. In order to compare the snapshot data to the template data, the relevant historical data are divided into data windows that are the same size as the snapshot data. The historical data sets are then organized by placing windows side-by-side along the time axis, which results in equal-length, nonoverlapping segments of data. The historical data window moves one observation at a time, with each old observation getting replaced by new one. The historical data windows with the largest values of similarity factors/smallest values in dissimilarity factors with the snapshot data are collected in a candidate pool and are called records to be analyzed by the process engineer. Pool accuracy, pattern matching efficiency, and pattern matching algorithm efficiency are important metrics that quantify the performance of the proposed pattern-matching technique.

N_P: The size of the candidate pool; it is the number of template data windows that have been labeled "similar/dissimilar" to the snapshot data by a pattern-matching technique. The data windows collected in the candidate pool are called records.
N_1: Number of correctly identified records (as per the algorithm).
N_2: Number of records in the candidate pool that are not correctly identified (as per the algorithm). Hence, $N_P = N_1 + N_2$.
N_{DB}: The total number of template windows that are actually similar to the current snapshot. In general, $N_{DB} \neq N_P$.

Pool accuracy, $P = \left(\frac{N_1}{N_P}\right) \times 100$

Pattern matching efficiency, $H = \left[1 - \left(\frac{(N_P - N_1)}{N_{DB}}\right)\right] \times 100$

Pattern matching algorithm efficiency, $\xi = \left(\frac{N_P}{N_{DB}}\right) \times 100 \left(\frac{(N_p)}{(N_{DB})}\right) \times 100$

A large value of pool accuracy is important in the case of detection of a small number of specific previous situations from a small pool of records without evaluating incorrectly identified records. A large value of pattern-matching efficiency is required in the case of detection of all of the specific previous situations from a large pool of records. The proposed method is completely data driven and unsupervised; no process models or training data are required. The user should specify only the relevant measured variables. The proposed pattern matching algorithm is as follows:

1. *Specify the template data windows (variables and time period)*
2. *Specify the snapshot data (variables and time period)*
3. *Specify the threshold values of similarity/dissimilarity factor*
   a. *For j = 1 to m (m = snapshot data windows)*
      N_P = []
      For i = 1 to n (n = number of template data windows)
      i. *Calculation of similarity/dissimilarity factors between snapshot and template data windows*
      ii. *Collection of template window records with the largest values of similarity factors/smallest values of dissimilarity factors in a pool*

```
end i
end j
```
N_P
Decide upon N_1, N_{DB}

Calculation of performance matrices: pool accuracy (P), pattern matching efficiency (H), and pattern matching algorithm efficiency (ξ).

Applications of moving window-based pattern matching are provided in Chapters 4 and 5.

2.4 Clustering

A *cluster* is a collection of objects which are "similar" and are "dissimilar" to the objects belonging to other collections. Clustering can be considered the most natural/primitive *unsupervised learning* problem; it deals with finding a *structure* in a collection of unlabeled data. The clustering technique is primitive in that no *a priori* assumptions are made regarding the group structures. Grouping can be made on the basis of similarities/dissimilarities or distances. Minimum distance cluster methods fall under partitioned clustering techniques. Partitioned clustering has the drawback that the scaling of the axis may greatly affect the performance of the cluster. The clustering method is further divided into hierarchical and nonhierarchical types. Because the possible decision is based on visualization being offered by hierarchical clustering, it is widely used in taxonomy, agriculture, remote sensing, and process control. A wide variety of clustering techniques are available and have been applied in clustering time series data as a part of chemometric applications [2,3,22,23]. Wang and McGreavy used clustering methods to classify abnormal behavior of refinery fluid catalytic cracking processes [24]. Johannesmeyer and Seborg [25] developed an efficient technique to locate and cluster similar records in the historical database using PCA similarity factors. Singhal and Seborg [26] used PCA and Mahalanobis distance similarity measures to locate and cluster similar operating conditions in a large multivariate database. Huang et al. [27] used PCA to cluster multivariate time series data by splitting large clusters into small clusters based on the percentage of variance explained by principal component analysis. Kavitha and Punithavalli [28] claimed that all traditional clustering or unsupervised algorithms are inappropriate for real-time data.

2.4.1 Hierarchical clustering

This variety of clustering techniques is processed either by a series of successive mergers or a series of successive divisions. Agglomerative hierarchical methods start with individual objects, ensuring as much the number of clusters as the objects initially. The most similar objects (or those with less inter-cluster distance) are grouped first, and these initial groups are merged according to their similarities. Eventually, as the similarities decrease (distance increases), all subgroups are fused into a single cluster. The divisive hierarchical method works in an opposite manner. An initial single group of objects is divided into two subgroups such that they are far from each other. This subdivision continues until there are as many subgroups as objects, that is until each object forms a group. The results of both the methods may be displayed in the form of a two-dimensional diagram called a dendrogram. Inter-cluster distances are expressed by single linkage, complete linkage, and average linkage [29,30].

2.4.2 Nonhierarchical clustering

In the nonhierarchical unsupervised K-means clustering method, the number of clusters K can be prespecified or can be determined iteratively as a part of the clustering procedure. The K-means clustering proceeds in three steps, which are as follows:

1. *Partition the items into K initial clusters.*
2. *Assign an incoming data item to the cluster whose centroid is nearest (distance is usually Euclidean). Recalculate the centroid for the cluster receiving the new item (data corresponding to a specific operating condition) and for the cluster losing that item.*
3. *Repeat step 2 until no more reassignment takes place or stable cluster tags are available for all the items.*

The mathematical representation of the *K*-means clustering algorithm is as follows:

- For a given assignment *C*, compute the cluster mean m_k:

$$m_k = \frac{\sum_{i:C(i)=k} x_i}{N_k}, \; k = 1 \ldots K \tag{2.30}$$

- For a current set of cluster means, assign each observation as

$$C(i) = \arg\min x_i - m_k^{\,2}, \; i = 1 \ldots . N, \; i < k < K \tag{2.31}$$

- Iterate on the above two steps until convergence.

K-means clustering has a specific advantage of not requiring the distance matrix as required in hierarchical clustering, hence ensuring a faster computation. The *K*-means algorithm has been applied to many engineering problems.

2.4.3 Modified K-means clustering using similarity factors

The time series data may be discriminated and classified using the following similarity-based modified *K*-means clustering algorithm.

1. *Specify database* $X = [X_1; X_2; \ldots . X_q; \ldots X_Q]^T$, *an initial guess for membership of each data set* X_q *and an assumed number of clusters K.*
2. *Compute the aggregate data set* $X_i \; (i = 1, 2, \ldots ., K)$ *for each of K clusters.*
   Let the jth *data set in the* i^{th} *cluster be defined as* X_{ij}.
   The aggregate data set X_i, *is defined as,*

$$X_i = [X_{i1}^T, \, X_{i2}^T, X_{i3}^T, \ldots \ldots \ldots , X_{ij}^T]$$

3. *Calculate the dissimilarity between an arbitrary data set* X_q *and each of the aggregate data sets* X_i *as* $d_{i,q} = 1 - SF_{i,q}$, *where* $SF_{i,q}$ *is the similarity between the* q^{th} *data set and* i^{th} *cluster.*
   Assign data set x_q *to the cluster with which it is least dissimilar or most similar.*
   Repeat this step for Q *data sets.*
4. *Calculate the average dissimilarity of each data set from its cluster as*

$$J(K) = \frac{1}{Q} \sum_{i=1}^{K} \sum_{x_q \in X_i} d_{i,q}$$

Check: If the value of $j(K)$ *has changed from a previous iteration, then go back to step 1. Otherwise stop.*

Selection of the number of clusters is crucial in a *K*-means clustering algorithm. The average dissimilarity factor $J(K)$ decreases as the value of *K* increases. The percentage in the average dissimilarity factor is defined as

$$dJ(K) = \frac{J(K+1) - J(K)}{J(K)} \times 100\%, \; K = 1, 2 \ldots \tag{2.32}$$

The value of $dJ(K)$ is plotted against the number of clusters K. The value of K for which $dJ(K)$ reaches a minimum, or is close to zero, is considered to be the near optimality (knee) in the plot of $J(K)$ versus K. $\psi(K)$ is proposed to be the sign of the difference of $dJ(K)$ with the changing value of K, which is used to locate these "knees."

$$\psi(K) = \text{sign}\,[dJ(K+1) - dJ(K)],\ \ K = 1, 2, 3\ldots \tag{2.33}$$

Thus, the quantity $\psi(K)$ is similar to the sign of the second derivative of $J(K)$. The values of K, where $\psi(K)$ changes from negative to positive (with increasing number of clusters, K) are the "knees" in the $J(K)$ versus K plot. Usually, the location of the first knee is selected to be the optimum number of clusters, but the user may choose other values if $J(K)$ decreases significantly after the first knee. If there is no value of K for which $\psi(K)$ changes sign, then the optimum number of clusters can be specified as the value of K for which $dJ(K)$ becomes very close to zero [26].

Some key definitions were introduced by Singhal and Seborg [26] to evaluate the performance of the clusters obtained using similarity factors. Among them, cluster purity is defined as characterizing each cluster in terms of how many data sets for a particular class are present in the ith cluster. *Cluster purity* is defined as

$$P_i = \left(\frac{\underset{j}{\max}\, N_{ij}}{N_{pi}} \right) \times 100\%,\ \ i = 1, 2, \ldots\ldots, K \tag{2.34}$$

where N_{ij} = the number of data sets of operating condition j in the ith cluster and N_{pi} is the number of data sets in the ith cluster.

Cluster efficiency measures the extent to which an operating condition is distributed in different clusters. This method is to penalize the large values of K when an operating condition is distributed into different clusters. Cluster efficiency is defined as

$$\eta_j = \left(\frac{\underset{i}{\max}\, N_{ij}}{N_{DBj}} \right) \times 100\%,\ \ j = 1, 2, \ldots\ldots, N_{op} \tag{2.35}$$

where N_{DBj} is the number of data sets for operating condition j in the database. A largenumber of data sets of a specific operating condition present in a cluster can be considered the dominant operating condition. *The modified K-means clustering algorithmis is applied in Chapter 3.* Program 2.5 presents the pseudocode for dissimilarity (one-similarity factor)- based modified *K*-means clustering algorithm.

Program 2.5 Pseudocode for dissimilarity (one-similarity factor)-based modified K-means clustering algorithm

```
Given data sets Xq (q=1, 2, 3, ……, Q), the number of samples (ns) in a
data set, number of columns (nc) in a data set (i.e. number of vari-
ables), an initial guess for membership of each data set Xq and weighting
```

factors α1 and α2. If no initial guess is provided, then each data set is randomly assigned to one of *K* clusters.

```
% Algorithm begins %
X=[X1;X2;X3;......XQ]T;
% the entire database, the size of Xq is ns×nc
JK=zeros(1, Q-1);
For K=2:1:Q
% k-means clustering algorithm repeats for each K value
J=1;
M=1;
Diq=zeros(1, K);
Diq1=zeros(K, Q);
while J<=Q
A=J*ns;
Xtest=X(M:A, nc);
for i=1:1:K
% computing aggregate data set for each of K clusters
Xi = [Xq1; Xq2; Xq3;........, Xqj]T;
% Xqj is the jth data set in ith cluster
[similarity_pca] = PCA_Similaity(Xtest, Xi);
% calculating PCA similarity factor
[similarity_distance] = Distance_Similarity(Xtest, Xi);
% calculating distance similarity factor
SF= α1* similarity_pca+ α2* similarity_distance;
% calculating combined similarity factor
Diq(i)= 1-SF;
% calculating dissimilarity value
end
[dmin, idx] = min(Diq);
Diq1(idx, J) = dmin;
M=J*ns+1;
J=J+1;
end
Sum1=0;
for kk=1:1:K
% calculating average dissimilarity of each data set from its cluster
Sum1=sum1+sum(Diq1(kk, :));
end
Sum1=sum1/Q;
JK(K-1)=sum1;
end
dJK=zeros(1, Q-2);
% calculating percentage change in JK
for w=1:1:Q-2
dJK(w)=[abs(JK(w+1)-JK(w))/JK(w)]*100;
end
psiK=zeros(1, Q-3);
for w1=1:1:Q-3
psiK(w1)=sign(dJK(w1+1)-dJK(w1));
end
% plotting the results to find optimum number of clusters
subplot(3,1,1), plot([2:1:Q],JK), ylabel('J(K)')
subplot(3,1,2), plot([2:1:Q-1], dJK), ylabel('dJ(K)%')
subplot(3,1,3),plot([2:1:Q-2], psiK), ylabel('\psi(K)')
```

2.5 *Partial least squares (PLS)*

Partial least squares (PLS) finds the latent variables (through principal component decomposition, PCA) from data by capturing the largest variance in the data and achieves the maximum correlation between the predictor X variables and response Y variables. PLS was not originally designed to be a tool for statistical discrimination, rather it is a tool for regression. PLS is preferred to PCA when discrimination is the goal and dimension reduction is needed.

Projections of the observed data to its latent structures by means of PLS were developed by Wold [31] and extended by several others [32–34]. Karp et al. [35] used partial least squares discriminant analysis to classify two-dimensional gel studies in expression proteomics. The partial least squares technique was used by Andrade et al. [36] for predicting clean octane numbers of catalytic reformed naphthas. Nguyen and Rocke [37,38] used partial least squares to classify tumors using microarray gene expression data. Machelle [39] used partial least squares to reduce the dimension of hyperspectral reflectance data to derive components to be used in a logistic discrimination of salt marsh plants exposed to petroleum and heavy metal contamination. Oliver et al. [40] used PLS in chemometric analysis of proteomic profiles for detection of ovarian cancer. Rathore et al. [41] used PLS in assessing comparability of a biotech process in different phases of its manufacturing, as well as other key activities during product commercialization.

When dealing with nonlinear systems, the underlying nonlinear relationship between predictor variables X and response Y variables can be approximated by quadratic PLS (QPLS), or splines. Sometimes this approach may not function well when the nonlinearities cannot be described by a quadratic relationship. Qin and McAvoy [42,43] and Qin [44] suggested a new approach to replace the inner model in PLS with a neural network model followed by the focused research and development (R&D) activities taken up by several other researchers [45–49]. This approach of neural networks partial least squares (NNPLS) employs the neural network as an inner model, keeping the outer mapping framework as a linear PLS algorithm. Kaspar and Ray [50] demonstrated their approach for an identification and control problem using linear PLS models [50]. Lakshminarayanan et al. [51] proposed the autoregressive exogenous model (ARX)/Hammerstein model as the modified PLS inner relation, and used it successfully in identifying dynamic models and proposition of PLS-based feed forward and feedback controllers. For multivariable processes, a multivariable controller can be used as a series of SISO NN controllers within a PLS framework. There are not many references on NNPLS controllers available in the open literature, though PLS controllers are well documented. Damarla and Kundu [52] implemented an NNPLS control strategy for nonlinear complex (2 × 2), (3 × 3), and (4 × 4) distillation processes.

2.5.1 *Linear PLS*

X and Y matrices are scaled in the following way before they are correlated as a PLS algorithm:

$$X = XS_x^{-1} \text{ and } Y = YS_y^{-1} \tag{2.36}$$

where $S_x = \begin{bmatrix} S_{x1} & 0 \\ 0 & S_{x2} \end{bmatrix}$ and $S_Y = \begin{bmatrix} S_{y1} & 0 \\ 0 & S_{y2} \end{bmatrix}$. S_X and S_Y are scaled matrices.

The outer relationship for the input matrix and output matrix with predictor variables can be written as

$$X = t_1 p_1^t + t_2 p_2^t + \dots\dots\dots + t_n p_n^t + E = TP^T + E \tag{2.37}$$

$$Y = u_1 q_1^t + u_2 q_2^t + \dots\dots + u_n q_n^t + F = UQ^T + F \tag{2.38}$$

where T and U represent the matrices of scores of X and Y data while P and Q represent the loading matrices for X and Y. If all the components of X and Y are described, the errors E and F become zero. The inner model that relates X to Y is through the relation between the scores T and U:

$$U = TB \tag{2.39}$$

where B is the regression matrix. The response Y can now be expressed as

$$Y = TBQ^T + F \tag{2.40}$$

To determine the dominant direction of projection of X and Y data, the maximization of covariance within X and Y is used as a criterion. The first set of loading vectors p_1 and q_1 represent the dominant direction obtained by maximization of covariance within X and Y. Projection of X data on p_1 and Y data on q_1 resulted in the first set of score vectors t_1 and u_1, hence the establishment of the outer relation. The matrices X and Y can now be related through their respective scores, which is called the inner model, representing a linear regression between t_1 and $u_1 : \hat{u}_1 = t_1 b_1$. The calculation of the first two dimensions is shown in Figure 2.2. The residuals calculated at this stage are given by the following equations:

$$E_1 = X - t_1 p_1^t \tag{2.41}$$

$$F_1 = Y - u_1 q_1^t = Y - t_1 b_1 q_1^t \tag{2.42}$$

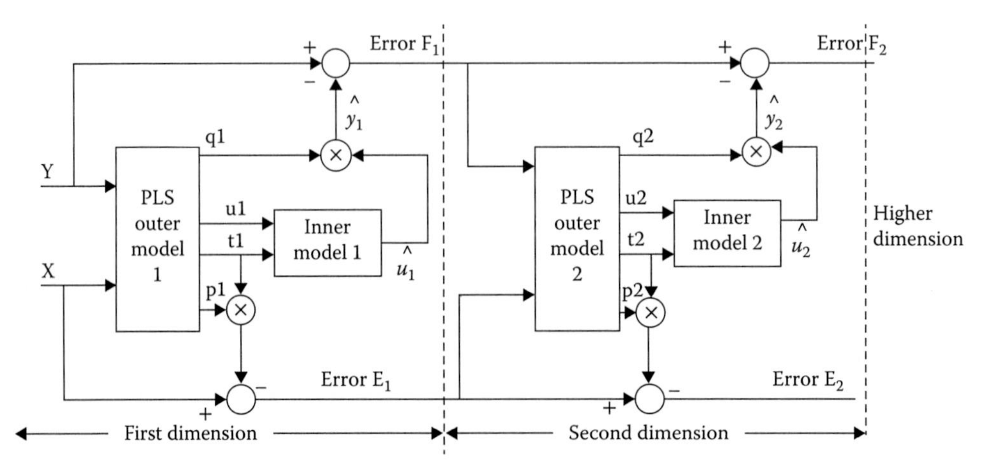

Figure 2.2 Schematic of linear PLS algorithm.

The procedure for determining the scores and loading vectors is continued by using the newly computed residuals until they are small enough, or the number of PLS dimensions required are exceeded. In practice, the number of PLS dimensions is calculated by percentage of variance explained and cross validation. For a given residual tolerance, the number of principal components can be much smaller than the original variable dimension. To get T and P matrices iteratively, deduction of $\left(tp^T\right)$ from X continues until the given tolerance gets satisfied. U and Q matrices are also found iteratively. The irrelevant directions originating from noise and redundancy are left as E and F. Y and X matrices, as shown in Equation 2.39. The multivariate regression problems are translated into several univariate regression problems in series or parallel on the application of PLS.

The linear PLS algorithm is as follows:

Assume X_0 and Y_0 are utoscaled X and Y matrices.

Assume $u = y_1$.

Repeat until weight w converges

1. *Calculate $w = X_0^T u$, and normalize w as $w = \frac{w}{\|w\|}$*
2. *Calculate $t = X_0 w$, and normalize t as $t = \frac{t}{\|t\|}$*
3. *Calculate $q = Y_0^T t$, and normalize q as $q = \frac{q}{\|q\|}$*
4. *Calculate $u = Y_0 q$*

After w converges, update the following:

5. *Calculate $t = X_0 w$, and normalize t as $t = \frac{t}{\|t\|}$*
6. *Calculate $p = X_0^T t$*
7. *Calculate $q = Y_0^T t$*
8. *Calculate $u = Y_0 q$*
9. *Calculate $b = \frac{u^T t}{(t^T t)}$*

where w, t, u, p, and q are the X weights, X scores, Y scores, X loadings, and Y loadings. b = coefficient of regression relating u and t.

10. *Calculate the following residual matrices:*

$$\widetilde{X_0} = X_0 - tp^T$$

$$\breve{Y}_0 = Y_0 - uq^T = Y_0 - tbq^T$$

11. *Repeat the aforesaid steps (1-10) for remaining k principal component dimensions.*

MATLAB code for PLS regression follows.

Program 2.6 PLS code to perform regression

```
function [T,P,U,Q,B,W] = pls (X,Y,tol2)
% PLS   Partial Least Squares Regrwssion
performs particial least squares regression % between the independent
variables, X and dependent Y as
% X = T*P' + E;
% Y = U*Q' + F = T*B*Q' + F1;
%
% Inputs:
% X     data matrix of independent variables
% Y     data matrix of dependent variables
% tol   the tolerance of convergence (defaut 1e-10)

Outputs:
% T     score matrix of X
% P     loading matrix of X
% U     score matrix of Y
% Q     loading matrix of Y
% B     matrix of regression coefficient
% W     weight matrix of X

% Input check
error(nargchk(1,3,nargin));
error(nargoutchk(0,6,nargout));

if nargin<2
  Y=X;
End

tol = 1e-10;
if nargin<3
  tol2=1e-10;

end

% Size of x and y
[rX,cX]  =  size(X);
[rY,cY]  =  size(Y);
assert(rX==rY,'Sizes of X and Y mismatch.');

% Allocate memory to the maximum size
n=max(cX,cY);
T=zeros(rX,n);
P=zeros(cX,n);
U=zeros(rY,n);
Q=zeros(cY,n);
B=zeros(n,n);
W=P;
k=0;
% iteration loop if residual is larger than specified
while norm(Y)>tol2 && k<n
    % choose the column of x has the largest square of sum as t.
    % choose the column of y has the largest square of sum as u.
    [dummy,tidx] =  max(sum(X.*X));
    [dummy,uidx] =  max(sum(Y.*Y));
```

```
    t1 = X(:,tidx);
    u = Y(:,uidx);
    t = zeros(rX,1);

    % iteration for outer modeling until convergence
    while norm(t1-t) > tol
        w = X'*u;
        w = w/norm(w);
        t = t1;
        t1 = X*w;
        q = Y'*t1;
        q = q/norm(q);
        u = Y*q;
    end

    % update p based on t
    t=t1;
    p=X'*t/(t'*t);
    pnorm=norm(p);
    p=p/pnorm;
    t=t*pnorm;
    w=w*pnorm;

    % regression and residuals
    b = u'*t/(t'*t);
    X = X - t*p';
    Y = Y - b*t*q';

    % save iteration results to outputs:
    k=k+1;
    T(:,k)=t;
    P(:,k)=p;
    U(:,k)=u;
    Q(:,k)=q;
    W(:,k)=w;
    B(k,k)=b;
    % uncomment the following line if you wish to see the convergence
%     disp(norm(Y))
end
T(:,k+1:end)=[];
P(:,k+1:end)=[];
U(:,k+1:end)=[];
Q(:,k+1:end)=[];
W(:,k+1:end)=[];
B=B(1:k,1:k);

end
```

2.5.2 *Dynamic PLS*

To incorporate the linear dynamic relationship (between time series data) in the PLS framework, the decomposition of the X block is given by Equation 2.37, and the dynamic analog of Equation 2.38 is as follows:

$$Y = G_1(t_1)q_1^T + G_2(t_2)q_2^T + \ldots G_n(t_n)q_n^T + F = Y_1^{\exp} + Y_2^{\exp} + .. + Y_n^{\exp} + F \tag{2.43}$$

where G_i denotes the linear dynamic model, which can be identified at each time instant by ARX (autoregressive with exogenous terms) and FIR (finite impulse response model) models as well. $G_i(t_i)q_i^T$ is a measure of Y space explained by the ith PLS dimension in latent subspaces. Y_1^{exp} is the experimental y data in the first dimension. G is the diagonal matrix comprising the dynamic elements identified at each of the n latent subspaces. Equation 2.44 represents the linear second-order ARX structure, which relates input and output signals of time series data.

$$y(k)+a_1y(k-1)+a_2y(k-2)=b_1x(k-1)+b_2x(k-2) \tag{2.44}$$

where $y(k)$ = output at k^{th} instant, $x(k)$ = input at k^{th} instant. The parameters of the ARX-based inner dynamic models relating the scores T and U are estimated by least squares regression. The ARX-based input matrix used in regression analysis is as follows:

$$X_{ARX}=\{U_{k-1}, U_{k-2}, T_{k-1}, T_{k-2}\} \tag{2.45}$$

The FIR model is also used for inner model development. The FIR-based input matrix is as follows:

$$X_{FIR}=\{T_{k-1}, T_{k-2}, T_{k-3}, T_{k-4}\} \tag{2.46}$$

T and U represent the matrices of scores of X and Y, respectively. The identified process transfer function in the discrete time domain is

$$G_p(z)=\frac{U(z)}{T(z)}=\frac{b_1z+b_2}{z^2+a_1z+a_2} \tag{2.47}$$

The post compensation of the U matrix (PLS inner dynamic model output) with loading matrix Q provides the PLS predicted output Y. The input matrix T to the PLS inner dynamic model is generated by post compensating the original X matrix with loading matrix P. Figure 2.3 represents the PLS-based identification of process dynamics. Prior to dynamic modeling, the order of the model should be selected. It is difficult to choose the model order. Autocorrelation signals render a good indication about order that depends on how many past input and past output values are taken in the input matrix for FIR and ARX models. The model parameters for both ARX and FIR models are estimated by a linear least squares technique.

For higher order, complex, and nonlinear processes, neural identification of process dynamics using the latent variables has been explored. For every input-output pair participating in training the neural networks, the use of the historical database eliminates the necessity of determining a cross-correlation coefficient. In a multiple-input multiple-output (MIMO) process control application, the cross-correlation coefficient would have guided the selection of the most effective inputs and their corresponding targets to be

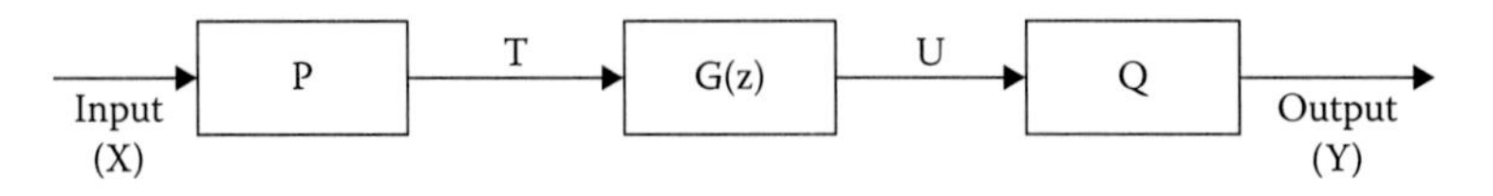

***Figure* 2.3** Schematic of PLS based dynamics.

used in training the neural networks. The PLS-identified process is somewhat decoupled owing to the orthogonality of the input scores, and the rotation of the input scores is correlated with the output scores. The scores corresponding to all the time series data are generated using the principal component decomposition. Instead of the process being identified by a linear ARX-based model coupled with least squares regression, in the nonlinear PLS design, the relationship between the T and U scores is estimated by a feed-forward backpropagation neural network. The network input is arranged in an ARX structure using the process historical database. The input ($N1$) and output ($N3$) to the multilayer (three layers) feed-forward neural network (FFNN) representing the forward dynamics of the process regarding its training and simulation phase are as follows: the training algorithm used is gradient based. The convergence criterion is the mean squared error (MSE).

Training phase

$$N1 = \{U(t-3), U(t-2), T(t-3), T(t-2)\} \tag{2.48}$$

$$N3 = U(t-1) \tag{2.49}$$

Simulation phase

$$N1 = \{U(t-2), U(t-1), T(t-1), T(t)\} \tag{2.50}$$

$$N3 = U(t) \tag{2.51}$$

Application of static linear PLS in classification is presented in Chapter 6.

2.6 Cross-correlation coefficient

The cross-correlation function of two waveforms is a graph of the similarity between these waveforms as a function of the time shift between them. Cross-correlation is a well-known and elegant method for detecting common periodicities between two signals of interest. It detects a desirable signal in the presence of noise or other undesirable signals. Cross-correlation techniques have been used as feature detectors in architectural acoustics, spectral analysis, and speech coding [53,54]. In this book, the cross-correlation function between input (pulse voltammetric) and output (current response) signals of the electronic tongue is considered for feature extraction in the subspace instead of using the e-tongue output only (Chapter 6). The experimental uncertainties and physical properties of the system are being reflected in the cross-correlation coefficient. Since the cross-correlation coefficient-based method takes into account the characteristics of the liquid–electrode interface as well as the output signal pattern, one can expect the highest magnification in features. The following equations are used to calculate the cross-correlation coefficients (r_{xy}) of the time series x and y, each with N data points:

$$r_{xy}(k) = \frac{c_{xy(k)}}{\sqrt{c_{xx(0)}\, c_{yy}(0)}} \tag{2.52}$$

$$c_{xy(k)} = \sum_{t=1}^{N-k}(x_t - \overline{x})(y_t - \overline{y}) + \sum_{t=N-k+1}^{N}(x_t - \overline{x})(y_{t-N+k} - \overline{y}) + \sum_{t=1}^{N}(x_t - \overline{x})(y_t - \overline{y}) \text{ where, } k = 1, 2....N \quad (2.53)$$

If $k = 0$

$$C_{xx}(0) = \sum_{t=1}^{N}(x_t - \overline{x})^2 \quad (2.54)$$

$$C_{yy}(0) = \sum_{t=1}^{N}(y_t - \overline{y})^2 \quad (2.55)$$

k indicates a time shift of one signal (c) with respect to the other; $k = 0$ signifies the original two time series synchronized in time. Here, x and y denote the input and output signals. Cross-correlation data can be thought of as one kind of normalized series of output signals with respect to its reference input signals, but with enhanced and unbiased features. The cross-correlation was, therefore, calculated using $k = 0$ to (length of the series − 1). *Application of cross-correlation-based feature extraction is presented in Chapter 6.* Program 2.7 presents the MATLAB code for finding cross-correlated series.

Program 2.7 MATLAB code for cross-correlated timed series

```
function [X1,X2,X3]=crosscorr(X)
% X is nx6 matrix of which columns 2, 4, and 6 are output and 1, 3, and 5
are input signals. X1, X2, and X3 are cross-correlated series.

X=load(data,'-ascii');
X=zscore(X) ;

X1=xcorr(X(:,1) , X(:,2));
X2=xcorr(X(:,3),X(:,4)) ;

X3=xcorr(X(:,5) , X(:,6)) ;
N=size(X1);
T=0:1:N(1,1)-1;
T=T' ;
n=size(X);
t=0:1:n(1,1)-1;
t=t';

subplot(2,1,1)
plot(t,X(:,2), 'r' ,t,X(:,4) ,'g', t, X(:,6),'b')
grid on
ylabel('normal output')
xlabel('k')
axis auto

subplot(2,1,2)
plot(T,X1 , 'r' , T, X2, 'g' , T, X3,'b')
grid on
ylabel('cross correlated output')
xlabel('k')
axis auto

end
```

2.7 *Sammon's nonlinear mapping*

Clustering algorithms are central to the design of statistical classifiers. Clustering techniques depend on user-defined control parameters, namely similarity measures, similarity thresholds, the number of iterations required, and those thresholds that control cluster evolution or amalgamation. There are a few nonlinear projection methods like the Kohonen self-organization map (SOM), generative topographic maps (GTM), Sammon projection, and autoassociative feed-forward networks. The nonlinear mapping (NLM) algorithm does not require any control parameter based on the *a priori* knowledge of the data, and it is named after John Sammon, Jr. who initially proposed it in [55]. The user only has to set the number of iterations and the convergence criterion. Sammon's NLM technique has been widely applied for pattern recognition and protein sequencing projects [56,57]. Sammon's NLM can take its place under nonclassical multidimensional scaling techniques as well. Sammon's NLM theory is based on a point mapping of N number of L dimensional vectors from the L space to a lower d dimensional space such that the inherent data structure is preserved following the mapping process. This allows researchers to map the higher dimensional data to a reduced d-dimensional space through a minimization of the error or stress function. The significance of the stress function is that it expresses how well the distances in the reduced space fit the distances in the original dimension space, giving more weight to the small distances.

By assuming N vectors in an L-space are designated as X_i, $i = 1,2.3\ldots N$ and these N vectors in a d-space ($d = 2$ or 3) are designated as Y_i, $i = 1,2.3\ldots N$, the distance between the vectors X_i and X_j in the L-space can be defined by $d_{ij} = \text{dist}[X_i X_j]$ and the distance between the corresponding vectors Y_i and Y_j in the d-space is $D_{ij} = \text{dist}[Y_i Y_j]$:

$$D_{ij} = \sqrt{\sum_{k=1}^{d} [Y_{ik} - Y_{jk}]^2} \tag{2.56}$$

$$d_{ij} = \sqrt{\sum_{k=1}^{L} [X_{ik} - X_{jk}]^2} \tag{2.57}$$

$$E_{ij} = \frac{1}{\sum_{i<j}^{N} D_{ij}} \sum_{i<j}^{N} \frac{(d_{ij} - D_{ij})}{D_{ij}} \tag{2.58}$$

The data sets in the reduced dimension are mapped by minimizing the stress function "E" using the gradient descent method. The error is a function of the $d \times N$ variables (Y_{pq}, where $p = 1\ldots N$ and $q = 1\ldots d$). The NLM algorithm adjusts Y_{pq} variables or equivalently changes the d-space configuration to decrease the error. It is explained below. Let $E(m)$ be defined as the mapping error after the mth iteration

$$E_{ij}(m) = \frac{1}{C} \sum_{i<j}^{N} \frac{(d_{ij}(m) - D_{ij}(m))^2}{d_{ij}}, \text{ where } C = \sum_{i<k}^{N} [d_{ij}] \tag{2.59}$$

$$d_{ij}(m) = \sqrt{\sum_{k=1}^{d} [Y_{ik}(m) - Y_{jk}(m)]^2} \tag{2.60}$$

The new d-space configuration at time m+1 is given by

$$Y_{pq}(m+1) = Y_{pq}(m) - (LR)\Delta mk(m), \text{ where } \Delta mk(m) = \frac{\dfrac{\partial E(m)}{\partial Y_{pq}(m)}}{\left|\dfrac{\partial^2 E(m)}{\partial Y_{pq}^2(m)}\right|} \tag{2.61}$$

LR is the learning rate, which is determined to be 0.3 or 0.4:

$$\frac{\partial E}{\partial Y_{pq}} = -\frac{2}{C}\sum_{j=1,j\neq p}^{N}\left[\frac{d_{pj} - D_{pj}}{d_{pj}D_{pj}}\right](Y_{pq} - Y_{jq}) \tag{2.62}$$

$$\frac{\partial^2 E}{\partial Y_{pq}^2} = -\frac{2}{C}\sum_{j=1,j\neq p}^{N}\frac{1}{d_{pj}D_{pj}}\left[(d_{pj} - D_{pj}) - \frac{(Y_{pq} - Y_{jq})^2}{D_{pj}}\left(1 + \frac{(d_{pj} - D_{pj})}{D_{pj}}\right)\right] \tag{2.63}$$

The NLM is obtained by repeatedly selecting two points at random and updating their projected coordinates using Equation 2.61. When applied to a family of homologous sequences, the Sammon projection is able to capture the essential features of the distance matrix, and it provides a simple, intuitive, and faithful representation of sequence similarity, in whatever metric one may wish to define it. More importantly, these maps can restore important structural and/or functional relationships. Sammon's mapping algorithm, unlike for example PCA, does not yield a mathematical or algorithmic mapping procedure for previously unseen data points. In other words, when a new point has to be mapped, the whole mapping procedure must be repeated. MATLAB code (Program 2.8) for Sammon's mapping (sammon.m) has been adapted from *theoval.cmp.uea.ac.uk/matlab/default.html*. The dimensionality reduction could be done using the MATLAB function *mdscale*. *Application of Sammon's NLM is presented in Chapter 6.*

Program 2.8 MATLAB code for Sammon's mapping

```
function [y, E] = sammon(x, n, opts)

Y = SAMMON(X) applies Sammon's nonlinear mapping procedure on multivari-
ate data X, where each row represents a pattern and each column repre-
sents a feature. On completion, an n-dimensional output Y contains the
corresponding co-ordinates of each point of X on the map. By default, a
two-dimensional map is created. If X contains any duplicated rows, SAMMON
will fail (ungracefully). The function also returns the value of the cost
function in E (i.e.,the stress of the mapping). A set of optimization
options can also be specified using a third argument, where OPTS is a
structure with fields:
%
%       MaxIter          - maximum number of iterations
%       TolFun           - relative tolerance on objective function
%       MaxHalves        - maximum number of step halvings
%       Input            - {'raw','distance'} if set to 'distance', X is
%                          interpreted as a matrix of pairwise distances.
%       Display          - {'off', 'on', 'iter'}
%       Initialisation - {'pca', 'random'}
```

```
%
>>>>>>>>>>>>>>>>>>>>>>>>>>>>>>>>>>>>>>>>>>>>

if nargin < 3
   opts.Display        = 'iter';
   opts.Input          = 'raw';
   opts.MaxHalves      = 20;
   opts.MaxIter        = 500;
   opts.TolFun         = 1e-9;
   opts.Initialisation = 'random';
end

% the user has requested the default options structure

if nargin == 0
   y = opts;
   return;
end

% create a two-dimensional map unless dimension is specified

   if nargin < 2
   n = 2;
end

% set level of verbosity

if strcmp(opts.Display, 'iter')

   display = 2;

elseif strcmp(opts.Display, 'on')

   display = 1;

else

   display = 0;

end

% create distance matrix unless given by parameters

if strcmp(opts.Input, 'distance')

   D = x;

else

   D = euclid(x, x);

end

% remaining initialisation

N     = size(x, 1);
scale = 0.5/sum(D(:));
D     = D + eye(N);
Dinv  = 1./D;

if strcmp(opts.Initialisation, 'pca')

   [UU,DD] = svd(x);
   y       = UU(:,1:n)*DD(1:n,1:n);
```

```
else
   y = randn(N, n);
end
one   = ones(N,n);
d     = euclid(y,y) + eye(N);
dinv  = 1./d;
delta = D - d;
E     = sum(sum((delta.^2).*Dinv));
% get on with it
for i=1:opts.MaxIter
    % compute gradient, Hessian and search direction
    delta    = dinv - Dinv;
    deltaone = delta*one;
   g        = delta*y - y.*deltaone;
   dinv3    = dinv.^3;
   y2       = y.^2;
   H        = dinv3*y2 - deltaone -2*y.*(dinv3*y) + y2.*(dinv3*one);
   s        = -g(:)./abs(H(:));
   y_old    = y;
   % use step-halving procedure to ensure progress is made
   for j=1:opts.MaxHalves
     y(:) = y_old(:) + s;
     d     = euclid(y,y) + eye(N);
     dinv  = 1./d;
     delta = D - d;
     E_new = sum(sum((delta.^2).*Dinv));
     if E_new < E
        break;
     else
        s = 0.5*s;
     end
end
% bomb out if too many halving steps are required
if j == opts.MaxHalves
     if display
        fprintf(1, 'Warning : MaxHalves exceeded.\n');
     end
     break;
end
 % evaluate termination criterion
 if abs((E - E_new)/E) < opts.TolFun
   if display
       fprintf(1, 'Optimisation terminated - TolFun exceeded.\n');
   end
```

```
    break;
end
% report progress
E = E_new;
if display > 1
    fprintf(1, 'epoch = %d : E = %12.10f\n', i, E*scale);
    end
end
% fiddle stress to match the original Sammon paper
E = E*scale;
% end sammon
function d = euclid(x,y)
d = sqrt(sum(x.^2,2)*ones(1,size(y,1))+ones(size(x,1),1)*sum(y.^2,2)'-2*(
x*y'));
```

2.8 Moving window-based PCA

Wang et al. [57] pointed out that, despite its tremendous success, conventional PCA could suffer major drawbacks in monitoring. The normal operating data may be insufficient when the process monitoring is started. Since the confidence limits of monitoring statistics are obtained through a statistical manner, the higher the number of normal observations, the better. It is thus desirable that the new observation, once found normal, is augmented into the normal data set to modify the confidence limits, hence making them consistent [58]. Once static PCA is built; almost settled mean, variance, and covariance exist among variables. It can't deal with the time varying behavior of the process. In real industrial processes, alteration happens frequently due to varying market demand of products, and variations in raw material quality and supply and in utility prices. Multivariate statistical process control (MSPC) is difficult to apply to above such circumstances having entrenched nonstationary and time-varying behavior. Therefore, a new method is required to ensure adaptive monitoring by a model with a changing process condition [59]. Wold [60] proposed the use of an exponentially weighted moving average (EWMA) filter for updating of PCA and PLS models. A straightforward approach is to use a moving (rectangular) time window, on which the statistical models are operational, for example, moving window PCA (MWPCA); a still more sophisticated way, for example, is by recursive PCA (RPCA). Li et al. [61] proposed two recursive PCA algorithms for sample-wise and block-wise recursions. In moving window-based methods, as the window slides along the data, a new process model is generated using the recent sample and estranging the old ones. Recursive techniques, on the other hand, update the model using the old and newer data samples without any disposal of data. Such recursive approaches have been applied in an adaptive statistical process control [62–64]. Industrial processes can demonstrate slow time-varying behaviors as well, such as decreasing catalyst activity, wear and tear in equipment, sensor and process wandering, and maintenance and cleaning. In such cases, the older samples are not revealing the most recent process status. Hence, RPCA is not the appropriate technique here, indulging a reduced execution speed with ever increasing data size. Although introduction of a forgetting factor can down-weight older samples, its selection is not a trivial task without *a priori* knowledge [58]. Wang et al. [57] proposed a MWPCA algorithm for adaptive monitoring. They incorporated the concept of recursive adaptation within a moving window to (i) update the mean and variance of the process variables, (ii) update the correlation matrix, and (iii) adjust the PCA model by recomputing the decomposition. In the MWPCA, the use of a constant number of samples in the window implies a constant model adaptation speed, which is not suitable for handling large data samples indicating wide range of variation in the process (used in model adaptation, hence monitoring the process). In this case, the computational speed of MWPCA drops significantly. A smaller window size recovers computational speed, however, and smaller data within the window may not be the actual representation of the underlying relationships among the process variables. In addition to that, process anomalies may be adapted as the process trends due to quick model adaptation speed. Wang et al. [57] also introduced an N-step-ahead horizon predictor into MWPCA. This implied that the PCA model, identified N-steps earlier, could be used to analyze the current observations. This proposition seemed useful for detecting slowly developing drifts. Jeng [65] proposed an adaptive monitoring technique combining RPCA and MWPCA. Liu et al. [66] proposed a moving window kernel PCA for adaptive monitoring of nonlinear processes [66]. Elshenawy et al. [67] proposed two new RPCA algorithms. The major advantages of these algorithms are low computation cost and simplicity of online implementation. The first algorithm is based on first-order perturbation theory (FOP), which is a rank-one update of the eigenpairs of the data covariance matrix [68,69]. The second one is based on the data projection method (DPM), which serves as a simple and reliable approach for adaptive subspace tracking [70].

2.8.1 Mathematical postulates of recursive PCA

A complete recursive PCA scheme is as follows:

1. *Recursive update correlation matrix with time varying mean and variance*
2. *Application of efficient algorithm for the principal component calculation of both sample-wise and block-wise augmented data matrix.*
3. *Recursive determination of the confidence limits for SPE and T^2 to facilitate adaptive monitoring / discrimination.*

Step I

Rarely, the covariance matrix is used to derive a PCA model, where the data is scaled to zero mean, but the variance is unscaled. The variance scaling affects the relative weighting of all variables. In RPCA, a recursive update of the mean, variance, and correlation matrix is essential to build a PCA model. Since a PCA model is constructed from the correlation matrix of the original process data, its update is achieved efficiently by calculating the current correlation matrix from the previous one, rather than by using the old process data. In the conventional batch-wise PCA, the raw data matrix $X_k^0 \in R^{k \times m}$ of k samples (rows) and m variables (columns) is first normalized to a matrix X_k with zero mean and unit variance. Its mean, standard deviation, and correlation matrix are given by b_k:

$$\sum\nolimits_k = \mathrm{diag}(\sigma_k(1) \cdots \sigma_k(m))$$

$$R_k \approx \frac{1}{k-1} X_k^T X_K \tag{2.64}$$

When a new measurement x_{k+1}^0 becomes available (sample-wise update), the new data is expected to augment the data matrix and calculate the correlation matrix recursively. The task for recursive calculation is to calculate b_{k+1}, X_{k+1}, and R_{k+1}. The augmented data matrix is given by

$$X_{k+1}^0 = \left| \begin{matrix} X_k^0 \\ \left(x_{k+1}^0\right)^T \end{matrix} \right| \tag{2.65}$$

For all the $k+1$ samples, the mean vector b_{k+1} and standard deviation are updated as follows:

$$b_{k+1} = \frac{k}{k+1} b_k + \frac{1}{k+1}\left(x_{k+1}^0\right) \tag{2.66}$$

$$\sigma_{k+1}(i)^2 = \frac{k-1}{k}\sigma_{k+1}(i)^2 + (\Delta b_{k+1}(i))^2 + \frac{\left(x_{k+1}^0(i) - b_{k+1}(i)\right)^2}{k} \tag{2.67}$$

where $\Delta b_{k+1} = b_{k+1} - b_k$. Given $\sum_{k+1} = \text{diag}\left(\sigma_{k+1}(1)\ldots.\sigma_{k+1}(m)\right)$, the new sample x_{k+1}^0 is scaled as

$$x_{k+1} = \sum_{k+1}^{-1} (x_{k+1}^0 - b_{k+1}) \tag{2.68}$$

Utilizing Δb_{k+1}, $\sum_{k+1}$, $\sum_k$, x_{k+1}, and the old R_k, the new correlation matrix R_{k+1} is given by

$$R_{k+1} = \frac{k-1}{k} \sum_{k+1}^{-1} \sum_k R_k \sum_k \sum_{k+1}^{-1} + \sum_{k+1}^{-1} \Delta b_{k+1} \Delta b_{k+1}^T \sum_{k+1}^{-1} + \frac{1}{k} x_{k+1} x_{k+1}^T \tag{2.69}$$

The eigenvector decomposition of R_{k+1} then provides the most recent RPCA model. While the change in mean is only a rank one modification to the correlation matrix, the change in variance $\left(\sum_{k+1}\right)$ completely changes the eigenstructure. For the case of a covariance-based PCA, there is no need to scale the variance at all. *A MATLAB program for a recursive update of the mean vector and correlation matrix is presented in Chapter 5.*

Step II

There are two numerically efficient algorithms to update the PCA model: (i) use of rank-one modification and (ii) use of Lanczos tridiagonalization presented the block-wise update of the model in RPCA and summarized rank-one modification algorithm [71].

Based on the known eigenstructure of R_k, one can compute the eigenpair of R_{k+1} by applying the rank-one modification algorithm twice. A major drawback of the rank-one modification algorithm is that all the eigenpairs have to be calculated, although only a few principal eigenpairs of the correlation matrix are needed. This is an excessive computational burden. Moreover, if one needs to update the model block-wise instead of sample-wise, a costly sequence of rank-one modifications (depending on the rank of the matrix; $x_{n_{k+L}} x_{n_{k+L}}^T$) needs to be applied. In view of this, the RPCA by the Lanczos tridigonalization algorithm received much attention, and this book has adapted this approach.

Lanczos Tridigonalization

Applying the Lanczos procedure to the symmetric correlation matrix $R_k \in R^{m\times m}$ yields a symmetric tridiagonal matrix, Γ. The extreme eigenvalues of the two matrices R_k and Γ are approximately the same.

$$\Gamma^{m_1 \times m_1} = L^{'}RL = \begin{pmatrix} \alpha_1 & \beta_1 & 0 & \cdots & 0 \\ \beta_1 & \alpha_2 & \beta_2 & 0 & \vdots \\ 0 & \beta_2 & \ddots & \ddots & 0 \\ \vdots & \cdots & \ddots & \vdots & \beta_{m_1-1} \\ 0 & \cdots & 0 & \beta_{m_1-1} & \alpha_{m_1} \end{pmatrix}$$

$$\Gamma = L^T R_k L \tag{2.70}$$

$\Gamma = R^{m_1 \times m_1}$ is a matrix containing the orthogonal Lanczos vectors and $m_1 < m$. The Lanczos algorithms for symmetric matrices were introduced in the 1950s and are described in Golub and Loan [72] where an algorithm to compute the diagonal elements and the subdiagonal elements is provided. The main advantages of the Lanczos tridiagonalization methods originate from the following:

The original matrix is not overwritten; it is modified.
Little storage is required since only matrix ± vector products are computed.
Only a few (l_k) of the largest eigenvalues are calculated. Usually l_k is much smaller than the dimension of the matrix R_k.

Because of a round of errors, a loss of orthogonality among the computed Lanczos vectors occurs after some eigenpairs have already been calculated, leading to several undesirable phenomena. To cope with this problem, a reorthogonalization procedure has been introduced. This is a two-step procedure.

1. The tridiagonalization of the correlation matrix R_k using the Lanczos algorithm with a reorthogonalization procedure
2. The calculation of the principal eigenpairs of the resulting tridiagonal matrix

The eigenvectors, V, and eigenvalues, D, of $\Gamma^{m_1 \times m_1}$ can be computed according to Equation 2.71. The projection of Γ on eigenvectors yields a PCA loading matrix, which is holding important current information about process variables.

$$[V, D] = \text{eig}(\Gamma) \tag{2.71}$$

$$P = \Gamma V \tag{2.72}$$

The unknown/new sample (measurement vector) is projected onto a recursively updated PCA model. The corresponding score and residual vectors are calculated using the following equations:

$$t_{\text{new}} = x_{\text{new}} P (P'P)^{-1} \tag{2.73}$$

$$e_{\text{new}} = x_{\text{new}} - t_{\text{new}} P' \tag{2.74}$$

Termination of the Lanczos Algorithm

A properly selected m_1 should guarantee that the eigenvalues of Γ are approximately the same as the m_1 largest eigenvalues of R_k. If Γ includes all significant eigenvalues of R_k, trace Γ will be very close to trace R_k. It is known that trace R_k is equal to the sum of all the eigenvalues of R_k.

$$\frac{\text{trace }(\Gamma)}{\text{trace }(R_k)} \geq \epsilon \tag{2.75}$$

where ϵ is a selected threshold. For example $\epsilon = 0.995$ means that 99.5% of the variance in R_k is represented by the m_1 eigenvalues. With the computed Γ, one can calculate its first few largest eigenvalues $(\lambda_1, \lambda_2, \lambda_3, \ldots \lambda_{lk})$, which are approximately equal to the l_k principal eigenvalues of R_k, and the associated eigenvectors $|C_{lk} = C_1, C_2, \ldots C_{lk}|$. Consequently, the PCA loading matrix for R_k is $Pl_k = LC_{lk}$. The *MATLAB code for the Lanczos algorithm is provided in Chapter 5.*

Step III

With the recursively updated scores and residual available:

Calculate the T^2 and Q statistics by Equations 2.8 and 2.9.

- Calculate the threshold limits of T^2 and Q statistics by the statistical metrics in Equations 2.10 through 2.14.
- If the T^2 and Q statistics of any unknown sample class do not violate a threshold limit prescribed against a specific class, the unknown sample belongs to the specific class. At this stage, augment the data matrix with a new measurement to repeat the same.

- If the T^2 and Q statistics of any unknown liquid sample violates a threshold limit prescribed against a specific class, the augmentation of the data matrix is not allowed and the algorithm terminates.

Chapter 5 presents RPCA-based classification.

2.9 Discriminant function and hyperplane

There are two approaches to develop statistical classifiers: one is a parametric approach, in which *a priori* knowledge of data distributions is assumed, and the other is a nonparametric approach, in which no *a priori* knowledge of data distributions is assumed. Neural networks, fuzzy systems, linear discriminant analysis (LDA, a generalization of Fisher's linear discriminant, FLD), and support vector machines (SVMs) are typical nonparametric classifiers. The basic idea behind LDA-based classification is to find a linear transformation that best discriminates among classes, and the classification is then performed in the transformed space based on some metrics such as Euclidean distance. In an SVM classifier, an optimal hyperplane is determined to maximize the distance of the hyperplane from each of the feature classes to be separated. Or one can say, in a support vector machine, the optimal hyperplane is determined in order to maximize the generalization ability. Through training using input-output pairs, classifiers acquire decision functions that classify an input datum into one of the given classes. Formally, given a set *xi*'s, one needs to determine the corresponding *yi*'s with the knowledge of the training patterns and their associated membership into either category available. This process is called *binary classification*. A straight line separates the patterns on both sides of it in the two-dimensional space. If it is a three-dimensional problem, the separating line becomes a plane. And in a problem of more than three dimensions, it becomes a hyperplane. In fact, one can draw infinitely many such lines/planes/hyperplanes between linearly separable data. For a binary classification problem, the class boundary given by a direct decision function corresponds to the curve where the function vanishes, while the class boundary given by two indirect decision functions corresponds to the curve where the two functions give the same values.

Multiclass classification can be roughly divided into two groups in its approaches. The first group consists of those algorithms like *nearest neighborhoods* [73] and *CART* [74]. The second group consists of methods that involve reduction of multiclass classification problems into binary ones. Depending on the reduction techniques that are used, the second group can be further divided into one-versus-the-rest method [75,76], pairwise comparison [77,78], decision directed acyclic graph (DDAG) [79], error-correcting output coding [80,81], and multi-class objective functions [82]. In practice, the choice of reduction method from multiclass to binary is problem dependent and not a trivial task since each reduction method has its own limitations [81]. In multiclass classification problems, the choice of approach will depend on constraints (required accuracy, application modality, whether online or offline), CPU (central processing unit) time available for development and training, and the type of the classification problem. In this book, classification using the pairwise comparison and decision directed acyclic graph (DDAG) method have been implemented in Fisher discriminant analysis (FDA) and SVM-based classifier development.

2.9.1 *Linear discriminant analysis (LDA)*

Linear discriminant analysis (LDA)/FDA is a well-known method for dimensionality reduction and classification that projects high-dimensional data onto a low-dimensional space where the data achieves maximum class separability [73,83,84]. LDA has been applied successfully in many applications including face recognition and microarray gene expression data analysis. PCA is the most popular unsupervised technique for dimensionality reduction without having any level information assigned to the projected data. It searches for directions in the data that have the largest variance and subsequently projects the data onto it without ensuring the maximum separability among the data. LDA is a supervised learning technique that relies on class labels or informative projections. The main disadvantage of LDA/FDA is that the number of variables used as the input for an LDA/FDA model has to be smaller than the total number of samples (observations/runs here). The scores of the most significant principal components are used as inputs into an LDA model. The LDA/FDA model developed on the basis of training data can be a basis of classifying unknown data and thus for the design of a LDA/FDA-based classifier.

Consider a set of n-dimensional samples $x_1,\ldots,x_n$. N_1 is the number of samples in the subset D_1 labeled as 1 and N_2 in the subset D_2 labeled as 2. One can search for a linear combination of features that maximizes the separability among them. In other words, one can find a scalar y by projecting the samples x onto a line such that $y = w^T x$, where w is the transform. Of all the possible lines, it is desirable to select the one that maximizes the separability of the scalars. The following are the steps for an LDA classifier.

Step I

In order to find a good projection vector, a measure of separation needs to be defined. The mean vector of each class in x-space and y-space is

$$\mu_i = \frac{1}{N_i}\sum_{x\in\omega_i} x \text{ and } \breve{\mu}_1 = \frac{1}{N_i}\sum_{y\in\omega_i} y = \frac{1}{N_i}\sum_{x\in\omega_i} w^T\ x = w^T\mu_i \tag{2.76}$$

Step II

The distance between the projected means can be chosen as the objective function:

$$J(2) = |\breve{\mu}_1 - \breve{\mu}_2| = \left|w^T(\mu_1 - \mu_2)\right| \tag{2.77}$$

Step III

However, the distance between projected means is not a good measure since it does not account for the standard deviation within classes. Fisher suggested maximizing the difference between the means, normalized by a measure of the within-class scatter. For each class, one can define the scatter, an equivalent of the variance, as

$$S_i^2 = \sum_{y\in\omega_i} (y - \breve{\mu}_1)^2 \tag{2.78}$$

where the quantity $S_1^2 + S_2^2$ is called the within-class scatter of the projected examples. The Fisher linear discriminant is defined as the linear function $w^T x$ that maximizes the criterion function:

$$J(w) = \frac{|\breve{\mu}_1 - \breve{\mu}_2|^2}{S_1^2 + S_2^2} \tag{2.79}$$

Step IV

Therefore, a desirable projection is where samples from the same class are projected very close to each other, and, at the same time, the projected means are as far apart as possible. To find the optimum w^*, one must express $J(w)$ as a function of w. A measure of the scatter in feature space x is as follows:

$$S_i = \sum_{x \in \omega_i} (x - \mu_i)(x - \mu_i)^T \tag{2.80}$$

$$S_1 + S_2 = S_w \tag{2.81}$$

where S_w is called the within-class scatter matrix. The scatter of the projection y can be expressed as a function of the scatter matrix in feature space x.

$$S_i^2 = \sum_{y \in \omega_i} (y - \breve{\mu}_1)^2 = \sum_{x \in \omega_i} (w^T x - w^T \mu_i)^2$$

$$= \sum_{x \in \omega_i} w^T (x - \mu_i)(x - \mu_i)^T w = w^T S_i w \tag{2.82}$$

$$\breve{S}_1^2 + \breve{S}_2^2 = w^T S_w w \tag{2.83}$$

Similarly, the difference between the projected means can be expressed in terms of the means in the original feature space.

$$(\breve{\mu}_1 - \breve{\mu}_2)^2 = (w^T \mu_1 - w^T \mu_2)^2 = w^T (\mu_1 - \mu_2)(\mu_1 - \mu_2)^T w = w^T S_B w \tag{2.84}$$

The matrix S_B is called the between-class scatter. Since S_B is the outer product of two vectors, its rank is at most one. One can finally express the Fisher criterion in terms of S_w and S_B in the following way:

$$J(w) = \frac{w^T S_B w}{w^T S_w w} \tag{2.85}$$

Step V

To find the maximum of w, Equation 2.85 is derived with respect to w and equated to zero :

$$\frac{d}{dw} J(w) = \frac{d}{dw}\left(\frac{w^T S_B w}{w^T S_w w}\right) = 0 \tag{2.86}$$

$$w^T S_w w \frac{d}{dw}(w^T S_B w) - w^T S_B w \frac{d}{dw}(w^T S_w w) = 0 \tag{2.87}$$

$$\left[w^T S_w w\right] 2S_B w - \left[w^T S_B w\right] 2S_w w = 0 \tag{2.88}$$

Dividing by $w^T S_w w$:

$$\left[\frac{w^T S_w w}{w^T S_w w}\right] S_B w - \left[\frac{w^T S_B w}{w^T S_w w}\right] S_w w = 0 \tag{2.89}$$

$$S_B w - J S_w w = 0 \tag{2.90}$$

$$S_w^{-1} S_B w - Jw = 0 \tag{2.91}$$

Solving the generalized eigenvalue problem $S_w^{-1} S_B w = Jw$ yields

$$w^* = \arg\max \frac{w^T S_B w}{w^T S_w w} = S_w^{-1}(\mu_1 - \mu_2) \tag{2.92}$$

This is known as Fisher's linear discriminant, although it is not a discriminant but rather a specific choice of direction for the projection of the data down to one dimension. After an optimal solution of w is been found, setting a threshold criterion on $y = w^t x$ makes FLD a true binary classifier. The optimal w can be easily found when $J(w)$ is expressed in a matrix-oriented way in which the expression of $J(w)$ is known as the generalized Rayleigh quotients.

Step VI

Once the transformation w is found, the classification is then performed in the transformed space based on some distance metric, such as. Euclidean distance $d(x,y) = \sqrt{\sum_i (x_i - y_i)^2}$,

Mahalanobis distance < and cosine measure $(x,y) = \left(1 - \frac{\sum_i x_i y_i}{\sqrt{\sum_i x_i^2}\sqrt{\sum_i y_i^2}}\right)$.

On the arrival of new instance, z, it is classified to $\{\arg\min d(zw, \overline{x_k} w)\}$, where, $\overline{x_k}$ is the centroid of the kth class. LDA will fail when the discriminatory information is not in the mean but rather in the variance of the data. The computational complexity of training time consists of evaluating the intra- and interclass covariance matrices, eigenvalue decomposition, and selecting discriminating features. If n and m are the number of instances and the number of features respectively, the time of selecting discriminating features is almost linear to m. Therefore, the total time is about $O(nm \min(n, m))$. Since we only need the largest $n - 1$ eigenvalues and corresponding eigenvectors, the complexity could be even lower if the feature vectors are sparse. *LDA is applied in Chapter 5.* The MATLAB function *classify* has been used for LDA-based discrimination.

2.9.2 Support vector machine (SVM)

The SVM is a supervised linear classifier by virtue of a separating hyperplane as its output. Given labeled inputs as training data, it can develop a predictive model to assign an unknown test data to either of the level classes. The data points that are closest to the hyperplane are termed "support vectors" and have maximum influence on the orientation and position of the separating hyperplanes. The selected decision boundary/optimal hyperplane will be one that leaves the greatest margin between the two classes, where margin is defined as the sum of the distances to the hyperplane from the closest points of the two classes. The trade-off between margin and misclassification error is controlled by a user-defined constant [85]. The support vector machine finds an optimal decision boundary in the case of linearly separable data. SVM maps the patterns into a higher dimensional feature space such that the troubled nonlinear separation in the original space becomes facile by the linear separation in the higher dimensional feature space. The use of kernel functions allows this transformation. Originally the SVM was devised for binary classification, or classifying data into two types. They are accurate, robust, and quick to apply to test instances. However, the elegant theory behind the use of large-margin hyperplanes cannot be easily extended to multiclass classification problems. An SVM constructs a hyperplane $g(x) = 0$ that acts as a decision surface in such a way that the margin of separation between the two classes is maximized. Like neural classifiers, applications of SVMs to any classification problem require the determination of several user-defined parameters, such as choice of an appropriate kernel and related parameters, determination of a suitable value of regularization parameter, and a suitable optimization technique. A number of methods have been proposed to implement SVMs to produce multiclass classification. Most of the research in generating multiclass support vector classifiers can be divided in two categories. One approach involves constructing several binary classifiers and combining their results, while another approach considers all data in one optimization formulation.

2.9.2.1 Determination of decision function in SVM

Consider an input pattern X, which is being transformed into feature vector $\Phi(X)$ that belongs to class C1 or C2. A linear discriminant function $g(X) = W^T\Phi(X) + b$ (where w is a weight vector and b is a bias) represents a hyperplane in D-dimensional space. W is a vector perpendicular to the hyperplane that represents the orientation and b is the position of the hyperplane in a D-dimensional space.

For any input pattern X_1 (corresponding feature vector $\Phi(X_1)$):

$$\begin{cases} \text{If } g(X_1) = W^T\Phi(X_1) + b > 0, \text{ then } X_1 \text{ belongs to C1} \\ \text{If } g(X_1) = W^T\Phi(X_1) + b < 0, \text{ then } X_1 \text{ belongs to C2} \\ \text{If } g(X_1) = W^T\Phi(X_1) + b = 0, \text{ then } X_1 \text{ is lying on the hyperplane} \end{cases} \tag{2.93}$$

A pattern X is given a belongingness $y = \pm 1$.

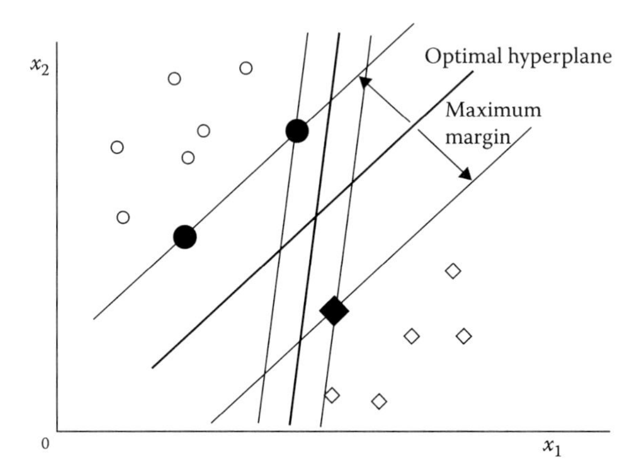

Figure 2.4 The optimal hyperplane with maximum margin determined with a few support vectors (solid symbols).

$$W^T\Phi(X)+b\}\begin{cases} >0 \text{ for } y_i=1 \\ <0 \text{ if } y_i=-1 \end{cases} \tag{2.94}$$

$g(X)=W^T\Phi(X)+b$ can act as a classifier if it is trained in an iterative manner with training set $(X_1, y_1) \cdots (X_n, y_n)$ to adjust values of W and b. Irrespective of $X_i \in C1/C2$, $y_i(W^T\Phi(X)+b)>0$, which can be the basis of design of the classifier.

The Euclidean distance from a training datum X to the separating hyperplane is given by

$$\frac{((W^T\Phi(X)+b))}{\|W\|} \text{ and } \left\{\frac{((W^T\Phi(X)+b))}{\|w\|}\right\}>\delta \tag{2.95}$$

$$(W^T\Phi(X)+b)\geq\|W\|\ \delta=1 \tag{2.96}$$

$$\|W\|\delta=1$$

Thus, to control separability, instead of Equation 2.94, the following inequality can be used:

$$W^T\Phi(X)+b\begin{cases} \geq 1 \text{ for } y_i=1 \\ \leq -1 \text{ if } y_i=-1 \end{cases} \tag{2.97}$$

Here, +1 and −1 on the right-hand sides of the inequalities can be a constant α and $-\alpha$, respectively. Equation 2.97 is equivalent to

$$y_i(W^T\Phi(X)+b)\geq 1 \text{ for } i=1\ldots m \text{ features} \tag{2.98}$$

$y_i(W^T\Phi(X)+b)=1$, the equality is true for support vector $\Phi(X)$. Support vectors have a tremendous influence on the design of the hyperplane $(W^T\Phi(X)+b)$. The distance between the separating hyperplane and the training datum nearest to the hyperplane is called the margin. The generalization ability depends on the location of the separating hyperplane, and the hyperplane should be distanced as much as possible from both the feature classes (Figure 2.4).

2.9.2.2 Determination of optimal separating hyperplane in SVM

There is no training data that exists between $W^T\Phi(X)+b=+1$ and $W^T\Phi(X)+b=-1$. Distance between these two hyperplanes or margins is $\frac{2}{\|W\|}$. In order to be qualified as a better classifier, this margin should be as large as possible. The distance between the hyperplane and the closest pattern is $\frac{1}{\|W\|}$, a separating hyperplane that maximizes separation between classes is an optimal hyperplane. For this purpose we need to solve the quadratic optimization problem with a linear equality constraint. Therefore, the optimal separating hyperplane is the solution of the following optimization problem:

$$\min\Phi(W)=\frac{1}{2}W^TW \text{ subjected to } \forall_i\, y_i(W^T\Phi(X_i)+b)=1 \tag{2.99}$$

Since the support vector machine classifier depends on support vectors, the constraint Equation 2.99 is an equality constraint. The mathematical tool of choice for simplifying this problem is the formation of Lagrangian function:

$$L(W,b)=\frac{1}{2}W^TW-\sum\alpha_i\left[y_i\left(\frac{1}{2}W^TW\right)-1\right]=\frac{1}{2}W^TW-\sum y_i\big[(\alpha)\big]_i\, b-\sum y_i\big[(\alpha)\big]_i\,(w.\Phi(X_i))+\sum\big[(\alpha)\big]_i \tag{2.100}$$

where α_i is the Lagrangian multiplier. The conditions of optimality are

$$\frac{\partial L}{\partial b}=0\Rightarrow\sum_{i=1}^{m}y_1\alpha_i=0,\ m \text{ is the number of features} \tag{2.101}$$

$$\frac{\partial L}{\partial W}=0\Rightarrow w=\sum_{i=1}^{m}y_i\alpha_i\ X_i=0,\ m \text{ is the number of features} \tag{2.102}$$

Putting the value of Equations 2.101 and 2.102 in Equation 2.100:

$$\begin{aligned} L(W,b)&=\frac{1}{2}W^TW-\sum y_i\alpha_i\ b-\sum y_i\alpha_i(W.\Phi(X_i))+\sum\alpha_i \\ &=\sum_{i=1}^{m}\alpha_i-\frac{1}{2}\sum\alpha_i\alpha_j\,y_i\ y_j\ \Phi(X_i)^T\Phi(X_j) \end{aligned} \tag{2.103}$$

This approach leads to solving the following dual problem:

$$\max L(\alpha)\sum_{i=1}^{m}\alpha_i = -\frac{1}{2}\sum_{i,\,j=1}^{m} y_i\alpha_i y_j\alpha_j \Phi(X_i)^T \Phi(X_j)$$

$$\text{subjected to } \forall_i\ \alpha_i \geq 0, \text{ and } \sum_i y_i\ \ \alpha_i = 0 \tag{2.104}$$

The direction W^* of the optimal hyperplane is then recovered from a solution α^* of the optimization problem (Equation 2.104):

$$W^* = \sum_i \alpha_i^* y_i \Phi(X_i) \tag{2.105}$$

The bias b^* becomes

$$b = \frac{1}{2}\left[\min\left\{\sum \alpha_i y_i \left(\Phi(X_i)^T \Phi(X_i)\right)\right\}_{(i|y_i=+1)} + \max\left\{\sum \alpha_i y_i \left(\Phi(X_i)^T \Phi(X_i)\right)\right\}_{(i|y_i=-1)}\right] \tag{2.106}$$

The linear discriminant function can then be written as

$$\hat{y}(x) = W^{*T}X + b^* = \sum_{i=1}^{m} y_i\alpha_i^* \Phi(X_i)^T \Phi(X) + b^* \tag{2.107}$$

For an unknown test vector Z, the decision class is determined by the sign of the following expression:

$$D(Z) = \text{sign}\left(\sum_{j=1}^{m} y_i\alpha_i^* \Phi(X_j)\cdot Z + b^*\right) \tag{2.108}$$

The optimization problem in Equation 2.103 and the linear discriminant function in Equation 2.106 only involve the patterns X through the computation of dot products in feature space. There is no need to compute the features $\Phi(X)$ but only the dot products directly. Boser et al. [86] proposed a kernel function $K(x, x')$ that represents a dot product $\Phi(X)^T \Phi(X)$ in some unspecified high-dimensional space. There are several kinds of kernel functions as described in Table 2.1.

The above discussion on SVM classification is a linear and binary one. For multiclass classification, one needs several binary classifications. The MATLAB functions svmtrain and svmclassify are used for SVM-based binary classification. *SVM is utilized in Chapter 6.*

Table 2.1 Commonly used kernel functions

Name of the kernel function	Expression
Linear	$K(u,v) = u^T v$
Polynomial of degree n	$K(u,v) = \left(u^T v + 1\right)^n$
Gaussian radial basis function (RBF)	$K(u,v) = e^{-\frac{1}{2}\left\{(u-v)^T \sum^{-1} (u-v)\right\}}$
Sigmoid	$K(u,v) = \tan\text{h}\left[u^T v\right] + b$

2.10 Multiclass decision function

There are different methods to determine the decision functions for a multiclass problem, which are as follows.

2.10.1 One against the rest approach

If the data set is to be classified into M classes, M binary classifiers may be created where each classifier is trained to distinguish one class from the remaining M-1 classes. For example, a class one binary classifier is designed to discriminate between class one data vectors and the data vectors pertaining to the rest of the classes. Other classifiers are constructed in the same manner. During the testing or application phase, data vectors are classified by finding the margin from the linear separating hyperplane. The final output is the class that corresponds to the SVM/LDA with the largest margin. However, if the outputs corresponding to two or more classes are very close to each other, those points are labeled as unclassified, and are prone to causing a subjective decision to be made. This multiclass method has an advantage in the sense that the number of binary classifiers to be constructed equals the number of classes. However, there are some drawbacks. Since the binary classifiers are obtained by training on different binary classification problems, it must be ensured that they are on comparable scales [87]. During the training phase, the memory requirement is very high and amounts to the square of the total number of training samples. This may cause problems for large training data sets, and may lead to computer memory problems, if there exist M classes and each have an equal number of training samples. During the training phase, the ratio of training samples of one class to rest of the classes will be 1:(M–1). This ratio, therefore, shows that training sample sizes will be unbalanced. Because of these limitations, the *against the rest* approach of multiclass classification had been proposed.

2.10.2 One against one approach

In this method, multiclass classifiers for all possible pairs of classes are created [73,88]. Therefore, for M classes, there will be $\frac{M(M-1)}{2}$ binary classifiers. The output from each classifier in the form of a class label is obtained. Given a test instance, the multiclass classification is then executed by evaluating all $\frac{M(M-1)}{2}$ binary classifiers and assigning the unknown/test instance to the class, which gets the highest number of resulted outputs (decision class) of the binary classifiers. In case of a tie, the strategy is to randomly select one of the class labels that are tied. The number of classifiers created by this method is generally much larger than the previous method. However, the number of training data vectors required for each classifier, and the ratio of training data vector size for one class against another, are much smaller. Therefore, this method is considered more symmetric than the *one against the rest* method. Moreover, the memory required to create the kernel matrix is much smaller. However, the main disadvantage of this method is the increase in the number of classifiers as the number of classes increases. For example, for 5 classes of interest, 10 classifiers need to be created.

2.10.3 Decision directed acyclic graph (DDAG)-based approach

Platt et al. [79] proposed a multiclass classification method called *directed acyclic graph SVM* (DAGSVM) based on the *decision directed acyclic graph* (DDAG) structure that forms a tree-like structure. The DDAG method in essence is similar to pairwise classification such that, for an M class classification problem, the number of binary classifiers is equal to $\frac{1}{2}M(M-1)$ and each classifier is trained to classify two classes of interest. Each classifier is treated as a node in the graph structure. Nodes in DDAG are organized in a triangle with the single root node at the top and increasing thereafter in an increment of one in each layer until the last layer that will have M nodes. The DDAG evaluates an input vector X starting at the root node and moves to the next layer based on the output values. For instance, it exits to the left edge if the output from the binary classifier is negative, and it exits to the right edge if the output from the binary classifier is positive. The binary classifier of the next node is then evaluated. The path followed is called the evaluation path. The DDAG method basically eliminates one class out from a list. Initially, the list contains all classes. Each node evaluates the first class against the last class in the list. For example, the root node evaluates class 1 against class M. If the evaluation results in one class out of two classes, the other is eliminated from the list. The process then tests the first and the last class in the new list. It is terminated when only one class remains in the list. The class label of the nodes in the final layer of the evaluation path represents the classes that remained in the list. Although the number of binary classifiers still equals the pairwise classification method, the inputs are evaluated M–1 times instead of $\frac{1}{2}M(M-1)$ times as is the case with pairwise classification. *DDAGs are utilized for classifier development in Chapters 5 and 6.*

2.10.3.1 DDAG algorithm

There are two steps in the DDAG algorithm:

1. Creation of a triangular DDAG graph
2. Evaluation path, hence membership generation (for different classes)

```
Creation of triangular DDAG graph:
MAX = max number of classes = number of level = 6
Number of nodes n_L in level L = L,
For L = 1 to MAX-1
{
   For M = MAX to {MAX = (L-1)}
    {
   For j = L to 1
       {
       %% binary classifier framework is created.
       n_L = L
          Binary classifier names at level L = M versus j
        } % end j
    } % end M
} % end L
At L = MAX-1
For M = MAX to [MAX = {M-(L-1)}]
{
```

```
For j = L to (M-L).
    {
    n_L = M-1
    Binary classifier names = M vs j
    } % end j
} % end M
At level L = MAX
Number of empty nodes = M (no binary classifiers are to be assigned to
the nodes).
%%%%%%%% triangular structure is created
```

Evaluation path creation

- *The generated triangular graph can be thought of as consisting several diagonals, along which the binary classifiers are laid. Every diagonal culminates in an empty node.*
- *Start assigning the test sample to the binary classifiers at the highest level: level L = M until level L = M-1.*
- *If the result of a binary classification is positive, →starting of evaluation path, set evalpath = +ve.*
- *If the result of a binary classification is negative, no evaluation path, set evalpath = -ve.*
- *Evaluation path creation is complete.*

```
Membership generation for each class
Set MAX = 6 for six categories
for i = MAX to 1
{
for j = 1 to (i-1)
    {
    Perform authentication of unknown(X) with binary classifier C_ij
    Store the result into R(i,j)
    Result evalpath = +ve / R(i,j)= +1 (if not in j-th class)
    Or result evalpath = -ve / R(i,j)= -1 (if not in i-th class)
    } % end j
} % end i
Move along the diagonal D = MAX to 2. Diagonal number represents the (D
vs j binary classifiers along them, where j = (MAX-1) to 1
Do
While D < 2
  {
  Membership [D] = count [D] = 0
  For D = MAX to 2
    {
    If evalpath = +ve
       {
       count [D] = count [D]+1
   else
       count [D] = count [D]
} end (%%if)
Membership [D] = count [D]
} end (%% for)
```

```
Membership [D]
D = D-1
} end (%%while)
```

References

1. S. Wold (1995). Chemometrics; what do we mean with it, and what do we want from it? *Chemometrics and Intelligent Laboratory Systems*, vol. 30, 109–115.
2. A. Sudjianto, G.S. Wasserman (1996). A nonlinear extension of principal component analysis for clustering and spatial differentiation. *IIE Transactions*, vol. 28, 1023–1028.
3. A. Trouve, Y. Yu (2000). Unsupervised clustering trees by nonlinear principal component analysis. *In Proceedings of 5th International Conference on Pattern Recognition and Image Analysis: New Info*. Tech. Samara, Russia, pp. 110–114.
4. S.J. Qin (2003). Statistical process monitoring: Basics and beyond. *Journal of Chemometrics*, vol. 17, 480–502.
5. T. Kourti, J.F. MacGregor (1995). Process analysis, monitoring and diagnosis, using multivariate projection methods. *Chemometrics and Intelligent Laboratory Systems*, vol. 28, 3–21.
6. B.M. Wise, N.B. Gallagher (1996). The process chemometrics approach to process monitoring and fault detection. *Journal of Process Control*, vol. 6, 329–348.
7. J.E. Jackson (1991). *A User's Guide to Principal Components*. New York, NY: John Wiley & Sons.
8. M. Kano, K. Nagao, S. Hasebe, I. Hashimoto, H. Ohno, R. Strauss, B. Bakshi (2000). Comparison of statistical process monitoring methods: Application to the Eastman challenge problem. *Computers and Chemical Engineering*, vol. 24, 175–181.
9. J.E. Jackson, G.S. Mudholkar (1979). Control procedures for residuals associated with principal component analysis. *Technometrics*, vol. 21(3), 341–349.
10. J.V. Kresta, J.F. MacGregor, T.E. Marlin (1991). Multivariate statistical monitoring of process operating performance. *Canadian Journal of Chemical Engineering*, vol. 69, 35–47.
11. W.J. Krzanowski (1979). Between-groups comparison of principal components. *Journal of American Statistical Association*, vol. 74, 703–707.
12. P. Nomikos, J.F. MacGregor (1995). Multivariate SPC charts for monitoring batch processes. *Technometrics*, vol. 37(1), 41–59.
13. J.F. MacGregor, T. Kourti (1995). Statistical process control of multivariate processes. *Control Engineering Practice*, vol. 3(3), 403–414.
14. B.M. Wise (1991). Adapting multivariate analysis for monitoring and modelling of dynamic systems. PhD Dissertation, University of Washington, Seattle.
15. P.C. Mahalanobis (1936). On the generalized distance in statistics. *Proceedings National Institute of Science*, vol. 2(1), 49–55.
16. A. Singhal, D.E. Seborg (2002). Clustering of multivariate time-series data. *In Proceedings of the American Control Conference*, Anchorage, AK.
17. F. Fukunaga, W. Koontz (1970). Applications of the Karhunen-Loeve expansion to feature selection and ordering. *IEEE Transactions on Computers*, vol. 19(5), 311–318.
18. M.C. Johannesmeyer (1999). Abnormal situation analysis using pattern recognition techniques and historical data. MSc Thesis, University of California, Santa Barbara.
19. M.C. Johannesmeyer, A. Singhaland, D.E. Seborg (2002). Pattern matching in historical data. *AIChE Journal*, vol. 48(9), 2022–2038.
20. A. Singhal, D.E. Seborg (2001). Matching patterns from historical data using PCA and distance similarity factors. In *IEEE Proceedings of the 2001 American Control Conference*, Piscataway, NJ, pp. 1759–1764.
21. A. Singhal, D.E. Seborg (2002). Pattern matching in multivariate time series databases using a moving window approach. *Industrial and Engineering Chemistry Research*, vol. 41, 3822–3838.
22. R. Agrawal, J. Gehrke, D. Gunopulos, P. Raghavan (1998). Automatic subspace clustering of high dimensional data for data mining applications. *In Proceedings of ACM SIGMOD International Conference on Management of Data*, Seattle, WA, pp. 94–105.
23. M.R. Anderberg (1973). *Cluster Analysis for Applications*. New York, NY: Academic Press.

24. X.Z. Wang, C. Mcgreavy,(1998). Automatic classification for mining process operational data. *Industrial and Engineering Chemistry Research*, vol. 37(6), 2215–2222.
25. M.C. Johannesmeyer, D.E. Seborg (1999). Abnormal situation analysis using pattern matching techniques. *AICHE. Annual Meeting, Dallas*, TX.
26. A. Singhal, D.E. Seborg (2005). Clustering multivariate time series data. *Journal of Chemometrics*, vol. 19, 427–438.
27. Y. Huang, T.J. McAvoy, J. Gertler (2000). Fault isolation in nonlinear systems with structured partial principal component analysis and clustering analysis. *Canadian Journal of Chemical Engineering*, vol. 78, 569–577.
28. V. Kavitha, M. Punithavalli (2010). Clustering time series data stream-A literature survey. *International Journal of Computer Science and Information Security*, vol. 8, 289–294.
29. R.A. Johnson, D.W. Wichern (2014). *Applied Multivariate Statistical Analysis*. Upper Saddle River, NJ: Pearson.
30. B.S. Everitt, S. Landau, M. Leese, D. Stahl (2011). *Cluster Analysis*. UK: John Wiley & Sons.
31. H. Wold (1966). Estimation of principal components and related models by iterative least squares. In *Multivariate Analysis II*, P.R. Krishnaiah (Ed.). New York, NY: Academic Press, pp. 391–420.
32. H. Wold (1985) Partial least squares. In *Encyclopedia of the Statistical Sciences*, S. Kotz, N.L. Johnson (Eds.), Hoboken, NJ: John Wiley & Sons, vol. 6, pp. 581–91.
33. S. Wold (1992). Nonlinear partial least squares modeling II Spline inner relation. *Chemometrics and Intelligent Laboratory Systems*, vol. 14, 71–84.
34. P. Geladi, B.R. Kowalski (1986). Partial least-squares regression: A tutorial. *Analytica Chimica Acta*, vol. 185, 1–17.
35. N. Karp, J. Griffin, K. Lilley (2005). Application of partial least squares discriminant analysis to two-dimensional difference gel studies in expression proteomics. *Proteomics*, vol. 5(1), 81–90.
36. J.M. Andrade, S. Muniategui, D. Prada (1997). Prediction of clean octane numbers of catalytic reformed naphthas using FT-m.i.r. and PLS. *Fuel*, vol. 76(11), 1035–1042.
37. D.V. Nguyen, D.M. Rocke (2002a). Tumor classification by partial least squares using microarray gene expression data. *Bioinformatics*, vol. 18, 39–50.
38. D.V. Nguyen, D.M. Rocke (2002b). Multi-class cancer classification via partial least squares with gene expression profiles. *Bioinformatics*, vol. 18, 1216–1226.
39. D. Machelle (2004). Classification of contamination in salt marsh plants using hyperspectral reflectance. *IEEE Transactions on Geoscience and Remote Sensing*, vol. 42(5), 1088–1095.
40. P. Oliver, M.E. Whelehan, E.J. Earll, T. Marianne, L. Eriksson (2006). Detection of ovarian cancer using chemometric analysis of proteomic profiles. *Chemometrics and Intelligent Laboratory Systems*, vol. 84(1–2), 82–87.
41. A.S. Rathore, S. Mittal, M. Pathaka, V. Mahalingam (2014). Chemometrics application in biotech processes: Assessing comparability across processes and scales. *Journal of Chemical Technology and Biotechnology*, vol. 89, 1311–1316.
42. S.J. Qin, T.J. McAvoy (1992). Nonlinear PLS modeling using neural networks. *Computers and Chemical Engineering*, vol.16(4), 379–391.
43. S.J. Qin, T.J. McAvoy (1996). Nonlinear FIR modeling via a neural net PLS approach. *Computers and Chemical Engineering*, vol. 20(2), 147–159.
44. S.J. Qin (1993). A statistical perspective of neural networks for process modelling and control. In *Proceedings of the 1993 Internation Symposium on Intelligent Control*, Chicago, IL, pp. 559–604.
45. D.J.H. Wilson, G.W. Irwin, G. Lightbody (1997). Nonlinear PLS using radial basis functions. *Transactions of the Institute of Measurement and Control*, vol. 19(4), 211–220.
46. T.R. Holcomb, M. Morari (1992). PLS/neural networks. *Computers and Chemical Engineering*, vol. 16(4), 393–411.
47. E.C. Malthouse, A.C. Tamhane, R.S.H. Mah (1997). Nonlinear partial least squares. *Computers and Chemical Engineering*, vol. 21(8), 875–890.
48. S.J. Zhao, J. Zhang, Y.M. Xu, Z.H. Xiong (2006). Nonlinear projection to latent structures method and its applications. *Industrial and Engineering Chemistry Research*, vol. 45, 3843–3852.
49. D.S. Lee, M.W. Lee, S.H. Woo, Y. Kim, J.M. Park (2006). Nonlinear dynamic partial least squares modeling of a full-scale biological wastewater treatment plant. *Process Biochemistry*, vol. 41, 2050–2057.

50. M.H. Kaspar, W.H. Ray (1993). Dynamic modeling for process control. *Chemical Engineering Science*, vol. 48(20), 3447–3467.
51. S. Lakshminarayanan, L. Sirish, K. Nandakumar (1997). Modeling and control of multivariable processes: The dynamic projection to latent structures approach. *AIChE Journal*, vol. 43, 2307–2323.
52. S.K. Damarla, M. Kundu (2011). Identification and control of distillation process using partial least squares based artificial neural network. *International Journal of Computer Applications*, vol. 29(7), 29–35.
53. J.P. Lewis (1995). Fast template matching. *Vision Interface*, vol. 95, 120–123.
54. R.B. Blackman, J.W. Tukey (1958). *The Measurement of Power Spectra*. New York, NY: Dover.
55. J.W. Sammon (1969). A nonlinear mapping for data structure analysis. *IEEE Transactions on Computers*, vol. 18(5), 401–409.
56. K.A. Dimitris (1997). A new method for analyzing protein sequence relationships based on Sammon maps. *Protein Science*, vol. 6(2), 287–293.
57. X. Wang, U. Kruger, G.W. Irwin (2005). Process monitoring approach using fast moving window PCA. *Indian Engineering Chemical Research*, vol. 44(15), 5691–5702.
58. X. Wang, U. Kruger, B. Lennox (2003). Recursive partial least squares algorithms for monitoring complex industrial processes. *Control Engineering Practice*, vol. 11, 613–632.
59. V.B. Gallagher, R.M. Wise, S.W. Butler, D.D. White, G.G. Barna (1997). Development and benchmarking of multivariate statistical process control tools for a semiconductor etch process; Improving robustness through model updating. *In Proceedings of ADCHEM 97*, Banff, Canada, pp. 78–83.
60. S. Wold (1994). Exponentially weighted moving principal component analysis and projection to latent structures. *Chemometrics and Intelligent Laboratory Systems*, vol. 23, 149–161.
61. W. Li, H.H. Yue, S. Valle-Cervantes, S.J. Qin (2000). Recursive PCA for adaptive process monitoring. *Journal of Process Control*, vol. 10, 471–486.
62. H.D. Jin, Y.-H. Lee, G. Lee, C. Han (2006). Robust recursive principal component analysis modeling for adaptive monitoring. *Indian Engineering Chemical Research*, vol. 45, 696–703.
63. S.W. Choi, E.B. Martin, A.J. Morris, I. Lee (2006). Adaptative multivariate statistical process control for monitoring time-varying processes. *Industrial and Engineering Chemistry Research* , vol. 45(9), 3108–3188.
64. X.B He, Y.P Yang (2008). Variable MWPCA for adaptive process monitoring. *Industrial and Engineering Chemistry Research*, vol. 47, 419–427.
65. J.C. Jeng (2010). Adaptive process monitoring using efficient recursive PCA and moving window PCA algorithms. *Journal of the Taiwan Institute of Chemical Engineers*, vol. 41, 475–481.
66. X. Liu, U. Kruger, T. Littler, L. Xie (2009). Moving window kernel PCA for adaptive monitoring of nonlinear processes. *Chemometrics and Intelligent Laboratory Systems*, vol. 96(2), 132–143.
67. L.M. Elshenawy, S. Yin, A.S. Naik, S.X. Ding (2010). Efficient recursive principal component analysis algorithms for process monitoring. *Indian Engineering Chemical Research*, vol. 49(1), 252–259.
68. B. Champagne (1994). Adaptive eigendecomposition of data covariance matrices based on first-order perturbations. *IEEE Transactions on Signal Processing*, vol. 42(10), 2758–2770.
69. T. Willink (2008). Efficient adaptive SVD algorithm for MIMO applications. *IEEE Transactions on Signal Processing*, vol. 56(2), 615–622.
70. X.G. Doukopoulos, G.V. Moustakides (2008). Fast and stable subspace tracking. *IEEE Transactions on Signal Processing*, vol. 56(4), 1452–1465.
71. W. Li, H.H. Yue, S. Valle-Cervantes, S.J. Qin (2000). Recursive PCA for adaptive process monitoring. *Journal of Process Control*, vol. 10, 471–486.
72. G. Golub, C.V. Loan (1996). *Matrix Computations*. London: Johns Hopkins Press.
73. T. Hastie, R. Tibshirani, J. Friedman (2001). *The Elements of Statistical Learning: Data Mining, Inference, Prediction*. New York, NY: Springer Series in Statistics.
74. L. Breiman, J.H. Friedman, R.A. Olshen, C.G. Stone (1984). *Classification and Regression Trees*. New York, NY: Chapman & Hall/CRC.
75. B. Scholkopfand, A.J. Smola, (2002). *Learning with Kernels: Support Vector Machines, Regularization, Optimization, and Beyond*. Cambridge, MA: MIT Press.

76. L. Bottou, C. Cortes, J.S. Denker, H. Drucker, I. Guyon, L.D. Jackel, (1994). Comparison of classifier methods: A case study in handwriting digit recognition. In International Conference on Pattern Recognition, *IEEE Computer Society Press*, pp. 77–87.
77. U.H.G. Kreeel, (1999). Pairwise classification and support vector machines. In *Advances in Kernel Methods*, Cambridge, MA: MIT Press.
78. T. Hastie, R. Tibshirani (1998). Classification by pairwise coupling. *Annals of Statistics*, vol. 26(2), 451–471.
79. J.C. Platt, N. Cristianini, J. Shawe-Taylor (2004). Large margin DAGs for multiclass classification. In *Advances in Neural Information Processing Systems*, S. Solla, T. Leen, K.R. Muller (Eds.), Cambridge, MA: MIT Press, vol. 12, pp. 547–553.
80. T.G. Dietterichand, G. Bakiri (1995). Solving multiclass learning problems via error-correcting output codes. *Journal of Artificial Intelligence Research*, vol. 2, 263–286.
81. E.L. Allwein, R.E. Schapire, Y. Singer, P. Kaelbling (2001). Reducing multiclass to binary: A unifying approach for margin classifiers. *Journal of Machine Learning Research*, vol. 1, 113–141.
82. J. Weston. C. Watkins (1998). Multi-class support vector machines. Technical Report CSD-TR-98-04, Department of Computer Science, Royal Holloway, University of London, UK.
83. R Duda, P. Hart, D. Stork (2012). *Pattern Classification*. Hoboken, NJ: John Wiley & Sons.
84. K. Fukunaga (1990). *Introduction to Statistical Pattern Recognition*. New York, NY: Academic press.
85. C. Cortes, V.N. Vapnik (1995). Support vector networks. *Machine Learning*, vol. 20(3), 1–25.
86. B.E. Boser, I.M. Guyon, V.N. Vapnik (1992). A training algorithm for optimal margin classifiers. *In Proceedings of the 5th Annual ACM Workshop on Computational Learning Theory*, D. Haussler (Ed.), Pittsburgh, PA, pp. 144–152.
87. B. Scholkopfand, A.J. Smola (2002). *Learning with Kernels: Support Vector Machines, Regularization, Optimization, and Beyond*. Cambridge, MA: MIT press.
88. S. Kner, L. Personnaz, G. Dreyfus (1990) Single-layer learning revisited: A stepwise procedure for building and training neural network. In *Neurocomputing: Algorithms, Architectures and Applications*, F.F. Soulie, J. Herault (Eds.), Berlin: Springer-Verlag, pp. 41–50.

chapter three

Classification among various process operating conditions

In process industries, changes in process operating conditions frequently occur because of variations in product demand, fluctuations in raw material quality and quantity, changes in utility prices, etc. It is necessary to find out the best, or Pareto optimum, operating conditions for a process (apart from detecting an abnormal operating condition), keeping in mind optimal usage of utility/energy/material leading to economic and efficient plant operation. Optimizing the design and operating conditions in a process ensures its product quality and is related to the overall plant economy. The desire for profitable plant operation necessitates the building of a robust strategy to identify various process operating conditions so that the process optimizer can function efficiently.

It is important for any monitoring algorithm to be able to distinguish between change of operating conditions and disturbances, which, alternatively, can be stated as identifying the processes in various operating conditions. In adaptive process monitoring, a routine model update without identifying the change in process operating conditions or process disturbance may lead to wrong conclusions. The process may get accustomed to the long disturbances, so that the adaptive algorithm allows the continuous and automatic model update without even recognizing this as a disturbance. A process maintained at a steady state puts unnecessary burdens on the algorithm because of futile updates. There may be sudden, frequent, and large process changes apart from the process manifesting slow time-varying behavior (catalyst deactivation, equipment aging, sensor and process drifting, and preventive maintenance and cleaning). It is prudent to choose an appropriate monitoring algorithm for a specific kind of process, which is supposed to be capable of identifying the change in various process operating conditions, and to be able to distinguish process disturbances.

Successful recognition of various process operating conditions and process disturbance allow the monitoring algorithms to function one step ahead and detect the process faults; it is important to reduce the false annunciations. A rigorous fault detection algorithm should be complemented by a robust and efficient process operating condition detection/classifying mechanism. The database regarding the detected range of process operating conditions may be used as an offline resource for fault detection algorithms/fault tolerant systems to function efficiently. The present chapter utilizes a modified *K-means* clustering algorithm using dissimilarity (one combined similarity) for classifying various operating conditions in a double effect evaporator, yeast fermentation bioreactor, and continuous crystallization process.

3.1 Yeast fermentation bioreactor process

3.1.1 Modeling and dynamic simulation of yeast fermentation bioreactor

Fermenting sugars with yeast is one of the important processes used for ethanol production. Yeast fermentation is a nonlinear and dynamic process that suffers very much from multivariable interactions, changeable parameters, effect of unknown disturbances, etc. Realizing optimum conditions such as pH, temperature, agitation speed, and dissolved oxygen concentration, etc., in a bioreactor is complex task. The monitoring and control of such a process is also a challenging one [1]. The present study on the classification of various operating conditions in the fermentation process is carried out using data obtained by simulating the process model. The following steps are carried out to generate data pertaining to various operating conditions and classify them.

3.1.1.1 The process description

Figure 3.1 illustrates ethanol production with a yeast fermentation bioreactor with constant substrate inflow and outflow [2–4]. The reaction volume inside the bioreactor is maintained at constant levels. Modeling of a bioreactor is similar to a continuously stirred tank with constant reaction volume.

Initially a biomass solution, which is a suspension of yeast, is fed to the bioreactor. Since the process is carried out in the presence of oxygen, a baker's yeast (saccharomyces cerevisiae) is used. There is a continuous supply of substrate, a solution of glucose, which feeds the microorganisms, to the bioreactor. For the formation of coenzymes, inorganic salts are added together with the biomass. The equilibrium concentration of oxygen in the liquid phase is influenced by inorganic salts because of the salting-out effect. The effect of dissolved inorganic salts, as well as temperature, on the equilibrium concentration of oxygen in the liquid phase is explained by Equations 3.1 through 3.20. Yeast hydrolyses glucose into ethanol and carbon dioxide. The contents of the reactor are biomass, substrate, and product. The product is continuously withdrawn along with biomass and substrate.

3.1.1.2 Mathematical model

The following equations are used to calculate molar concentrations and ionic strength of ions in the reaction medium.

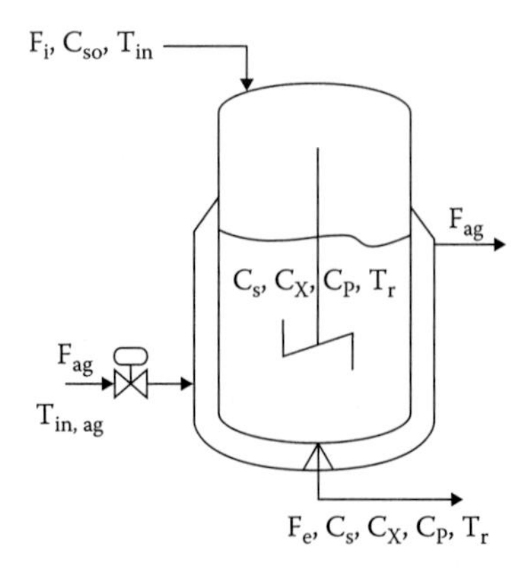

Figure 3.1 Bioreactor process.

$$C_{Na} = \frac{m_{NaCl}}{M_{NaCl}} \frac{M_{Na}}{V} \tag{3.1}$$

$$C_{Ca} = \frac{m_{CaCO_3}}{M_{CaCO_3}} \frac{M_{Ca}}{V} \tag{3.2}$$

$$C_{Mg} = \frac{m_{MgCl_2}}{M_{MgCl_2}} \frac{M_{Mg}}{V} \tag{3.3}$$

$$C_{Cl} = \left[\frac{m_{NaCl}}{M_{NaCl}} + 2\frac{m_{MgCl_2}}{M_{MgCl_2}} \right] \frac{M_{Cl}}{V} \tag{3.4}$$

$$C_{CO_3} = \frac{m_{CaCO_3}}{M_{CaCO_3}} \frac{M_{CO_3}}{V} \tag{3.5}$$

$$C_H = 10^{-pH} \tag{3.6}$$

$$C_{OH} = 10^{-(14-pH)} \tag{3.7}$$

$$I_{Na} = 0.5C_{Na}(1)^2 \tag{3.8}$$

$$I_{Ca} = 0.5C_{Ca}(2)^2 \tag{3.9}$$

$$I_{Mg} = 0.5C_{Mg}(2)^2 \tag{3.10}$$

$$I_{Cl} = 0.5C_{Ca}(-1)^2 \tag{3.11}$$

$$I_{CO_3} = 0.5C_{CO_3}(-2)^2 \tag{3.12}$$

$$I_H = 0.5C_H(1)^2 \tag{3.13}$$

$$I_{OH} = 0.5C_{OH}(-1)^2 \tag{3.14}$$

The global effect of the ionic strengths is given as

$$\sum H_i I_i = H_{Na}I_{Na} + H_{Ca}I_{Ca} + H_{Mg}I_{Mg} + H_{Cl}I_{Cl} + H_{CO_3}I_{CO_3} + H_H I_H + H_{OH}I_{OH} \tag{3.15}$$

Equation 3.16 gives the dependence of the equilibrium concentration of oxygen with temperature in distilled water [4]:

$$C^*_{O_{2,o}} = 14.6 - 0.3943T_r + 0.007714T_r^2 - 0.0000646T_r^3 \tag{3.16}$$

The equilibrium concentration of oxygen in the liquid phase is obtained from the following Setchenov-type equation:

$$C^*_{O_{2,}} = C^*_{O_{2,o}} 10^{-\sum H_i I_i} \tag{3.17}$$

Equation 3.18 given below relates the mass transfer coefficient of oxygen with temperature [5]:

$$(k_1 a) = (k_1 a)_0 (1.204)^{T_r - 20} \tag{3.18}$$

The rate of oxygen consumption is

$$r_{O_2} = \mu_{O_2} \frac{1}{Y_{O_2}} c_X \frac{c_{O_2}}{K_{O_2} + c_{O_2}} \tag{3.19}$$

Equation 3.20 gives the expression of the maximum specific growth rate:

$$\mu_X = A_1 e^{\frac{-Ea_1}{R(T_r+273)}} - A_2 e^{\frac{-Ea_2}{R(T_r+273)}} \tag{3.20}$$

The nonlinear dynamic model containing ordinary differential equations (Equations 3.21 through 3.27) represents the actual process [6]. There are five input variables: substrate feed concentration ($C_{S,in}$), substrate inlet flow rate (F_I), substrate inlet temperature (T_{in}), coolant inlet flow rate (F_{ag}), and coolant inlet temperature ($T_{in,ag}$). And there are six states: biomass concentration (C_X), substrate concentration (C_S), ethanol concentration (C_P), dissolved oxygen concentration (C_{O_2}), reactor temperature (T_r), and cooling jacket temperature (T_{ag}). There is constant inflow and outflow of substrate so that the volume of the reaction medium is not considered a state. The values of parameters used in the model are provided in Table 3.1 [7]. Table 3.2 presents the values of input and steady state values.

$$F_i = F_e \tag{3.21}$$

$$\frac{dc_X}{dt} = \mu_X c_X \left(\frac{c_S}{K_S + c_S} \right) e^{-K_P c_P} - \frac{F_e}{V} c_X \tag{3.22}$$

$$\frac{dc_P}{dt} = \mu_P c_X \left(\frac{c_S}{K_{S1} + c_S} \right) e^{-K_{P1} c_P} - \frac{F_e}{V} c_P \tag{3.23}$$

$$\frac{dc_S}{dt} = \frac{1}{R_{SX}} \mu_X c_X \left(\frac{c_S}{K_S + c_S} \right) e^{-K_P c_P} - \frac{1}{R_{SP}} \mu_P c_X \left(\frac{c_S}{K_{S1} + c_S} \right) e^{-K_{P1} c_P} + \frac{F_i}{V} c_{S,in} - \frac{F_e}{V} c_S \tag{3.24}$$

Table 3.1 Parameters used in the bioreactor process model

$A_1 = 9.5 \times 10^8$	$H_{Cl} = 0.844$	$R_{SP} = 0.435$
$A_2 = 2.55 \times 10^{33}$	$H_{CO_3} = 0.485$	$R_{SX} = 0.607$
$A_T = 1 m^2$	$H_{HO} = 0.941$	$V = 1000$ L
$C_{heat,ag} = 4.18$ J g^{-1} K^{-1}	$(k_l a)_0 = 38$ h^{-1}	$V_j = 50$ L
$C_{heat,r} = 4.18$ J g^{-1} K^{-1}	$K_{O_2} = 8.86$ mg/L	$Y_{O_2} = 0.970$ mg/mg
$E_{a1} = 55000$ J/mol	$K_P = 0.139$ g/L	$\Delta H_r = 518$ KJ/mol O_2
$E_{a2} = 220000$ J/mol	$K_{P1} = 0.070$ g/L	$\mu_{O_2} = 0.5$ h^{-1}
$H_{Na} = -0.550$	$K_S = 1.030$ g/L	$\mu_P = 1.790$ h^{-1}
$H_{Ca} = -0.303$	$K_{S1} = 1.680$ g/L	$\rho_{ag} = 1000$ g/L
$H_{Mg} = -0.314$	$K_T = 3.6 \times 10^5$ Jh^{-1} m^{-2} K^{-1}	$\rho_r = 1080$ g/L
$m_{MgCl_2} = 100$ g	$m_{NaCl} = 500$ g	$P_H = 6$
$H_H = -0.774$	$m_{CaCO_3} = 100$ g	

Source: F. Godia et al.: Batch alcoholic fermentation modeling by simultaneous integration of growth and fermentation equations. *J. Chem. Technol. Biotechnol.* 1988. 41, 155–165. Copyright Wiley-VCH Verlag GmbH & Co. KGaA. Reproduced with permission.

Table 3.2 Values of process variables

Inputs	States
Glucose inlet flow rate, $F_I = 51$ Lh^{-1}	Biomass concentration, $C_X = 0.8539$ g/L
Glucose outlet flow rate, $F_e = 51$ Lh^{-1}	Ethanol concentration, $C_P = 12.1722$ g/L
Glucose inlet temperature, $T_{in} = 25$°C	Glucose concentration, $C_S = 33.42$ g/L
Glucose inlet concentration, $C_{S,in} = 60$ g/L	Oxygen concentration, $C_{O_2} = 4.5939$ g/L
	Reactor temperature, $T_r = 28.7188$°C
Coolant inlet temperature, $T_{in,ag} = 15$°C	Coolant outlet temperature, $T_{ag} = 26.3472$°C
Coolant inlet flow rate, $F_{ag} = 18$ Lh^{-1}	

$$\frac{dc_{O_2}}{dt} = (k_1 a)\left(c_{O_2}^* - c_{O_2}\right) - r_{O_2} - \frac{F_e}{V} c_{O_2} \tag{3.25}$$

$$\frac{dT_r}{dt} = \frac{F_i}{V}(T_{in} + 273) - \frac{F_e}{V}(T_r + 273) + \frac{r_{O_2}\Delta H_r}{32\rho_r C_{heat,r}} + \frac{K_T A_T (T_r - T_{ag})}{V_j \rho_{ag} C_{heat,ag}} \tag{3.26}$$

$$\frac{dT_{ag}}{dt} = \frac{F_{ag}}{V_j}\left(T_{in,ag} - T_{ag}\right) + \frac{K_T A_T (T_r - T_{ag})}{V_j \rho_{ag} C_{heat,ag}} \tag{3.27}$$

3.1.1.3 *Analysis of dynamic behavior of yeast fermentation bioreactor*

The set of nonlinear differential-algebraic equations derived in the previous section simulates the actual behavior of the industrial yeast fermentation continuous bioreactor. The mathematical model of continuous bioreactor was implemented as a MATLAB® Simulink® S-function whose Simulink block diagram is shown in Figure 3.2; its corresponding MATLAB codes are presented in Programs 3.1 and 3.2. Figure 3.3 presents the response of the continuous bioreactor to step decrease in the substrate inlet temperature (from 25 to

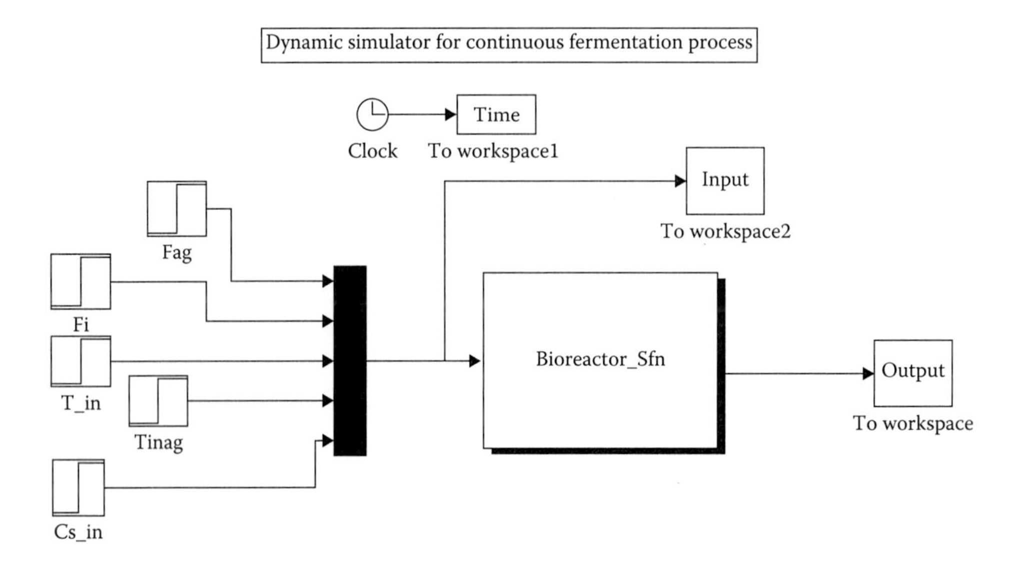

Figure 3.2 Simulink model of a yeast fermentation bioreactor.

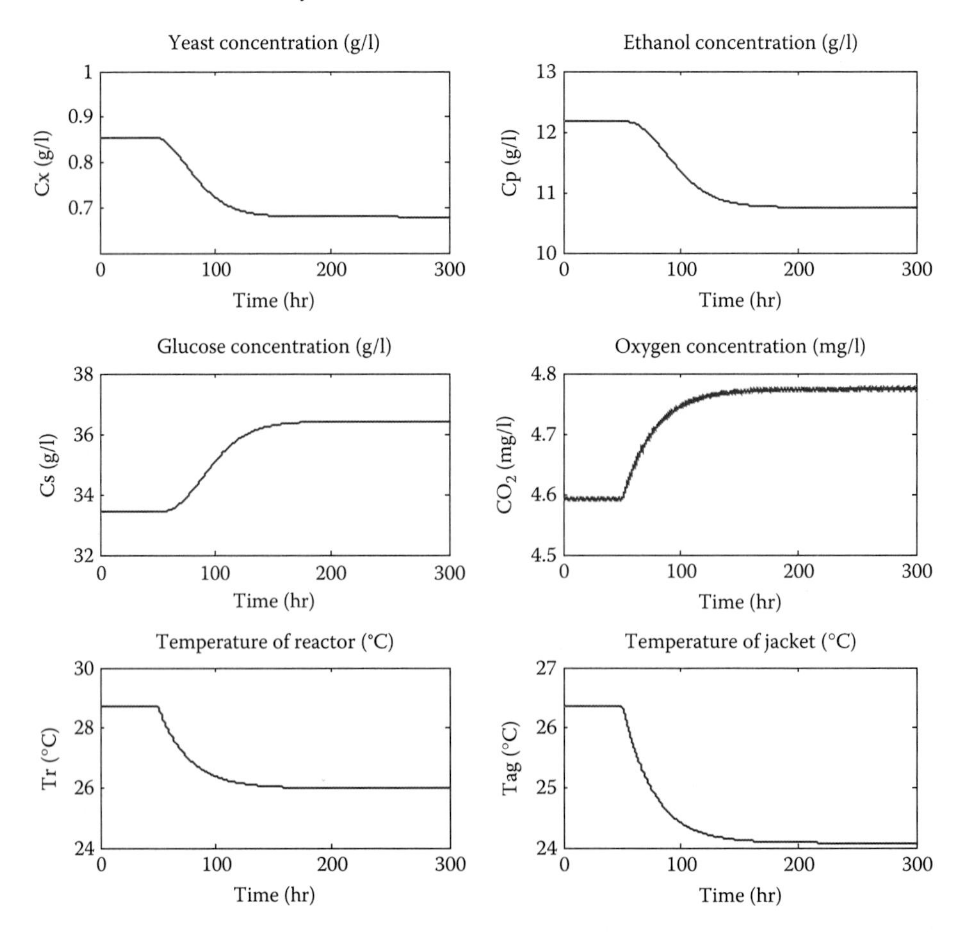

Figure 3.3 Response of bioreactor to step decrease in substrate inlet temperature (25–23°C).

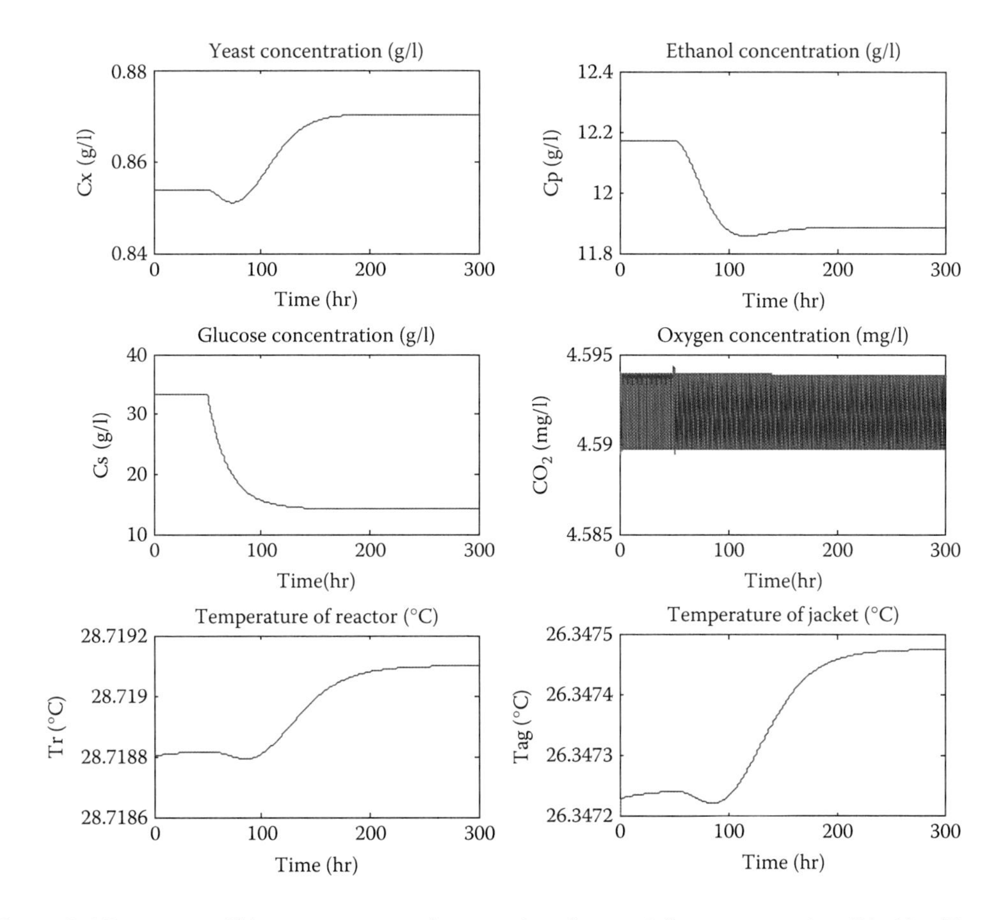

Figure 3.4 Response of bioreactor to step decrease in substrate inlet concentration (60–40 g/L).

23°C). When the substrate inlet temperature suddenly decreased from its nominal value to 23°C and settled at this new steady state, the temperature of the reactor and the jacket initially decreased, and finally achieved new steady states lower than their respective nominal values. The decrease in the substrate inlet temperature caused the product (ethanol) concentration to decrease and reach the new steady state. The substrate inlet concentration (Figure 3.4) has almost negligible impact on the reactor temperature, the jacket temperature, and the concentration of oxygen but can significantly change the concentration of yeast, glucose (i.e., substrate) and ethanol. A unit step increase in the substrate inlet flow rate (Figure 3.5) greatly amplified the yeast concentration, the oxygen concentration, and the product concentration but induced the glucose concentration to decay. For the increased flow rate of substrate, the mathematical model accurately predicts the increased product, hence, the model is a perfect representative of an industrial yeast fermentation bioreactor. For the abrupt increase in the coolant inlet flow rate (Figure 3.6), the jacket temperature dropped immediately, but within a short period of time, the jacket started gaining enough heat so that the jacket temperature was increasing slowly until reaching the new steady state. Because of increased reactor and jacket temperature, the product concentration increased. The substrate inlet flow rate can serve as a manipulated variable to regulate the product concentration, and the remaining input variables can be treated

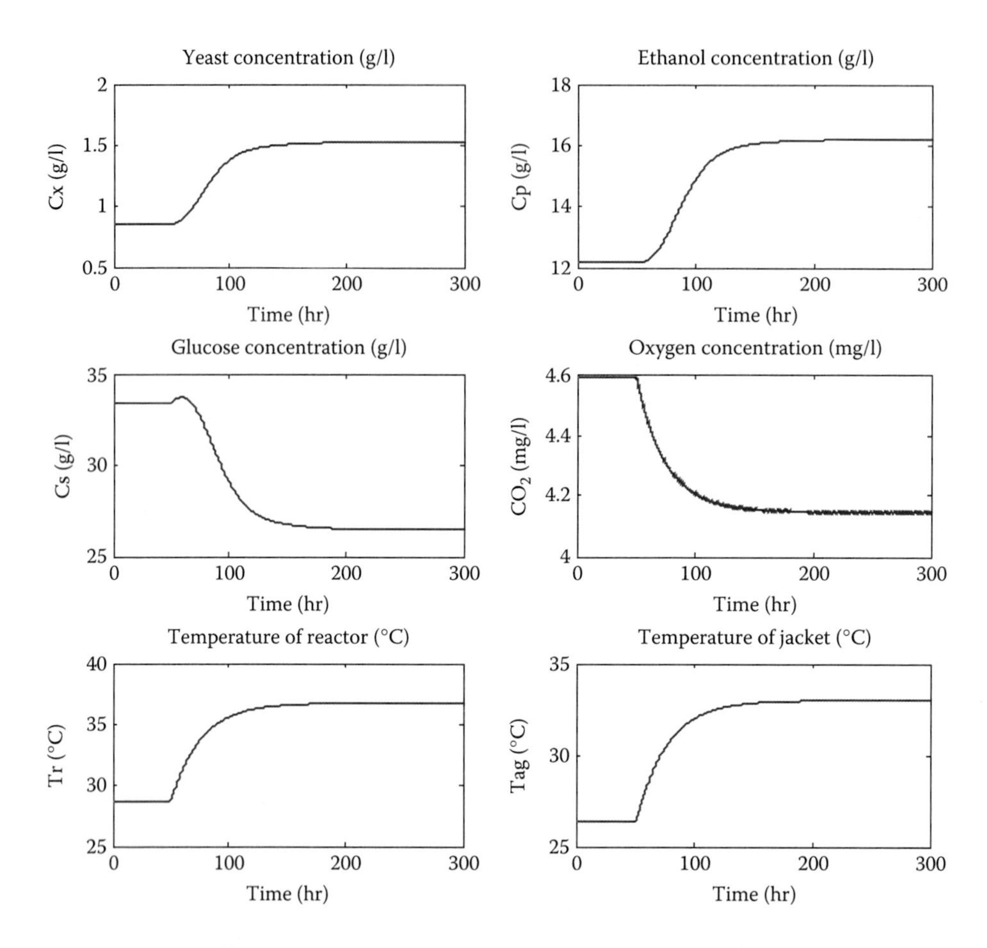

Figure 3.5 Response of bioreactor to unit step increase in substrate inlet flow rate (51–52 l/h).

as disturbances. Since the product concentration is a secondary measurement, the ethanol (product) concentration can be controlled through controlling reactor temperature by manipulating the inlet flow rate of substrate [8].

Program 3.1 MATLAB code for yeast fermentation bioreactor S-function

```
function [sys,x0,str,ts]=Bioreactor_Sfn(t,x,u,flag)
switch flag,
case 0,
  [sys,x0,str,ts]=mdlinitializesizes;
case 1,
  sys=mdlderivatives(t,x,u);
case 3,
  sys=mdloutputs(x);
case {2,4,9},
  sys=[];
otherwise,
```

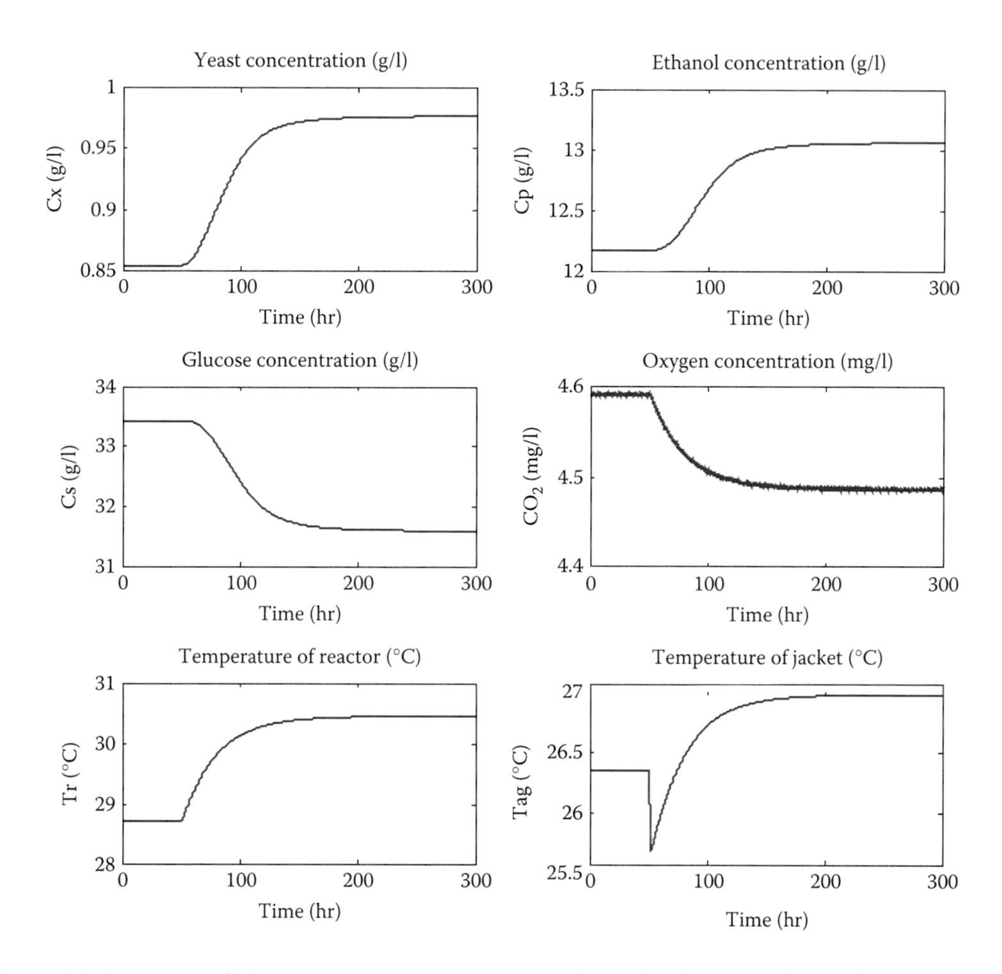

Figure 3.6 Response of bioreactor to step increase in coolant inlet flow rate (18–25 l/h).

```
  error(['Unhandled flag = ',num2str(flag)]);
end
function [sys,x0,str,ts]=mdlinitializesizes
sizes=simsizes;
sizes.NumContStates=6;
sizes.NumDiscStates=0;
sizes.NumOutputs=6;
sizes.NumInputs=5;
sizes.DirFeedthrough=1;
sizes.NumSampleTimes=1;
sys=simsizes(sizes);
x0=[0.8539,12.1722,33.424,4.592,28.7188,26.3472];
str=[];
ts=[0 0];
function sys=mdlderivatives(t,x,u)
Fag=u(1);
Fi=u(2);
Tin=u(3);
Tinag=u(4);
```

```
Csin=u(5);
sys=Yeast_Ferment_Model(t,x,Fag,Fi,Tin,Tinag,Csin);
function sys=mdloutputs(x)
sys=x;
%%%%%%%%%%%%%%%%%%%%%%%%%%%%%%%%%%%%%%%%%%%%%
```

Program 3.2 MATLAB code for model of bioreactor

```
function dy= Yeast_Ferment_Model(t,x,Fag,Fi,Tin,Tinag,Csin)
dy=zeros(6,1);
% Parameter values
A1=(9.5)*(10^8);
A2=(2.55)*(10^33);
At=1;
Chag=4.18;
Chr=4.18;
Ea1=55000;
Ea2=220000;
Hna=-0.55;
Hca=-0.303;
Hmg=-0.314;
Hh=-0.774;
Hcl=0.844;
Hco3=0.485;
Hoh=0.941;
klao=38;
Ko2=8.86;
Kp=0.139;
Kp1=0.070;
Ks=1.030;
Ks1=1.680;
Kt=(3.6)*(10^5);
Rsp=0.435;
Rsx=0.607;
V=1000;
Vj=50;
Yo2=0.970;
deltaHr=518;
muo2=0.5;
mup=1.790;
rhoag=1000;
rhor=1080;
R=8.314;
% inorganic salts for coenzymes
mnacl=500;
mcaco3=100;
mmgcl2=100;
pH=6;
% inputs
Fe=51;
```

```
% molar concentrations of ions
Cna=(mnacl/58.443)*(23/V);
Cca=(mcaco3/100.0869)*(40.078/V);
Cmg=(mmgcl2/95.211)*(24.3080/V);
Ccl=((mnacl/58.443) + ((2*mmgcl2)/95.211))*(35.453/V);
Cco3=(mcaco3/100.0869)*(60/V);
Ch=10^(-pH);
Coh=10^(-14-pH);
% ionic strength
Ina=0.5*Cna*(1)^2;
Ica=0.5*Cca*(2)^2;
Img=0.5*Cmg*(2)^2;
Icl=0.5*Ccl*(-1)^2;
Ico3=0.5*Cco3*(-2)^2;
Ih=0.5*Ch*(1)^2;
Ioh=0.5*Coh*(-1)^2;
% global effect of ionic strength
sumHiIi=Hna*Ina + Hca*Ica + Hmg*Img + Hcl*Icl + Hco3*Ico3 + Hh*Ih +
Hoh*Ioh;
% equilibrium conc. of oxygen fn of temp.
Co2o=(14.6-0.3943*x(5) +
0.00771*(x(5)^2)-0.0000646*(x(5)^3))*10^(-sumHiIi);
C1o2=Co2o*10^(-sumHiIi);
% mass transfer coefficient for oxygen
k1a=k1ao*(1.204)^(x(5)-20);
% rate of oxygen consumption
ro2=muo2*(1/Yo2)*x(1)*(x(4)/(Ko2 + x(4)));
mux=A1*(exp(-Ea1/(R*(x(5) + 273))))-A2*(exp(-Ea2/(R*(x(5) + 273))));
% Component and energy balances
dy(1)=(mux*x(1)*x(3)*exp(-Kp*x(2)))/(Ks + x(3))-(Fe*x(1))/V;
dy(2)=(mup*x(1)*x(3)*exp(-Kp1*x(2)))/(Ks1 + x(3))-(Fe*x(2))/V;
dy(3)=(mux*x(1)*x(3)*exp(-Kp*x(2)))/(Rsx*(Ks + x(3)))-
(mup*x(1)*x(3)*exp(-Kp1*x(2)))/(Rsp*(Ks1 + x(3))) +
(Fi*Csin)/V-(Fe*x(3))/V;
dy(4)=k1a*(C1o2-x(4))-ro2-((Fe*x(4))/V);
dy(5)=((Fi*(Tin + 273))/V)-((Fe*(x(5) + 273))/V) + ((ro2*deltaHr)/
(32*rhor*Chr)) + ((Kt*At*(x(5)-x(6)))/(V*rhor*Chr));
dy(6)=((Fag*(Tinag-x(6)))/Vj) + ((Kt*At*(x(5)-x(6)))/(Vj*rhoag*Chag));
end
%%%%%%%%%%%%%%%%%%%%%%%%%%%%%%%%%%%%%%%%%%%%%%
```

3.1.2 *Generation of process historical database for yeast fermentation process*

To create the historical database, four operating modes (normal operation) as named in Table 3.3 were selected. The values of model parameters and the initial conditions were maintained at their respective base values while simulating the bioreactor model in every operating mode. The simulated continuous bioreactor was operated for 300 hours in each operating mode to generate a single data set. Each data set contains 150,004 measurements of six state variables (c_X, c_P, c_S, c_{O_2}, T_r, and T_{ag}). Eighteen data sets belonging to four operating conditions were generated.

Table 3.3 Operating conditions for yeast fermentation bioreactor

Op. cond.	Description	No. of data sets
1	Perturbations in the steady state value of substrate flow rate (random noise added)	6
2	Ramp increase in substrate inlet concentration (slope = 0.5,1,1.5, initial output = steady state value)	3
3	Step change in coolant flow rate (up to ±50% from nominal value)	6
4	Oscillation in coolant inlet temperature (high frequency oscillation–30 cycles/sec, medium frequency oscillation–20 cycles/sec, low frequency oscillation–10 cycles/sec)	3

3.1.3 *Application of modified K-means clustering algorithm on historical database for yeast fermentation process*

Using the pseudocode given in Chapter 2, the dissimilarity (one-SF/combined similarity) factor-based modified *K*-means clustering technique was implemented (using different combinations of weighting factors α_1 and α_2) on the historical database generated in the preceding section for the yeast fermentation process. The algorithm was repeated for an assumed number of clusters, $K = 2$ through 8 for every combination of weighting factors. The results are reported in Figures 3.7 through 3.10 and in Tables 3.4 through 3.11. For all the combinations chosen, the optimum number of clusters found by the modified *K*-means clustering algorithm is four ($K = 4$), which is precisely equal to the number of operating conditions in the database. The performance indices of the clustering technique are computed and tabulated in Table 3.12. It is evident that each operating condition occurs in only a single cluster. The developed dissimilarity (one-SF/combined similarity) factor-based *K*-means clustering algorithm exhibited good performance with 100% cluster purity and 100% clustering efficiency.

3.2 *Commercial double-effect evaporator*

3.2.1 *Modeling and dynamic simulation of double effect evaporator*

A multiple effect evaporation process is widely used in agro-based industries (like tomato juice concentration), sugar manufacturing plants, paper industries, and many others. The present study on classification of various operating conditions in a double-effect evaporation process, is carried out using data obtained by simulating the process model. These mathematical models are based on linear and nonlinear mass, energy balance equations, and relationships pertaining to the physical properties of the solution. The following steps are carried out to generate data pertaining to various operating conditions and to classify them.

3.2.1.1 *The process description*

The industrial equipment, schematically shown in Figure 3.11, consists of two effects named (from left to right) Tank 1 and Tank 2 in which the tomato juice is concentrated. Because of steam economy (water evaporated / steam given), and to avoid high viscosity fluids at low

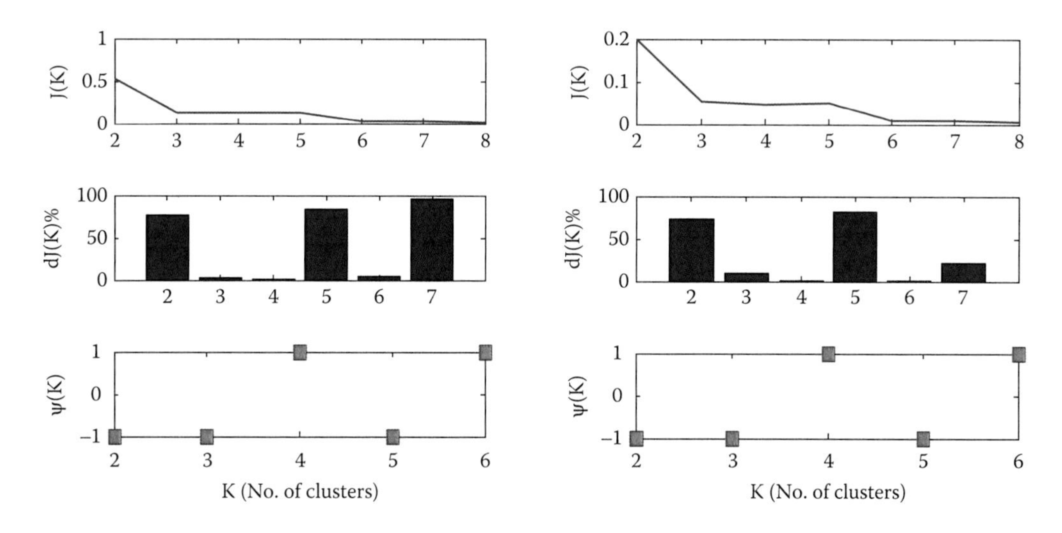

Figure 3.7 Clustering performance for $\alpha_1 = 1$ and $\alpha_2 = 0$ (left column), and $\alpha_1 = 0.37$ and $\alpha_2 = 0.63$ (right column).

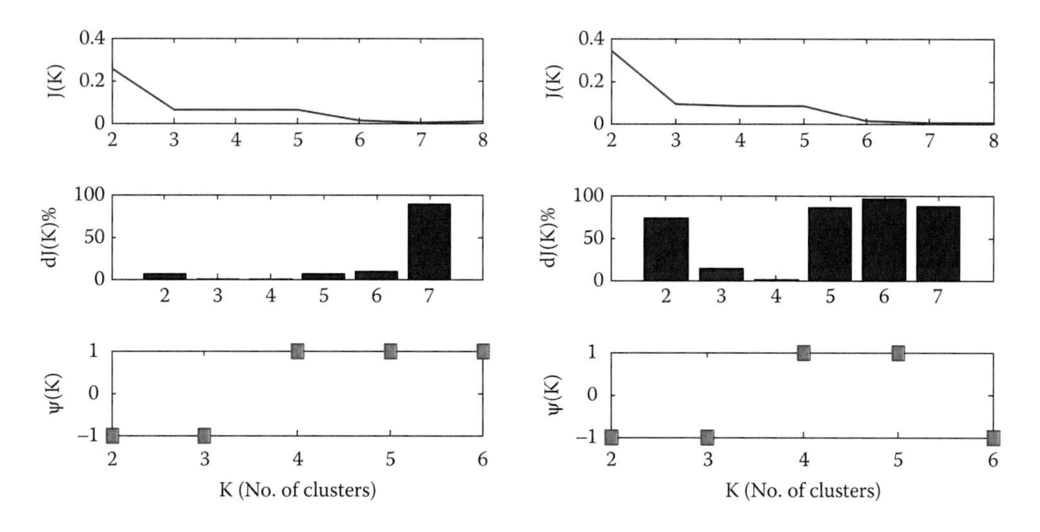

Figure 3.8 Clustering performance for $\alpha_1 = 0.5$ and $\alpha_2 = 0.5$ (left column), and $\alpha_1 = 0.63$ and $\alpha_2 = 0.37$ (right column).

temperatures, the operation mode of the evaporator system is chosen as backward feeding. The raw tomato juice is a binary solution of soluble solids and water, both considered inert in a chemical sense. The fresh tomato juice having flow rate F and temperature T_f enters effect 2. The concentration of soluble solids in the feed is indicated by X_f. The concentrated tomato juice in Tank 2 leaves from the bottom and enters Tank 1 with flow rate P_2 and temperature T_2. The concentration of soluble solids in the product of Tank 2 is given by X_2. The final product leaves Tank 1 from the bottom with flow rate P_1, temperature T_1

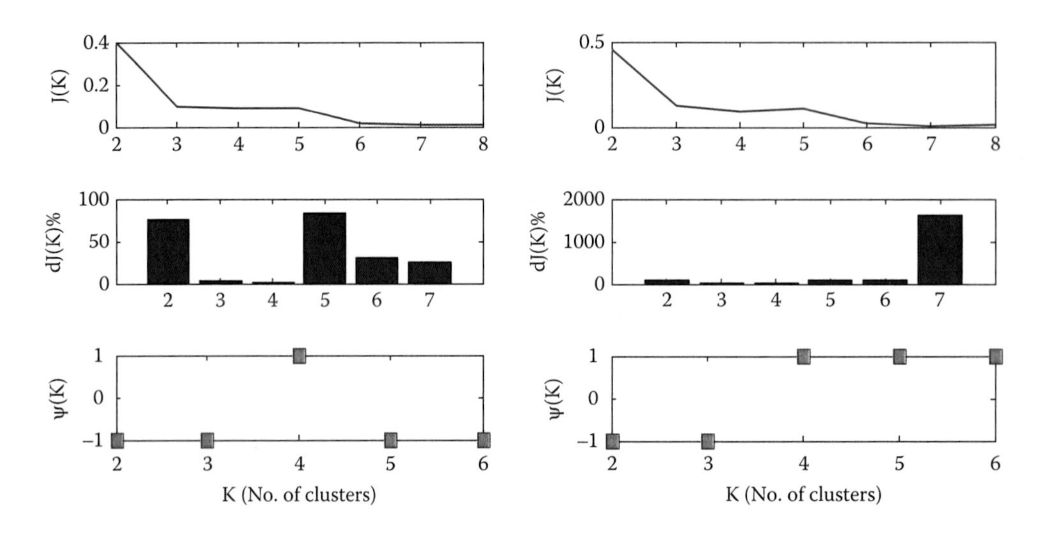

Figure 3.9 Clustering performance for $\alpha_1 = 0.75$ and $\alpha_2 = 0.25$ (left column), and $\alpha_1 = 0.85$ and $\alpha_2 = 0.15$ (right column).

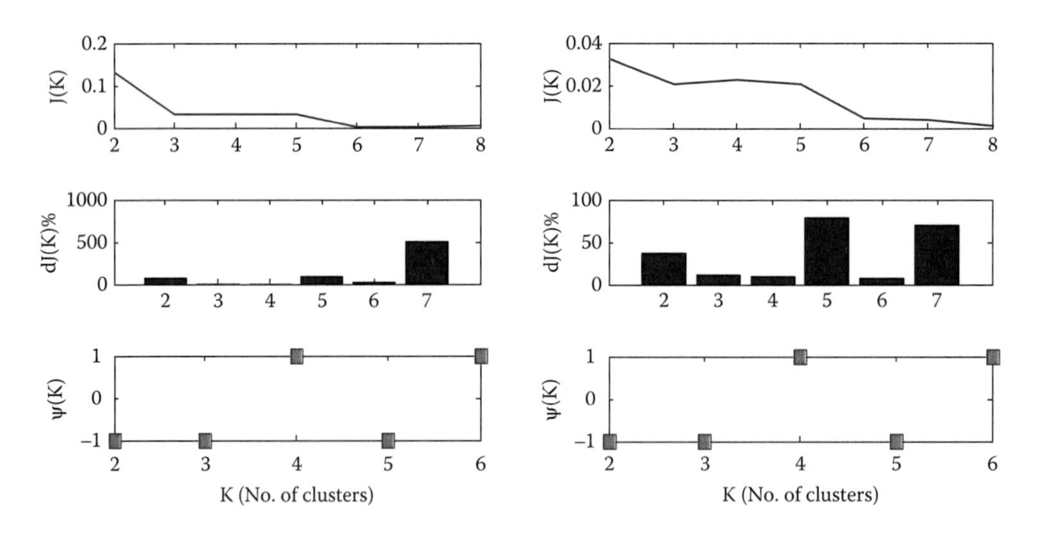

Figure 3.10 Clustering performance for $\alpha_1 = 0.25$ and $\alpha_2 = 0.75$ (left column), and $\alpha_1 = 0.15$ and $\alpha_2 = 0.85$ (right column).

and concentration of soluble solids X_1. The mass holdups in the two tanks are represented by M_1 and M_2. To avoid low levels and to get an adequate heat transfer coefficient, the level of tomato concentrate in Tank 1 and Tank 2 can be controlled by adjusting the flow rate of the product of Tank 2 and the flow rate of feed, respectively. The heat required to evaporate the water in the juice is supplied by the steam entering Tank 1. The temperature and flow rate of the steam are denoted by T_S and S, respectively. The vapor generated in the first effect leaves Tank 1 from the top and is sent to the second effect to heat the tomato juice

Table 3.4 Dissimilarity factor for $\alpha_1 = 1$ and $\alpha_2 = 0$

Cluster Data set	1	2	3	4
X_1	8.2862698e-08	0.729254394	0.110654006	0.645802736
X_2	3.0696541e-08	0.729342621	0.110693038	0.645786322
X_3	4.9635985e-09	0.729427936	0.110730963	0.645770429
X_4	2.6023222e-09	0.729510783	0.110767954	0.645754964
X_5	2.1239728e-08	0.729591505	0.110804143	0.645739862
X_6	5.8988593e-08	0.729670371	0.110839633	0.645725070
X_7	0.729495998	2.9340853e-08	0.567355735	0.999999943
X_8	0.729469385	6.7002137e-10	0.567296899	0.999999956
X_9	0.72945162	5.0159204e-09	0.567257834	0.999999965
X_{10}	0.09789312	0.460449190	0.053982895	0.864043920
X_{11}	0.094837846	0.467317183	0.049314538	0.867744352
X_{12}	0.09185918	0.47446268	0.044877522	0.871190357
X_{13}	0.11099888	0.43525950	0.074951942	0.846508542
X_{14}	0.11456152	0.429497662	0.080781315	0.841377857
X_{15}	0.118248151	0.423928868	0.086843507	0.835920081
X_{16}	0.646994370	0.999999962	0.944274965	7.522282e-11
X_{17}	0.645567163	0.999999955	0.944058166	1.780686e-11
X_{18}	0.645115017	0.999999950	0.943960474	1.388849e-10

Table 3.5 Dissimilarity factor for $\alpha_1 = 0.37$ and $\alpha_2 = 0.63$

Cluster Data set	1	2	3	4
X_1	5.4493191e-05	0.270262140	0.041002704	0.239284544
X_2	3.1287575e-05	0.270317843	0.041040302	0.239301555
X_3	9.2716648e-06	0.270371295	0.041076313	0.239317585
X_4	1.1739939e-05	0.270422843	0.041110981	0.239332781
X_5	3.1897800e-05	0.270472751	0.041144497	0.239347257
X_6	5.1314791e-05	0.270521230	0.041177007	0.239361103
X_7	0.271203904	0.000306612	0.213294129	0.376070316
X_8	0.270042534	1.34048396e-07	0.210006267	0.370028423
X_9	0.270017722	1.24907907e-06	0.209978099	0.370025590
X_{10}	0.037286073	0.1719229733	0.021154163	0.321152810
X_{11}	0.036976760	0.1752830801	0.020247500	0.323341366
X_{12}	0.036764178	0.1788138383	0.019494717	0.325503841
X_{13}	0.042707617	0.1621875989	0.029254134	0.314451036
X_{14}	0.044578639	0.1606079653	0.031963708	0.313105052
X_{15}	0.046453956	0.1590584167	0.034717901	0.311596654
X_{16}	0.239781636	0.3701007056	0.349659812	1.6366593e-07
X_{17}	0.239253380	0.3701008914	0.349579408	2.5176468e-08
X_{18}	0.239085973	0.3701010030	0.349543148	1.3851948e-07

Table 3.6 Dissimilarity factor for $\alpha_1 = 0.5$ and $\alpha_2 = 0.5$

Cluster Data set	1	2	3	4
X_1	4.32656636e-05	0.364974827	0.055375195	0.323169250
X_2	2.48377432e-05	0.365037242	0.055413089	0.323179364
X_3	7.35948839e-06	0.365097268	0.055449494	0.323188807
X_4	9.31794953e-06	0.365155275	0.055484642	0.323197676
X_5	2.53200977e-05	0.365211542	0.055518709	0.323206048
X_6	4.07381973e-05	0.365266291	0.0555518346	0.323213985
X_7	0.365772114	0.000243349	0.2863544608	0.5048176995
X_8	0.364844900	1.0652587e-07	0.2837329059	0.500022549
X_9	0.364821543	9.9236763e-07	0.2837024891	0.500020303
X_{10}	0.049792290	0.2314601291	0.0279283466	0.433177959
X_{11}	0.048916349	0.235544085	0.026245460	0.435678490
X_{12}	0.048132989	0.239820743	0.024732439	0.438105820
X_{13}	0.056799467	0.218535769	0.038683841	0.424240680
X_{14}	0.059019552	0.216093141	0.042037182	0.422113726
X_{15}	0.061268632	0.213714224	0.045473978	0.419790377
X_{16}	0.323809660	0.500079917	0.472358177	1.29909122e-07
X_{17}	0.323095907	0.500080063	0.472249628	1.998499810e-08
X_{18}	0.322869744	0.5000801509	0.472200692	1.099647592e-07

Table 3.7 Dissimilarity factor for $\alpha_1 = 63$ and $\alpha_2 = 0.37$

Cluster Data set	1	2	3	4
X_1	3.20381354e-05	0.459687515	0.069747686	0.407053957
X_2	1.83879111e-05	0.459756640	0.069785876	0.407057173
X_3	5.44731194e-06	0.459823242	0.069822676	0.407060029
X_4	6.89595925e-06	0.459887707	0.069858303	0.407062571
X_5	1.87423946e-05	0.459950332	0.069892922	0.407064840
X_6	3.01616030e-05	0.460011352	0.069926662	0.407066867
X_7	0.4603403243	0.0001800862	0.359414792	0.633565082
X_8	0.4596472669	7.90033526e-08	0.357459544	0.630016675
X_9	0.4596253648	7.35656186e-07	0.357426878	0.630015015
X_{10}	0.0622985071	0.290997284	0.0347025294	0.545203109
X_{11}	0.0608559384	0.295805091	0.0322434208	0.548015615
X_{12}	0.0595018009	0.300827648	0.0299701611	0.550707800
X_{13}	0.0708913168	0.274883940	0.0481135475	0.534030324
X_{14}	0.0734604662	0.271578316	0.0521106573	0.531122400
X_{15}	0.0760833071	0.268370031	0.0562300559	0.527984100
X_{16}	0.4078376851	0.630059129	0.5950565424	9.61523083e-08
X_{17}	0.4069384340	0.630059235	0.5949198484	1.47935284e-08
X_{18}	0.4066535154	0.630059298	0.5948582357	8.14100319e-08

Table 3.8 Dissimilarity factor for $\alpha_1 = 0.75$ and $\alpha_2 = 0.25$

Cluster Data set	1	2	3	4
X_1	2.16742631e-05	0.547114611	0.083014601	0.484485993
X_2	1.24342199e-05	0.547189931	0.083053063	0.484482843
X_3	3.6822259e-06	0.547262602	0.083090228	0.484479618
X_4	4.6602759e-06	0.547333029	0.083126298	0.484476320
X_5	1.2670668e-05	0.547401524	0.083161426	0.484472955
X_6	2.0398592e-05	0.547468331	0.083195733	0.484469528
X_7	0.547634056	0.000121689	0.426855098	0.752408821
X_8	0.547157143	5.3597947e-08	0.425514902	0.750011252
X_9	0.547136584	4.9869177e-07	0.425480161	0.7500101344
X_{10}	0.073842706	0.345954659	0.040955621	0.6486109399
X_{11}	0.071877097	0.351430634	0.037779999	0.6517114218
X_{12}	0.069996088	0.357141714	0.034804981	0.6546480892
X_{13}	0.083899177	0.326897636	0.056817891	0.6353746113
X_{14}	0.086790540	0.322795401	0.061409249	0.6317457919
X_{15}	0.089758391	0.318821546	0.066158742	0.6278552294
X_{16}	0.485402015	0.750039940	0.708316571	6.49921724e-08
X_{17}	0.484331535	0.750040009	0.708153897	1.00014024e-08
X_{18}	0.483992381	0.750040050	0.708080583	5.50518221e-08

Table 3.9 Dissimilarity factor for $\alpha_1 = 0.85$ and $\alpha_2 = 0.15$

Cluster Data set	1	2	3	4
X1	1.3037702e-05	0.619970524	0.0940703635	0.54901269
X2	7.4728105e-06	0.620051007	0.0941090536	0.549004235
X3	2.2113210e-06	0.620128736	0.0941465226	0.548995942
X4	2.7972064e-06	0.620204131	0.0941829607	0.548987778
X5	7.6108971e-06	0.620277516	0.0942185131	0.548979718
X6	1.2262751e-05	0.620349147	0.0942532935	0.548971744
X7	0.6203788331	7.3025374e-05	0.4830553530	0.851445269
X8	0.6200820403	3.2426777e-08	0.4822277012	0.850006734
X9	0.6200626009	3.0122143e-07	0.4821912309	0.850006066
X10	0.0834628755	0.391752471	0.0461665310	0.734784132
X11	0.0810613970	0.397785254	0.0423938153	0.738124594
X12	0.0787413279	0.404070103	0.0388339977	0.741264996
X13	0.0947390621	0.370242383	0.0640715121	0.719828183
X14	0.0978989350	0.365476306	0.0691580758	0.715598618
X15	0.1011542958	0.360864475	0.0744326484	0.711081170
X16	0.5500389573	0.850023949	0.8026999289	3.9025392e-08
X17	0.54882578678	0.850023987	0.8025156054	6.0079642e-09
X18	0.5484414356	0.850024010	0.8024325399	3.3086647e-08

Table 3.10 Dissimilarity factor for $\alpha_1 = 0.25$ and $\alpha_2 = 0.75$

Cluster Data set	1	2	3	4
X_1	6.4857064e-05	0.182835044	0.027735790	0.16185250
X_2	3.7241266e-05	0.182884552	0.027773115	0.16187588
X_3	1.1036750e-05	0.182931935	0.027808760	0.161897996
X_4	1.3975623e-05	0.182977521	0.027842987	0.161919032
X_5	3.7969526e-05	0.183021560	0.027875993	0.161939142
X_6	6.1077801e-05	0.183064251	0.027907935	0.161958443
X_7	0.183910172	0.000365009	0.1458538235	0.257226577
X_8	0.182532658	1.5945380e-07	0.1419509092	0.250033845
X_9	0.182506502	1.4860434e-06	0.1419248164	0.250030471
X_{10}	0.025741872	0.116965598	0.0149010721	0.217744979
X_{11}	0.025955601	0.119657536	0.0147109212	0.219645559
X_{12}	0.026269890	0.122499772	0.0146598978	0.221563552
X_{13}	0.029699756	0.110173902	0.0205497905	0.213106749
X_{14}	0.031248565	0.109390880	0.0226651163	0.212481660
X_{15}	0.032778872	0.108606902	0.0247892142	0.211725524
X_{16}	0.162217305	0.250119894	0.2363997838	1.94826071e-07
X_{17}	0.161860279	0.250120117	0.2363453592	2.99685938e-08
X_{18}	0.161747107	0.250120251	0.2363208012	1.64877696e-07

Table 3.11 Dissimilarity factor for $\alpha_1 = 0.15$ and $\alpha_2 = 0.85$

Cluster Data set	1	2	3	4
X_1	7.3493624e-05	0.109979131	0.016680027	0.097325810
X_2	4.2202675e-05	0.110023476	0.016717125	0.097354494
X_3	1.2507655e-05	0.110065801	0.016752467	0.097381672
X_4	1.5838692e-05	0.110106419	0.016786324	0.097407574
X_5	4.3029298e-05	0.110145567	0.016818906	0.097432379
X_6	6.9213643e-05	0.110183434	0.016850375	0.097456226
X_7	0.111165396	0.000413673	0.089653568	0.15819012
X_8	0.109607761	1.806249e-07	0.085238110	0.150038363
X_9	0.109580485	1.683513e-06	0.085213747	0.150034539
X_{10}	0.016121705	0.071167786	0.009690162	0.131571787
X_{11}	0.016771301	0.073302917	0.010097105	0.133232387
X_{12}	0.017524651	0.075571383	0.010630881	0.134946645
X_{13}	0.018859872	0.066829156	0.013296170	0.128653176
X_{14}	0.020140170	0.066709975	0.014916289	0.128628834
X_{15}	0.021382968	0.066563973	0.016515308	0.128499584
X_{16}	0.097580364	0.150135885	0.142016426	2.207928e-07
X_{17}	0.097366027	0.150136139	0.141983651	3.396203e-08
X_{18}	0.097298053	0.150136291	0.141968844	1.868428e-07

Table 3.12 Performance indices of modified *K*-means clustering algorithm for classification of various operating conditions in yeast fermentation process

α_1	α_2	CN	N_P	$P(\%)$	DOC	η	Operating Condition			
							1	2	3	4
1	0	1	6	100	1	100	6	0	0	0
		2	3	100	2	100	0	3	0	0
		3	6	100	3	100	0	0	6	0
		4	3	100	4	100	0	0	0	3
0.37	0.67	1	6	100	1	100	6	0	0	0
		2	3	100	2	100	0	3	0	0
		3	6	100	3	100	0	0	6	0
		4	3	100	4	100	0	0	0	3
0.5	0.5	1	6	100	1	100	6	0	0	0
		2	3	100	2	100	0	3	0	0
		3	6	100	3	100	0	0	6	0
		4	3	100	4	100	0	0	0	3
0.63	0.37	1	6	100	1	100	6	0	0	0
		2	3	100	2	100	0	3	0	0
		3	6	100	3	100	0	0	6	0
		4	3	100	4	100	0	0	0	3
0.75	0.25	1	6	100	1	100	6	0	0	0
		2	3	100	2	100	0	3	0	0
		3	6	100	3	100	0	0	6	0
		4	3	100	4	100	0	0	0	3
0.85	0.15	1	6	100	1	100	6	0	0	0
		2	3	100	2	100	0	3	0	0
		3	6	100	3	100	0	0	6	0
		4	3	100	4	100	0	0	0	3
0.25	0.75	1	6	100	1	100	6	0	0	0
		2	3	100	2	100	0	3	0	0
		3	6	100	3	100	0	0	6	0
		4	3	100	4	100	0	0	0	3
0.15	0.85	1	6	100	1	100	6	0	0	0
		2	3	100	2	100	0	3	0	0
		3	6	100	3	100	0	0	6	0
		4	3	100	4	100	0	0	0	3

*CN-Cluster Number, *DOC-Dominant Operating Condition

in Tank 2. The temperature of the first effect can be controlled by manipulating the steam flow rate, and the flow rate of the vapor leaving the second effect can act as a manipulated variable to control the temperature of Tank 2. Since the tomato juice is acidic and contains thermolabile compounds, the range of operation temperatures is considered [51 °C, 74 °C] and the evaporation starts at of 51 °C in order to minimize the risk of decomposition of organic substances.

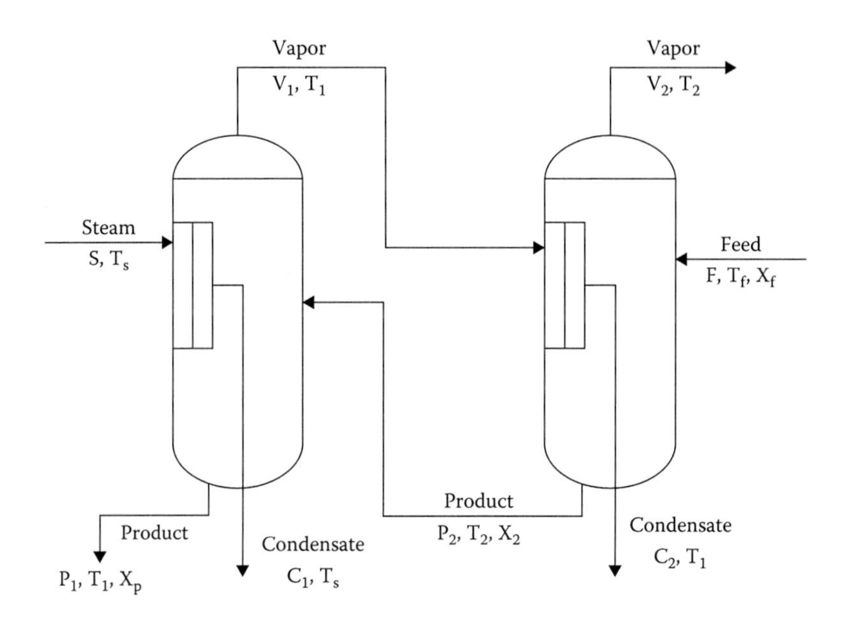

Figure 3.11 Schematic of double-effect evaporator.

3.2.1.2 *Mathematical model of double-effect evaporator*

The evaporation process involves heat and mass transfer [9]. The macroscopic mode of a double-effect evaporator is based on the conservative laws and the empirical relationships [10,11]. For modeling, only the juice phase is considered.

The assumptions made in the derivation of the mathematical model are listed below.

- Negligible heat losses to the surroundings
- Homogeneous composition and temperature inside each effect
- Variable liquid holdup and negligible vapor holdup
- Overhead vapors considered as pure steam
- Latent heat of vaporization or condensation varied with temperature
- No boiling point elevation of the solution

The model of the double-effect evaporator is a system of differential-algebraic equations formed from the following equations.

The total mass balance around Tank 1 and Tank 2 are

$$\frac{dM_1}{dt} = P_2 - P_1 - V_1 \tag{3.28}$$

$$\frac{dM_2}{dt} = F - P_2 - V_2 \tag{3.29}$$

The mass balance for soluble solids in the output of Tank 1 and Tank 2 are

$$\text{First effect: } \frac{d(M_1 X_P)}{dt} = P_2 X_2 - P_1 X_P \Rightarrow M_1 \frac{dX_P}{dt} + X_P \frac{dM_1}{dt} = P_2 X_2 - P_1 X_P \tag{3.30}$$

$$\Rightarrow M_1 \frac{dX_P}{dt} = P_2X_2 - P_1X_P - X_P \frac{dM_1}{dt} \tag{3.31}$$

From Equation 3.28,

$$M_1 \frac{dX_P}{dt} = P_2X_2 - P_1X_P - X_P(P_2 - P_1 - V_1) \Rightarrow \frac{dX_P}{dt} = \frac{P_2(X_2 - X_P) + X_PV_1}{M_1} \tag{3.32}$$

$$\frac{dX_2}{dt} = \frac{F(X_f - X_2) + X_2V_2}{M_2} \tag{3.33}$$

In addition, the energy balances are

$$\frac{d[M_1h(T_1, X_P)]}{dt} = P_2h(T_2, X_2) + S\lambda(T_S) - P_1h(T_1, X_P) - V_1H(T_1) \tag{3.34}$$

$$\Rightarrow M_1 \frac{dh(T_1, X_P)}{dt} + h(T_1, X_P)\frac{dM_1}{dt} = P_2h(T_2, X_2) + S\lambda(T_S) - P_1h(T_1, X_P) - V_1H(T_1) \tag{3.35}$$

where $S(T_S) = U_1A_1(T_S - T_1)$.
Using Equation 3.28,

$$\frac{dh(T_1, X_P)}{dt} = \frac{P_2[h(T_2, X_2) - h(T_1, X_P)] + U_1A_1(T_S - T_1) - V_1[H(T_1) - h(T_1, X_P)]}{M_1} \tag{3.36}$$

Second effect: $$\frac{d[M_2h(T_2, X_2)]}{dt} = Fh(T_f, X_f) + V_1\lambda(T_1) - P_2h(T_2, X_2) - V_2H(T_2) \tag{3.37}$$

where $V_1(T_1) = U_2A_2(T_1 - T_2)$.
Using Equation 3.29,

$$\frac{d[h(T_2, X_2)]}{dt} = \frac{F[h(T_f, X_f) - h(T_2, X_2)] + U_2A_2(T_1 - T_2) - V_2[H(T_2) - h(T_2, X_2)]}{M_2} \tag{3.38}$$

The enthalpy of the product is [11]

$$h(T, X) = (4.177 - 2.506X)T \tag{3.39}$$

and the pure solvent enthalpy is derived from the steam tables as

$$H(T) = 2495 + 1.958T - 0.002128T^2 \tag{3.40}$$

The pure solvent liquid enthalpy for the condensate streams is determined from the steam tables as

$$h(T) = 4.177T \tag{3.41}$$

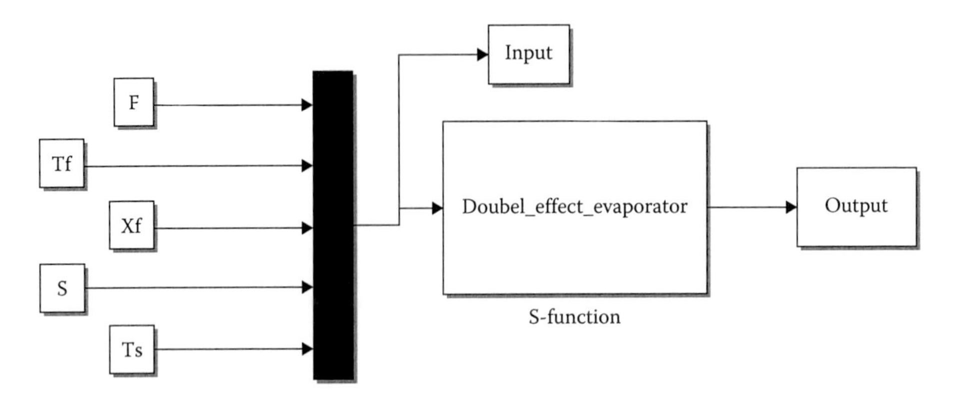

Figure 3.12 Simulink model of double-effect evaporator.

The latent heat of vaporization is

$$\lambda(T) = H(T) - h(T) = 2495 - 2.219T - 0.002128T^2 \tag{3.42}$$

By employing Equations 3.39 through 3.42 and rearranging, Equations 3.36 and 3.38 can be written as

$$\frac{dT_1}{dt} = \frac{P_2(4.177 - 2.506X_2)(T_2 - T_1) - U_2A_2(T_1 - T_2) + U_1A_1(T_S - T_1)}{M_1(4.177 - 2.506X_P)} \tag{3.43}$$

$$\frac{dT_2}{dt} = \frac{F(4.177 - 2.506X_f)(T_f - T_2) + U_2A_2(T_1 - T_2) + V_2[4.177T_2 - H(T_2)]}{M_2(4.177 - 2.506X_2)} \tag{3.44}$$

3.2.1.3 *Analysis of dynamic behavior of double-effect evaporator model*

The nonlinear differential-algebraic equations in Equations 3.28 through 3.44 describe the industrial double-effect evaporator. MATLAB (version 8.6 (R2015b)) was used to perform the dynamic simulations in order to analyze the transient behavior of the industrial double-effect evaporator described by the system of nonlinear differential-algebraic equations. The Simulink model for the double-effect evaporator is shown in Figure 3.12, and the associated MATLAB codes, which will be used while simulating the model, are provided in Programs 3.3 and 3.4. Using the steady-state operating point given in Table 3.13, the double-effect evaporator was simulated for 25 hours with a variable sample time. Figure 3.13 illustrates the impact of changes in the Tank 2 liquid product flow rate on Tank 1. The Tank 2 liquid product flow rate suddenly increased from its steady-state value to 15200 at time = 5 hours and then abruptly decreased from 15200 to its steady-state value at time = 10 hours. While P_2 remained constant at 15200 kg/hr for the period of 5 hours (from time = 5 to 10 hours), holdup M_1 kept on increasing and the concentration of soluble solids in Tank 1 liquid product monotonically decreased. When P_2 returned to its steady-state value, holdup M_1 became stable at the new steady state, and X_P came back to its original steady state. The influence of steam temperature on the temperature of Tank 1 and Tank 2 can be

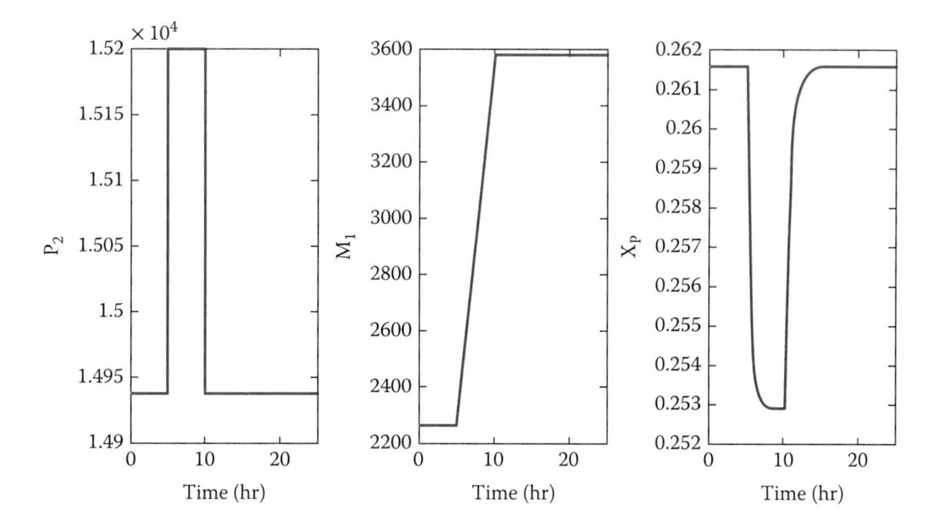

Figure 3.13 Effect of pulse input change in Tank 2 product flow rate.

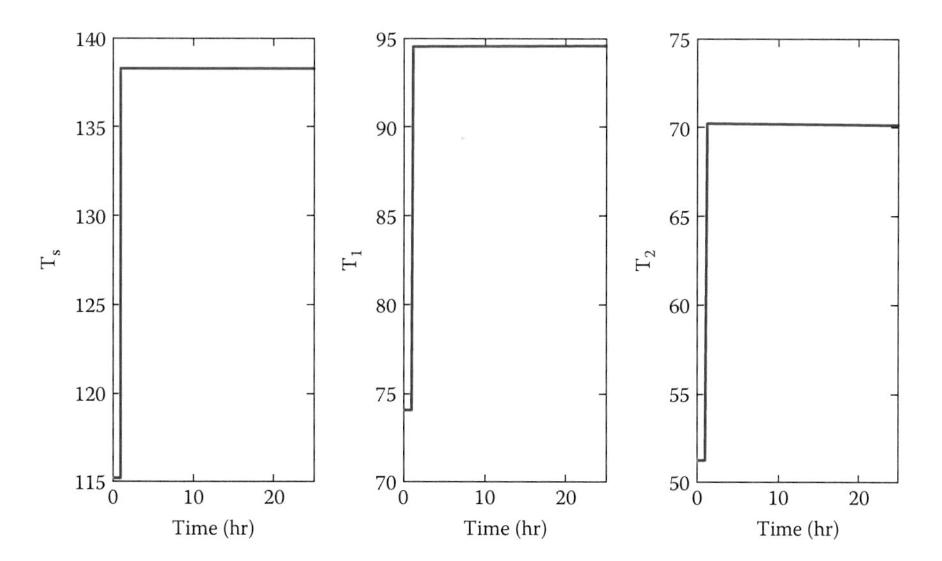

Figure 3.14 Effect of steam temperature on T_1 and T_2.

seen from Figure 3.14. An increase in the steam temperature increased the temperature of Tank 1 and Tank 2. As shown in Figure 3.15, when the Tank 1 vapor flow rate changed from its original base value to a new steady state at time = 1 hour, holdup M_1 monotonically increased level, and the product concentration rapidly decreased during the first few hours of operation and then settled at a new steady state lower than its original one. The dynamic simulations performed so far suggest that P_2 can be used to control holdup M_1, steam flow rate (S) to Tank 1 temperature (T_1), and Tank 1 vapor flow rate (V_1) to main product concentration (X_P). The step increase in feed flow rate (Figure 3.16) induced larger

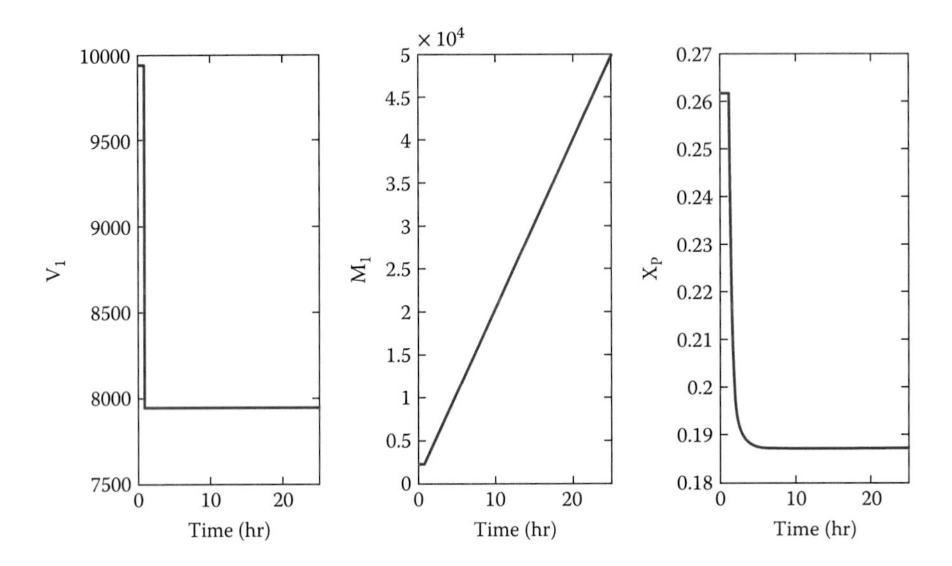

Figure 3.15 Effect of step change in Tank 1 vapor flow rate.

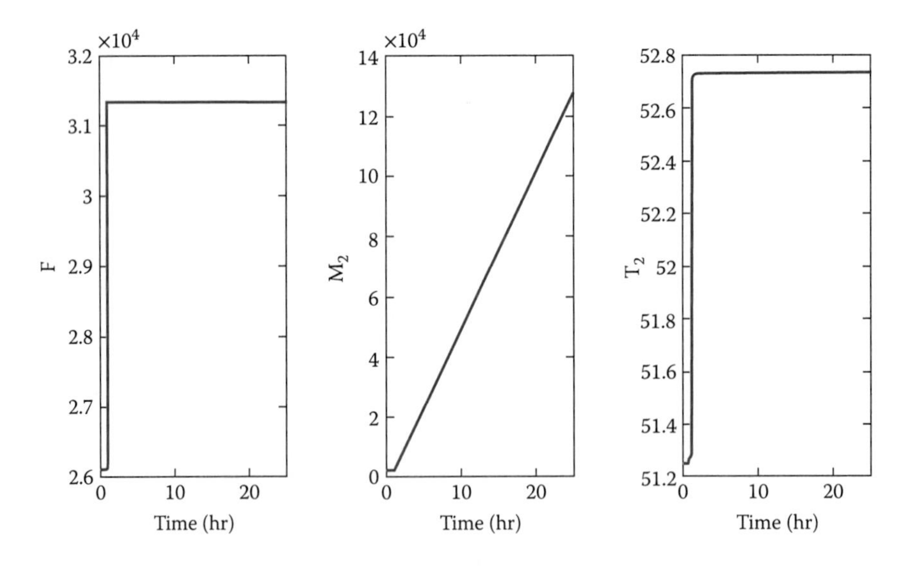

Figure 3.16 Influence of feed flow rate on second effect.

variations in the Tank 2 holdup (M_2) than it caused in the Tank 2 temperature (T_2), whereas the Tank 2 vapor flow rate, as Figure 3.17 exhibits, has greater influence on Tank 2 temperature than it has on Tank 2 holdup. Therefore, the feed flow rate can act as a suitable manipulated variable to control Tank 2 holdup, and the temperature of Tank 2 can be regulated by varying Tank 2 vapor flow rate. The temperature of the feed, the concentration of soluble solids in the feed, and the steam temperature can act as disturbance inputs.

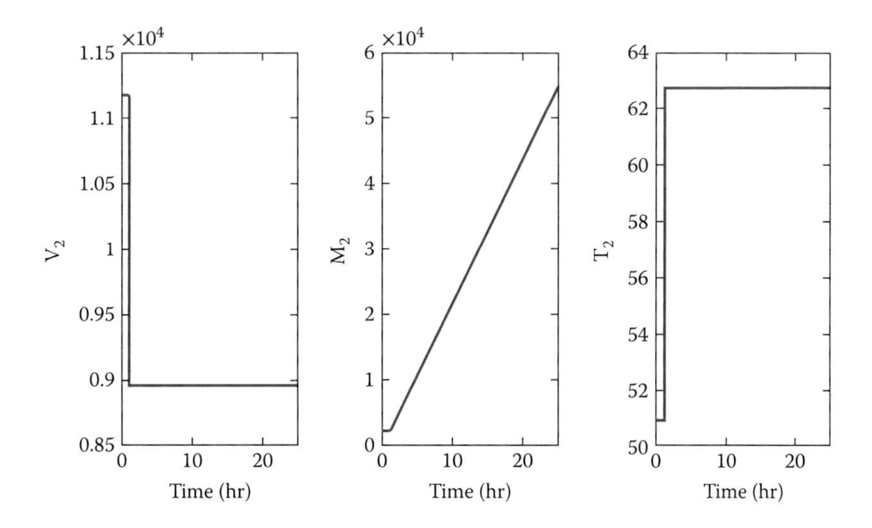

Figure 3.17 Influence of Tank 2 vapor product flow rate on second effect.

Program 3.3 MATLAB code for double-effect evaporator S-function

```
function [sys,x0,str,ts]=Double_Effect_Evaporator(t,x,u,flag)
switch flag,
case 0,
  [sys,x0,str,ts]=mdlinitializesizes;
case 1,
  sys=mdlderivatives(t,x,u);
case 3,
  sys=mdloutputs(x);
case {2,4,9},
  sys=[];
otherwise,
  error(['Unhandled flag = ',num2str(flag)]);
end
function [sys,x0,str,ts]=mdlinitializesizes
sizes=simsizes;
sizes.NumContStates=6;
sizes.NumDiscStates=0;
sizes.NumOutputs=6;
sizes.NumInputs=9;
sizes.DirFeedthrough=1;
sizes.NumSampleTimes=1;
sys=simsizes(sizes);
x0=[2268,2268,0.26155,0.08765,74.1,51.25];
str=[];
ts=[0 0];
function sys=mdlderivatives(t,x,u)
F=u(1);
```

Table 3.13 Values of model parameters and steady states of inputs and states [10]

Term	Abbreviation	Value
Tank 1 mass holdup	M_1	2268
Tank 2 mass holdup	M_2	2268
Input feed flow rate	F	26107.933
Input steam flow rate	S	1.1055e + 04
Tank 1 liquid product flow rate	P_1	5006
Tank 2 liquid product flow rate	P_2	14938
Vapor flow rate from Tank 1	V_1	9932
Vapor flow rate from Tank 2	V_2	11169.933
Feed composition	X_f	0.05015
Tank 1 composition	X_P	0.26155
Tank 2 composition	X_2	0.08765
Steam temperature	T_S	115.2337
Feed temperature	T_f	83.80344
Temperature in Tank 1	T_1	74.1
Temperature in Tank 2	T_2	51.25
Heat transfer area of Tank 1	A_1	102
Heat transfer area of Tank 2	A_2	412
Overall heat transfer coefficient for Tank 1	U_1	5826
Overall heat transfer coefficient for Tank 2	U_2	2453

Source: C.H. Runyon, T.R. Rumsey, K.L. McCarthy (1991). Dynamic Simulation of a Nonlinear Model of a Double Effect Evaporator. Journal of Food Engineering, vol. 14, 185–201. Reproduced with permission.

```
Tf=u(2);
Xf=u(3);
S=u(4);
Ts=u(5);
sys=model(t,x,F,Tf,Xf,S,Ts);
function sys=mdloutputs(x)
sys=x;
%%%%%%%%%%%%%%%%%%%%%%%%%%%%%%%%%%%%%%%%%%%%%%
```

Program 3.4 MATLAB code for mathematical model of double-effect evaporator

```
function dy=model(t,x,F,Tf,Xf,S,Ts)
dy=zeros(6,1);
% heat transfer area
A1=102;
A2=412;
% heat transfer coefficient
```

```
U1=5826;
U2=2453;
% Flowrate of vapor leaving Tank 1 and Tank 2
V1=9932;
V2=[11169.9336384439];
% Flowrate of product of Tank 1 and Tank 2
P1=5006;
P2=14938;
% state variables
M1=x(1);
M2=x(2);
Xp=x(3);
X2=x(4);
T1=x(5);
T2=x(6);
% Total mass balance around Tank 1 and Tank 2
dy(1)=P2-P1-V1;
dy(2)=F-P2-V2;
% Mass balance for soluble solids in the output of Tank 1 and Tank 2
dy(3)=(1/M1)*(P2*(X2-Xp) + Xp*V1);
dy(4)=(1/M2)*(F*(Xf-X2) + X2*V2);
% Latent heat of vaporization
lambdaTs=2495-2.219*Ts-0.002128*Ts^2;
% Energy balance
dy(5)=((4.177-2.506*X2)*P2*(T2-T1) + S*lambdaTs-U2*A2*(T1-T2))/
((4.177-2.506*Xp)*M1);
HT2=2495 + 1.958*T2-0.002128*T2^2; % pure solvent vapor enthalpy
dy(6)=((4.177-2.506*Xf)*F*(Tf-T2) + U2*A2*(T1-T2)-V2*(HT2-4.177*T2))/
((4.177-2.506*X2)*M2);
end
%%%%%%%%%%%%%%%%%%%%%%%%%%%%%%%%%%%%%%%%%%%%%
```

3.2.2 *Generation of historical database for double-effect evaporation process*

The simulated (open loop) commercial double-effect evaporator for concentrating the tomato juice was considered for the generation of the database. Five different operating conditions (normal operation), as explained in Table 3.14, were created. Twenty-one data sets, each data set consisting of 12504 samples of six state variables ($M_1, M_2, X_P, X_2, T_1, T_2$), were generated by operating the double-effect evaporator under each operating mode for 25 hours using a variable sample time.

3.2.3 *Application of modified K-means clustering algorithm on double-effect evaporation process database*

In this case study, a PCA similarity factor ($\alpha_1 = 1$ and $\alpha_2 = 0$) was only considered to verify whether the modified *K*-means clustering algorithm can perform well without a distance similarity factor. The modified *K*-means clustering algorithm was applied on the historical database generated in the previous section. The algorithm was repeated for an assumed number of clusters, $K = 2$ through 10 (using the weighting factors $\alpha_1 = 1$ and $\alpha_2 = 0$) and the results are displayed in Figure 3.18. As we can see from the plot of $\psi(K)$ versus K (subplot (c) of Figure 3.18), $\psi(K)$ changes from negative to positive at $K = 5$, which indicates the presence

Table 3.14 Generation of database for double-effect evaporation process

Op. cond.	Description	No. of data sets
1	Step increase in feed flowrate (F is varied in the range [26107.933, 3.9162e + 04] with step length of 2.6108e + 03)	6
2	Step increase in steam flowrate (S is varied in the range [1.1055e + 04, 1.6583e + 04] with step length of 1.1055e + 03)	6
3	Oscillations in feed flowrate (high frequency oscillation-30 cycles/sec, medium frequency oscillation-15 cycles/sec, low frequency oscillation-10 cycles/sec)	3
4	Oscillations in feed concentration (high frequency oscillation-35 cycles/sec, 30 cycles/sec, 25 cycles/sec)	3
5	Step increase in feed concentration (X_f is changed from its nominal value to 0.0552, 0.0652, and 0.0802)	3

of a knee (near optimality) in the plot of $J(K)$ versus K at $K = 5$. Therefore, the optimum number of clusters is five, which is exactly the number of operating conditions in the database. Thus, the developed, modified, K-means clustering algorithm is able to determine the exact number of clusters. Table 3.15 presents the dissimilarity (one-similarity factor) calculated based on a PCA similarity factor between each data set and each of five clusters, and the analysis of these five clusters is given in Table 3.16. The modified K-means clustering algorithm using only PCA similarity factor (followed by dissimilarity calculation as a one-similarity factor) did well in grouping the database very accurately.

3.3 Continous crystallization process

3.3.1 Modeling and dynamic simulation of continuous crystallization process

Crystallization from solution is a widely applied unit operation in both the batch crystallization (small scale) and continuous crystallization (large scale) modes for solid-liquid separations. Continuous crystallization is used for the bulk production of inorganic (e.g., potassium chloride, ammonium sulfate) and organic (e.g., adipic acid) materials. Batch

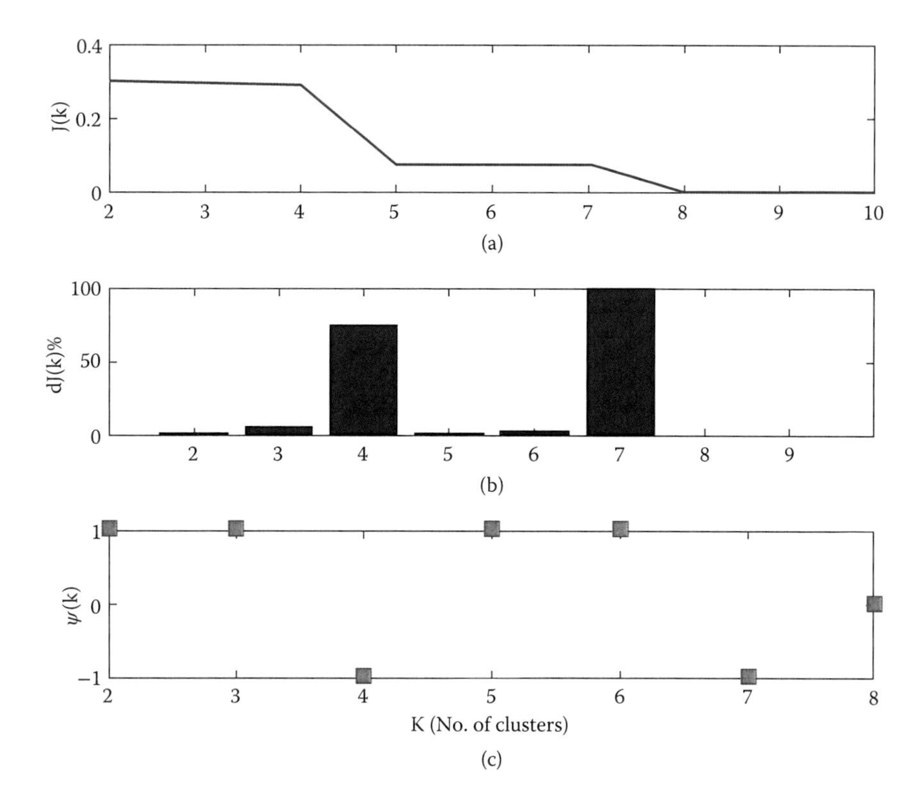

Figure 3.18 Clustering performance for the double-effect evaporator.

cooling crystallization is especially used in the production of value-added products, such as pharmaceuticals, photo materials, and fine chemicals. In a crystallizer, phenomena like primary nucleation, crystal growth, and attrition of crystals due to crystal–stirrer collisions are found on a microscopic scale and have been modeled [12]. On a macroscopic scale, crystallization processes are adequately described by population balance models (PBMS). The model related to microscopic phenomena are to be integrated into the PBMS. Population balance models are distributed parameter systems consisting of partial differential equations (PDEs) describing the evolution of the crystal size distribution (CSD), which is a function of time and crystal size, coupled with ordinary differential equations (ODEs) related to the concentration and temperature profile of the liquid phase. Mathematical modeling by population balance has become a standard tool for chemical engineers, with applications as diverse as crystallization, catalysis, particulate processes, and bioengineering. Hulbert and Katz [13] are generally credited with being the first authors to publish a general formulation of the theory. The standard reference on the subject is the textbook authored by Ramkrishna [14]. A population balance for a solids process, such as crystallization, with one property coordinate can be described by the following partial differential equation:

$$\frac{\partial F(L,t)}{\partial t} + \frac{\partial\big(G(L,F,t)F(L,t)\big)}{\partial L} = \sum \dot{F}_i(F,L,t) \tag{3.45}$$

Table 3.15 Dissimilarity factor between data sets and clusters in double-effect evaporator

Cluster Data set	1	2	3	4	5
X_1	3.0713e-05	0.50016	0.000188	0.502061	0.516264
X_2	1.7042e-05	0.50012	0.000155	0.50203	0.516267
X_3	9.2917e-06	0.50010	0.000132	0.50201	0.516270
X_4	4.8518e-06	0.50008	0.000116	0.50199	0.516272
X_5	2.3542e-06	0.50007	0.000104	0.50198	0.516274
X_6	1.036e-06	0.500066	9.55491e-05	0.501978	0.516275
X_7	0.50005	0	0.500004	0.009843	0.500717
X_8	0.5000	4.44089e-16	0.500004	0.009843	0.500717
X_9	0.50005	4.440892e-16	0.500004	0.00984	0.500717
X_{10}	0.50005	6.661338e-16	0.5000041	0.00984	0.500717
X_{11}	0.50005	1.110223e-16	0.500004	0.009843	0.500717
X_{12}	0.50005	1.11022e-16	0.5000041	0.00984	0.500717
X_{13}	0.00207	0.50000	0.0016690	0.50186	0.528740
X_{14}	0.00019	0.500001	5.633e-05	0.50181	0.519334
X_{15}	9.41e-05	0.500004	1.68238e-06	0.50174	0.517734
X_{16}	0.50209	0.016596	0.501813	0.00108	0.49537
X_{17}	0.50193	0.009064	0.501740	2.0420e-05	0.499081
X_{18}	0.50190	0.005905	0.50177	0.000711	0.49999
X_{19}	0.51663	0.500849	0.517672	0.49879	2.480e-06
X_{20}	0.51663	0.500837	0.51767	0.498775	1.5411e-06
X_{21}	0.51663	0.500819	0.51767	0.498740	1.128e-06

Table 3.16 Performance indices of dissimilarity-based modified *K*-means clustering algorithm in classifying operating conditions for double-effect evaporator

Cluster			Dominant		Operating Condition				
No.	N_P	$P(\%)$	Op. Cond.		1	2	3	4	5
1	6	100	1		**6**	0	0	0	0
2	6	100	2		0	**6**	0	0	0
3	3	100	3		0	0	**3**	0	0
4	3	100	4		0	0	0	**3**	0
5	3	100	5		0	0	0	0	**3**
Average	**NA**	**100**	**NA**	$\eta =$	**100**	**100**	**100**	**100**	**100**
				$\eta_{avg} = 100$					

where F(L, t) is the population density function, G(F, L, t) is the linear growth rate of the crystals, and F_i are generation and loss terms that account for phenomena such as nucleation, attrition, flow into and out of the control volume, etc. In certain special cases, particularly batch processes, Equation 3.45 can be solved by the method of moments [15–18]. Historically, most researchers who have attempted to solve Equation 3.45 have employed the method of lines, in which the differential equation is discretized into a finite number of ordinary differential equations [19–25]. Other methods, including statistical methods, are reviewed by Ramkrishna [14]. Wulkow and coworkers have developed and marketed a commercial software package (Parsival) for the dynamic simulation of particle size distributions and flow sheets with solids processes [26,27]. The software is based on the Galerkin h–p method (Wulkow 1996), an adaptive finite element algorithm [28].

The present study on classification of various operating conditions in the continuous crystallization process is carried out using simulated data derived from a process model. The following steps are carried out to generate data pertaining to various operating conditions and to classify them.

3.3.1.1 The process description

As a final case study of this chapter, we consider the continuous crystallizer with fines removal (Figure 3.19) used for the continuous crystallization of potassium chloride from water (KCI) [29].

The solution (fresh feed) enters the crystallizer with volumetric flow rate, q, and concentration, c_{in}. The product with the particle size distribution, F, is withdrawn from the bottom at volumetric flow rate, q and the concentration of the product is denoted by c. Material is withdrawn from the reactor at a flow rate equal to Rq where R is the fines dissolution ratio. The fines dissolver is modeled by assuming that all particles smaller than the fines cutoff size L_f are dissolved, and other crystals pass through unchanged.

3.3.1.2 Mathematical model of continuous crystallization process

The mathematical model of a continuous crystallizer is derived by making the following assumptions.

- Ideal mixing
- Isothermal operation
- Constant overall volume (liquid + solid)

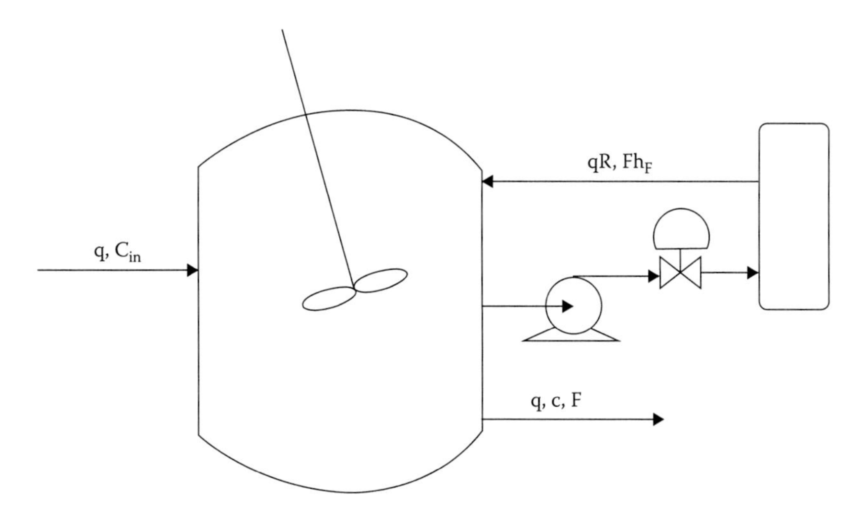

Figure 3.19 Continuous crystallizer.

- Nucleation of crystals of negligible size
- Size-independent growth rate
- No particle breakage, attrition or agglomeration

The population balance is

$$\frac{\partial F(L,t)}{\partial t}+G(t)\frac{\partial F(L,t)}{\partial t}=-\frac{q}{V}(1+h_f(L))F(L,t)\ ,\ F(0,t)=\frac{B(t)}{G(0,t)} \tag{3.46}$$

where V is the crystallizer volume.

The fines dissolution function is given as

$$h_f(L)=R\left(1-h\left(L-L_f\right)\right) \tag{3.47}$$

where h is the unit step function.

The following empirical power laws express the dependence of nucleation and growth rate on solute concentration, c:

$$G=k_g\left(c-c_{\text{sat}}\right)^g \tag{3.48}$$

$$B=k_b\left(c-c_{\text{sat}}\right)^b \tag{3.49}$$

The mass balance for solute in the liquid phase is

$$\frac{d(V\varepsilon cM_A)}{dt}=qc_{in}M_A-qc\varepsilon M_A-3Vk_v\rho G\int_0^\infty FL^2\,dL+qk_v\rho\int_0^\infty h_f(L)FL^3\,dL \tag{3.50}$$

where M_A is the molar mass of the solute, ρ is the density of the crystals, and ε is the void fraction given by

$$\varepsilon=1-k_v\int_0^\infty FL^3\,dL,\ k_v \text{ is the volume shape factor of the crystals} \tag{3.51}$$

Differentiating the above equation,

$$\frac{d\varepsilon(t)}{dt}=-k_v\int_0^\infty\frac{\partial F}{\partial L}L^3\,dL \tag{3.52}$$

Substituting Equation 3.45 and performing partial integration,

$$\frac{d\varepsilon(t)}{dt}=-3k_vG\int_0^\infty FL^2\,dL+\frac{q}{V}k_v\int_0^\infty\left(h_f(L)+1\right)FL^3\,dL \tag{3.53}$$

From Equations 3.50 and 3.53, one gets

$$M_A \frac{dc(t)}{dt} = \frac{q(\rho - M_A c)}{V} + \frac{\rho - M_A c(t)}{\varepsilon(t)} \frac{d\varepsilon(t)}{dt} + \frac{q M_A c_{in}}{V\varepsilon(t)} - \frac{q\rho}{V\varepsilon(t)} \tag{3.54}$$

3.3.2 Generation of historical database for continuous crystallization process

The simulated continuous crystallizer has three inputs: feed flow rate, feed solute concentration, and fines dissolution ratio. Among these three inputs, fines dissolution ratio can be considered a potential manipulated variable to control the third moment of the crystalize distribution, and the remaining inputs are disturbance variables. Ward and Yu [29] developed a Simulink block (PCSS) to model the transient evolution of a population density function for a physical system modeled by a population balance equation. They modeled the continuous crystallization process discussed in the previous section using the Simulink block (PCSS) and also designed a proportional integral controller to control the third moment of crystal size distribution through manipulation of fines dissolution ratio (R). The Simulink and MATLAB files written by Ward and Yu [29] for solving the mathematical model of a continuous crystallizer (Equations 3.45 through 3.54) are provided in Appendix A of Ward and Yu [29]. Their work was adapted here to create the historical database for the continuous crystallization process to evaluate the developed modified *K*-means clustering algorithm. Figure 3.20 displays the closed loop control of the simulated continuous crystallizer. The process data provided in Table 3.17 was employed for both open loop and closed loop simulations. As presented in Table 3.18, eight different normal operating conditions were created.

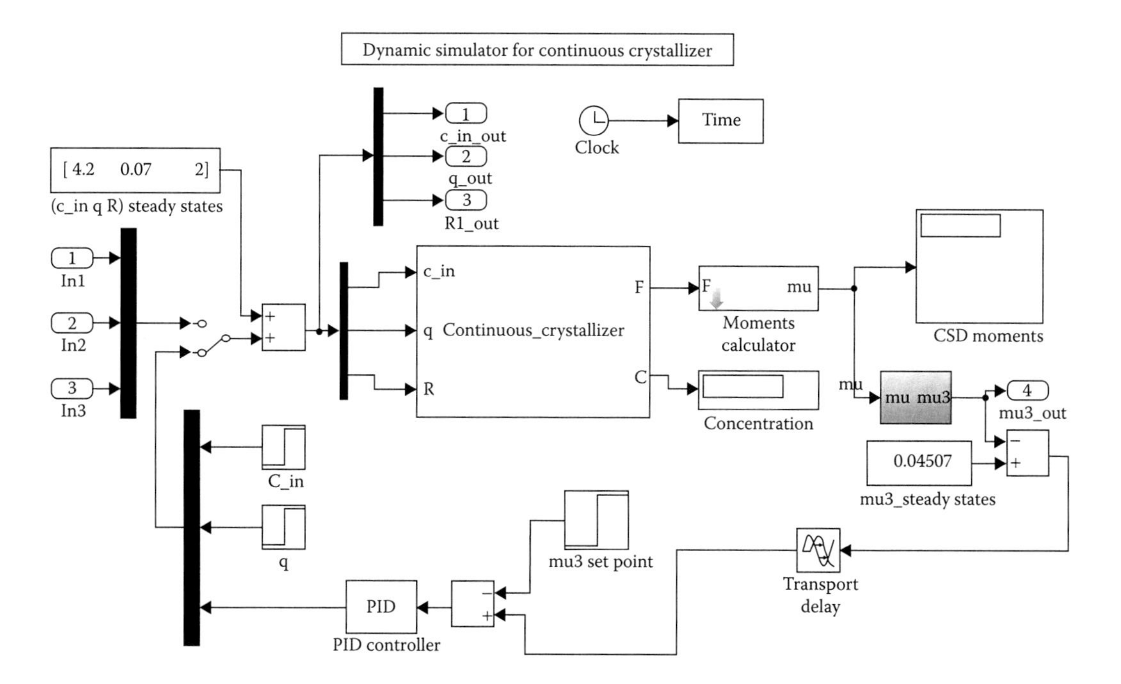

Figure 3.20 Simulink block diagram for closed loop control (PI) of continuous crystallizer.

The simulated continuous crystallizer was operated for 1000 minutes using variable sample time in each operating mode to generate a single data set consisting of measurements of product concentration, and zeroth, first, second, and third moments of crystal size distribution.

3.3.3 *Application of modified K-means clustering algorithm in continuous crystallizer*

Here, distance similarity factor was only considered in a modified *K*-means clustering algorithm to cluster the historical database generated in the former section. So the values of the weighting factors α_1 and α_2 were selected as 0 and 1, respectively. The number of clusters, *K*, was varied from 2 to 12 and the algorithm was run for every value of *K*. It is seen from Figure 3.21 that the sign of the difference of $dJ(K)$ $(\psi(K))$ changes multiple times from negative to positive, thus indicating the locations of multiple knees in the plot of $J(K)$ versus *K*. Therefore, the algorithm found four different optimum numbers of clusters. But the location of third knee, that is $K=8$, can be the exact optimum number of clusters since $dJ(K)$ has zero at $K=8$, supporting the actual groups in the database considered. So the algorithm was repeated again for $K=8$ and the results are shown in Tables 3.19 and 3.20. Table 3.19 presents dissimilarity factors between data sets and optimally computed clusters for the continuous crystallization process. Table 3.20 provides clustering performance. The modified *K*-means algorithm could not properly partition the database. Out of 35 data sets, three data sets were misclassified. Considering

Table 3.17 Values of parameter and steady states of inputs and outputs for the KCl crystallization process

Term	Abbreviation	Value	Units
Growth rate constant	k_g	3.0513e-02	Mm L/min mol
Nucleation rate constant	k_b	8.357e + 09	L^3/min mol^4
Growth rate exponent	g	1	
Nucleation rate exponent	b	4	
Saturation concentration	c_s	4.038	mol/L
Feed flow rate	q	0.035	L/min
Fines removal cut size	L_f	0.3	mm
Crystal density	ρ_c	1989	g/L
Molar mass KCI	M	74.551	g/mol
Crystallizer volume	V	10.5	L
Recycle ratio	R	2	–
Feed flow rate	q_{in}	4.2	L/min
Feed concentration	c_{in}	0.07	mol/L
Concentration of potassium chloride	c	4.0877085	mol/L
0th moment of crystal size distribution	μ_0	2.7630694	mm/(mm^3*mm)
1st moment of crystal size distribution	μ_1	0.2682994	m/mm^3
2nd moment of crystal size distribution	μ_2	0.0769161	m^2/mm^3
3rd moment of crystal size distribution	μ_3	0.0450665	m^3/mm^3

Table 3.18 Various operating conditions for continuous crystallizer

Op. cond.	Description	No. of data sets
1	Step increase in feed concentration (open loop) (c_{in} is varied in the range [4.41, 6.3] with step length of 0.21)	10
2	Oscillations in feed concentration (open loop) (high frequency oscillation-150, 100, 50 cycles/sec)	3
3	Ramp increase in feed concentration (open loop) (slope = {0.004;0.0045;0.005}, initial output = 4.2 mol/L)	3
4	Oscillations in feed flowrate (open loop) (high frequency oscillation-50 cycles/sec, 40 cycles/sec, 30 cycles/sec)	3
5	Fines removal cut size (open loop) (L_F = {0.6, 0.65, 0.7})	3
6	Sinusoidal change in feed concentration (closed loop) (amplitude = 1, frequency = {0.2,0.4,0.6,0.8,1})	5
7	Sinusoidal change in feed flow rate (closed loop) (amplitude = 1, frequency = {1.1,1.3,1.5,1.7,1.9})	5
8	Band-limited white noise added to third moment measurements (closed loop) (noise power = {10, 15, 20})	3

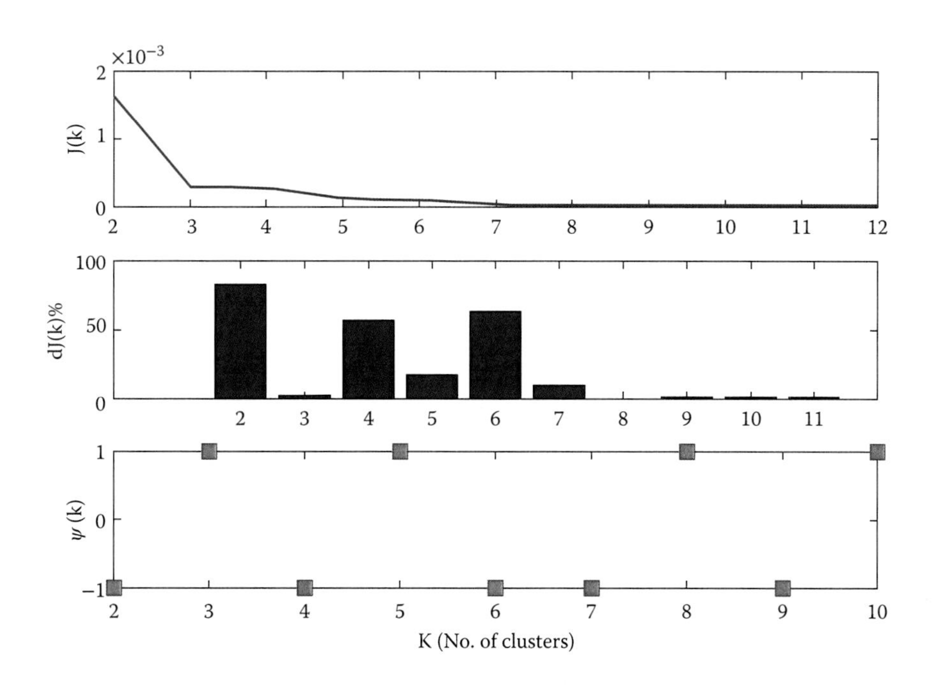

Figure 3.21 Clustering performance for continuous crystallization process

the average cluster purity (88%) and the average cluster efficiency (88%) obtained, it is reasonable to assert that the modified *K*-means algorithm worked satisfactorily on the larger database considered.

Table 3.19 Dissimilarity factors between data sets and optimally computed clusters for continuous crystallization process

Cluster Data set	1	2	3	4	5	6	7	8
X1	**0.0002596**	0.0003940	0.0196166	0.0005544	0.0020572	0.0017325	0.0010060	0.0006146
X2	**0.0001521**	0.0005205	0.0186708	0.0007129	0.0017296	0.0014277	0.0008323	0.0007984
X3	**9.68485e-05**	0.0006207	0.0180450	0.0008345	0.0015266	0.0012420	0.0007361	0.0009358
X4	**6.83247e-05**	0.0006896	0.0176488	0.0009169	0.0014047	0.0011311	0.0006812	0.0010280
X5	**5.29257e-05**	0.0007337	0.0173985	0.0009693	0.0013314	0.0010641	0.0006479	0.0010868
X6	**4.42688e-05**	0.0007602	0.0172409	0.0010008	0.0012877	0.0010236	0.0006267	0.0011224
X7	**3.93412e-05**	0.00077440	0.0171437	0.0010179	0.0012627	0.0009997	0.00061254	0.0011421
X8	**3.66451e-05**	0.00078010	0.0170871	0.0010250	0.0012501	0.0009867	0.00060290	0.0011509
X9	**3.53899e-05**	0.00077989	0.0170585	0.0010252	0.0012458	0.0009812	0.00059629	0.0011521
X10	**3.51315e-05**	0.00077554	0.0170497	0.0010205	0.0012474	0.00098091	0.00059184	0.00114799
X11	0.0009476	**5.98087e-06**	0.0222885	3.80921e-05	0.0035879	0.0030072	0.00150579	7.360516e-05
X12	0.0011447	**4.77380e-06**	0.0231594	1.03110e-05	0.0039702	0.0033485	0.00171431	3.463808e-05
X13	0.0014983	4.30431e-05	0.0245331	6.63672e-06	0.0046132	0.0039313	0.00209107	1.116684e-05
X14	0.0146451	0.02165439	**1.8112e-05**	0.0228239	0.0091549	0.00967189	0.01298225	0.0235773
X15	0.0155168	0.0226192	**4.04190e-08**	0.0238056	0.0098973	0.01041145	0.01376225	0.0245758
X16	0.0163046	0.0234791	**1.21380e-05**	0.0246793	0.0105790	0.0110882	0.01446718	0.0254638
X17	0.0013286	1.56571e-05	0.02374989	**1.00836e-07**	0.0042988	0.0036188	0.00184364	2.72246e-05
X18	0.0015444	5.04883e-05	0.0246516	9.566535e-06	0.0046864	0.00400482	0.00214515	8.87960e-06
X19	0.0015562	5.10136e-05	0.0246676	**8.986431e-06**	0.0047065	0.00401379	0.00213749	1.21545e-05
X20	0.0009437	0.0039340	0.0097950	0.0044855	**1.370540e-06**	0.00014350	0.00120787	0.0047619
X21	0.0010006	0.0040426	0.0096266	0.0046015	**3.955294e-06**	0.00015252	0.00124698	0.0048832
X22	0.0009519	0.0039388	0.0097206	0.0044906	**1.497334e-06**	0.0001384	0.00119493	0.0047704
X23	0.0006612	0.0031631	0.0104380	0.0036565	0.00012929	**0**	0.00056480	0.0039933
X24	0.0006612	0.0031631	0.0104380	0.0036565	0.00012929	**0**	0.00056480	0.0039933
X25	0.0006612	0.0031631	0.0104380	0.0036565	0.00012929	**0**	0.00056480	0.0039933
X26	0.0006612	0.0031631	0.0104380	0.0036565	0.00012929	**0**	0.00056480	0.0039933
X27	0.0006612	0.003163	0.0104380	0.0036565	0.00012929	**0**	0.00056480	0.0039933
X28	0.0005285	0.001855	0.0130472	0.0022061	0.00098867	0.00043765	**1.4744934e-05**	0.0025797
X29	0.0004855	0.001661	0.0135090	0.0019918	0.00108343	0.00051439	**2.0338933e-06**	0.0023517
X30	0.0004544	0.001483	0.0139713	0.0017936	0.00118689	0.00060035	**8.6814271e-07**	0.0021400
X31	0.0004352	0.001326	0.0144122	0.0016186	0.00129319	0.00069038	**9.7981490e-06**	0.00195254
X32	0.0004261	0.001195	0.0148118	0.0014712	0.00139562	0.00077838	**2.5841790e-05**	0.0017939
X33	0.0027900	0.0004688	0.0287787	0.0003140	0.00669069	0.00597413	0.003699925	**0.00020086**
X34	0.0014407	5.1678e-05	0.0243585	2.22187e-05	0.00449128	0.00388460	0.002150631	**7.83648e-07**
X35	0.0009012	1.21423e-05	0.0221553	5.015466e-05	0.00349677	0.00293947	0.001486694	8.17007e-05

In conclusion, this chapter presented three case studies to classify various operating conditions of processes chosen using a *K*-means clustering algorithm (which is either based on PCA similarity or distance similarity or a combination of both). The same task could have been accomplished by various other algorithms (for example, PLS, enhanced Sammon's *NLM*, dissimilarity (K-L expansion based), LDA, SVM) proposed in this book for various other purposes.

Table 3.20 Performance indices of modified *K*-means clustering algorithm for continuous crystallizer

CN	N_P	P(%)	DOC.		Operating Condition 1	2	3	4	5	6	7	8
1	10	100	1		**10**							
2	3	67	2			**2**						**1**
3	3	100	3				**3**					
4	3	67	4			**1**		**2**				
5	3	100	5						**3**			
6	5	100	6							**5**		
7	5	100	7								**5**	
8	3	67	8					**1**				**2**
AVG	**NA**	**88**	**NA**	$\eta =$	**100**	**67**	**100**	**67**	**100**	**100**	**100**	**67**
				η_{avg} = **88**								

CN: Cluster Number; DOC: Dominant Operating Condition.

References

1. S. Damarla, M. Kundu (2013). Control of yeast fermentation bioreactor in subspace. *International Journal of Computer Applications*, vol. 64(5), 13–20.
2. J.A. Roels (1982). Mathematical models and the design of biochemical reactors. *Journal of Chemical Technology and Biotechnology*, vol. 32, 59–72.
3. B. Volesky, L. Yerushalmi, J.H.T. Luong (1982). Metabolic-heat relation for aerobic yeast respiration and fermentation. *Journal of Chemical Technology and Biotechnology*, vol. 32, 650–659.
4. G.H. Cho, C.Y. Choi, Y.D. Choi, M.H. Han (1982). Ethanol production by immobilised yeast and its CO_2 gas effects in a packed bed reactor. *Journal of Chemical Technology and Biotechnology*, vol. 32, 959–967.
5. B. Sevella (1992). *Bioengeneering Operations*, Tankonykiado (Ed.). Budapest: Technical University of Budapest.
6. S. Aiba, M. Shoda, M. Nagatani (1968). Kinetics of product inhibition in alcoholic fermentation. *Biotechnology and Bioengineering*, vol. 10, 846–864.
7. F. Godia, C. Casas, C. Sola (1988). Batch alcoholic fermentation modeling by simultaneous integration of growth and fermentation equations. *Journal of Chemical Technology and Biotechnology*, vol. 41, 155–165.
8. Z. Nagy, S. Agachi (1996). Nonlinear model predictive control of a continuous fermentation reactor using artificial neural networks. *In Automatic Control and Testing Conference*, Cluj-Napoca, 23rd–24th May, Section A2, pp. 235–240.
9. D.M. Himmelblau, K.B. Bischoff (1968). *Process Analysis and Simulation: Deterministic System*. New York, NY: John Wiley & Sons.
10. C.H. Runyon, T.R. Rumsey, K.L. McCarthy (1991). Dynamic simulation of a nonlinear model of a double effect evaporator. *Journal of Food Engineering*, vol. 14, 185–201.
11. V. Miranda, R. Simpson (2005). Modeling and simulation of an industrial multiple effect evaporator: Tomato concentrate. *Journal of Food Engineering*, vol. 66, 203–210.
12. A. Gerstlauer, A. Mitrovic, S. Motz, E.-D. Gilles (2001). A population model for crystallization processes using two independent particle properties. *Chemical Engineering Science*, vol. 56, 2553–2565.
13. H.M. Hulbert, S. Katz (1964). Some problems in particle technology: A statistical mechanical formulation. *Chemical Engineering Science*, vol. 19, 555–574.
14. D. Ramkrishna (2000). *Population Balances*. San Diego, CA: Academic Press.

15. C.T. Chang, M.A.F. Epstein (1982). Identification of batch crystallization control strategies using characteristic curves. In *Nucleation, Growth and Impurity Effects in Crystallization Process Engineering*, M.A.F. Epstein (Ed.). New York, NY: AIChE.
16. S.H. Chung, D.L. Ma, R.D. Braatz (1999). Optimal seeding in batch crystallization. *Canadian Journal of Chemical Engineering*, vol. 77, 590–596.
17. A.G. Jones (1974). Optimal operation of a batch cooling crystallizer. *Chemical Engineering Science*, vol. 29, 1075–1087.
18. J.D. Ward, D.A. Mellichamp, M.F. Doherty (2006). Choosing an operating policy for seeded batch crystallization. *AIChE Journal*, vol. 52, 2046–2054.
19. P.J. Hill, K.M. Ng (1995). New discretization procedure for the breakage equation. *AIChE Journal*, vol. 41(5), 1204–1216.
20. M.J. Hounslow, R.L. Ryall, V.R. Marshall (1988). A discretized population balance for nucleation, growth, and aggregation. *AIChE Journal*, vol. 34(11), 1821–1832.
21. S. Kumar, D. Ramkrishna (1996a). On the solution of population balance equations by discretization. I. A fixed pivot technique. *Chemical Engineering Science*, vol. 51(8), 1311–1332.
22. S. Kumar, D. Ramkrishna (1996b). On the solution of population balance equations by discretization. II. A moving pivot technique. *Chemical Engineering Science*, vol. 51(8), 1333–13342.
23. S. Kumar, D. Ramkrishna (1997). On the solution of population balance equations by discretization. III. Nucleation, growth, and aggregation of particles. *Chemical Engineering Science*, vol. 52(24), pp. 4659–4679.
24. P. Marchal, R. David, J.P. Klein, J. Villermaux (1988). Crystallization and precipitation engineering. I. An efficient method for solving population balance in crystallization with agglomeration. *Chemical Engineering Science*, vol. 43(1), 59–67.
25. P.K. Pathath, A. Kienle (2002). A numerical bifurcation analysis of nonlinearoscillations in crystallization processes. *Chemical Engineering Science*, vol. 57, 4391–4399.
26. A. Gerstlauer, C. Gahn, H. Zhou, M. Rauls, M. Schreiber (2006). Application of population balances in the chemical industry—Current status and future needs. *Chemical Engineering Science*, vol. 61, 205–217.
27. M. Wulkow, A. Gerstlauer, U. Nieken (2001). Modeling and simulation of crystallization processes using Parsival. *Chemical Engineering Science*, vol. 56, 2575–2588.
28. M. Wulkow (1996). The simulation of molecular weight distributions in polyreaction kinetics by discrete Galerkin methods. *Macromolecular Theory and Simulations*, vol. 5(3), 393–416.
29. J.D. Ward, C.-C. Yu (2008). Population balance modeling in Simulink: PCSS. *Computers and Chemical Engineering*. vol. 32, 2233–2242.

chapter four

Detection of abnormal operating conditions in processes using moving window-based pattern matching

Chemical process plant design and associated control systems are becoming increasingly complex and instrument laden to meet product quality demands, as well as economic, energy, environmental, and safety constraints. It is important to supervise the process and detect faults, if any, while the plant is still under control so that abnormal situations can be promptly identified. It should be mentioned that fault detection and diagnosis can be seen as an integral part of optimal and safe plant operation. Deployment of advanced control strategies alone may not be the answer to dealing with abnormal process conditions. Venkatasubramanian et al. [1–3] discussed fault detection and diagnosis (FDI) in chemical processes and classified the available methods into quantitative and qualitative model-based methods and process history-based methods. Quantitative model-based FDI techniques, namely the parity space approach, the observer-based approach, and the parameter estimation approach make use of a mathematical model of the process to estimate measured outputs. Model-based methods are usually dependent on fairly accurate deterministic process models. The inconsistencies between the estimated and actual behaviors are *residuals,* or fault indicators, which reflect the abnormal situation of the supervised process. The residuals are then analyzed with the aim of localizing the fault. In process history-based methods, a large amount of historical process data is needed. Data can be presented as *a priori* knowledge to a diagnostic system, which is called feature extraction. This extraction process can be either qualitative or quantitative in nature. Two of the major methods that extract qualitative history information are expert systems and trend-modeling methods. Methods that extract quantitative information can be broadly classified as nonstatistical and statistical methods. Data-based methods for fault detection and diagnosis depend entirely on process measurements. For fault detection, thresholds/cutoff parameters are created from normal operating conditions based on historical measurements. Various faulty data are obtained from historical data. Process data under consideration in their prevailing state are compared with faulty/normal data from a historical database, and normal/abnormal operating conditions are detected, which is supposed to be the crucial step in the design of a model -free expert system dedicated to fault diagnosis.

This chapter is dedicated to the detection of abnormal/faulty operating conditions in three processes, including a pilot-scale process that uses a process historical database applying moving window-based pattern matching.

4.1 Detection of abnormal operating conditions in a fluid catalytic cracking unit

4.1.1 Introduction to the fluid catalytic cracking (FCC) process

Fluid catalytic cracking (FCC) is one of the major conversion technologies in the oil refinery and hydrocracking units. It is widely used to convert, or crack, the high-boiling, high-molecular weight oil to lighter fractions such as kerosene, gasoline, LPG, heating oil, and petrochemical feedstock. Apart from that, FCC produces an important fraction of propylene for the polymer industry. Currently more than 400 FCC units are in operation worldwide.

Commercial production of petroleum started in 1859 in Titusville, Pennsylvania (USA); however, it could not meet the appreciable conversion capability target. Thermal cracking (heated under pressure) was first introduced in 1913, by William M. Burton at Standard Oil of Indiana. Thermal cracking has been almost completely replaced by catalytic cracking because it produces more gasoline with a higher octane rating, along with many high-value products. In mid-1942, the first commercial FCC unit (PCLA-1) based on a reactor and regenerator unit using a clay-based catalyst got started [4]. In catalytic cracking, temperature ranges from 850° to 950° F at much lower pressures of 10–20 psi. The catalysts used in refinery cracking units are typically solid materials (zeolite, aluminum hydrosilicate, treated bentonite clay, Fuller's earth, bauxite, and silica-alumina) in the form of powders, beads, pellets, or extrudites. Three types of catalytic cracking processes are available: fluid catalytic cracking (FCC), moving-bed catalytic cracking, and thermoform catalytic cracking (TCC). The catalytic cracking process is very flexible, and operating parameters can be adjusted to meet changing product demand. In addition to cracking, isomerization, dehydrogenenation, hydrogen transfer, cyclization, condensation, alkylation, and dealkylation also happen in FCC process. There are three basic functions in the catalytic cracking process: (i) feedstock reacts with catalyst and breaks down into different hydrocarbons, (ii) regeneration of catalyst by burning off coke, and (iii) fractionation of hydrocarbon stream is separated into various products. Major feed stocks to FCC are vacuum gas oil (VGO), hydro-treated VGO, hydrocracker bottom, coker gas oil (CGO), deasphalted oil (DAO), reduced crude oil (RCO), and vacuum residue (VR). Although the FCC process has been in commercial operation over several decades now, the technology is continuously evolving in order to meet new challenges including processing complex feedstock and meeting more stringent environmental regulations [5]. The contributions of Avidan and Shinnar (1990), Avidan et al. (1990), Ewell and Gadmer (1978), and McDonald and Harkins (1987) are referred to for a detailed description of the FCC, its historic development, the different types of units in current use, and the design philosophy underlying them [6–9].

Numerous articles relating to the FCC process provide various aspects of mathematical modeling and simulation, stability, optimization, and optimal control. As the investigation of the dynamics and control of a system requires a model to work with, modeling and simulation are the subjects of many of the FCC-centric articles. Some articles include integrated models for the regenerator and reactor coupled [10–24]; some have only regenerator models [25–29]; and others have reactor or cracking models [30–33]. A rigorous mathematical model should consider the two-phase nature of the fluidized beds in both the reactor and the regenerator, together with their strong interaction using reliable kinetics schemes. A Model IV FCCU differs from other fluid catalytic cracking unit (FCCU) designs primarily in the relative positions of the reactor and regenerator, the configuration of catalyst circulation lines, and the method of manipulating catalyst circulation rate [7]. In other FCCU designs, slide valves can be manipulated to vary catalyst flow rate, while in a Model IV

FCCU, slide valves are available for emergency closure only (to prevent backflow of air into the reactor riser in an abnormal situation). Different models for type IV FCC units of varying degrees of sophistication have been developed [20].

The FCC process is a highly nonlinear, slow, and multivariable one possessing strong interactions among the variables, subject to a number of operational constraints, while working at high temperatures and pressures. These complex characteristics together with its economic significance makes the FCC unit a potential candidate and challenging problem for the application of advanced process engineering tools like advanced process control, optimization, and fault diagnosis and identification (FDI). Although numerous articles have been published concerning different modeling approaches, targeted towards advanced control and optimization of an FCC unit, few are directly devoted to the application of fault diagnosis and identification (FDI) in the FCC system. Among the few, Yang et al. [34] used an approach based on the neural network, Heim et al. [35] used causal and knowledge-based models, Huang et al. [36] used a heuristic-based extended Kalman filter, Sundarraman and Srinivasan [37] used a trend analysis-based approach, and Wang et al. [38] used a recursive partial least squares (PLS) algorithm to address FDI in an FCC process operating under conventional regulatory control. Pranatyasto and Qin [39] reported a PCA-based FDI system dedicated to proper functioning of a model predictive control (MPC) strategy in an FCC unit. Sotomayor et al. [40] designed a fault detection and isolation system to monitor malfunctions in sensors and actuators of an FCC unit MPC system using subspace identification methods.

In this context, the current chapter avails us of the opportunity to implement some simple yet effective fault detection algorithms in detecting the simulated abnormal operating conditions occurring in an FCC unit using a type IV FCC model.

4.1.2 FCC process description

A type IV industrial FCC unit consists of two interconnected gas-solid fluidized beds. The first one is the reactor, in which gas oil is catalytically converted to gasoline, coke, and light hydrocarbon gases. A schematic of the FCCU reactor/regenerator showing flow is shown in Figure 4.1 [23]. Preheated feed is mixed with hot slurry recycle (from the bottom of the main fractionator) and injected into the reactor riser, where it mixes with hot regenerated catalyst and vaporizes fully. The hot catalyst provides the sensible heat, heat of vaporization, and heat of reaction necessary for the endothermic cracking reactions. As a result of the cracking reactions, coke is deposited on the surface of the catalyst. Since coke poisons the catalyst, continuous regeneration of catalyst is done in the second fluidized bed, the regenerator, by using air to burn the coke deposited on the catalyst in the reactor. Separation of catalyst and gas occurs in the disengaging zone of the reactor. An entrained catalyst is removed through cyclones. The catalyst is returned to the stripping section of the reactor where steam is injected to remove entrained hydrocarbons. Reactor product gas is passed to the main fractionator for heat recovery and fractionates into various product streams. Wet gas from the overheads of the main column (C and lighter) is compressed and further separated in downstream fractionators. Spent catalyst is transported from the reactor to the regenerator through the spent catalyst U-bend. Air is injected into the bottom of the regenerator lift pipe to assist the circulation of the catalyst. Catalyst in the regenerator is fluidized with air flow provided by the lift and combustion air blowers. Coke on spent catalyst is usually 5–10% hydrogen, depending on the coking characteristics of the feedstock. Carbon and hydrogen on the catalyst react with oxygen to produce carbon monoxide, carbon dioxide, and water.

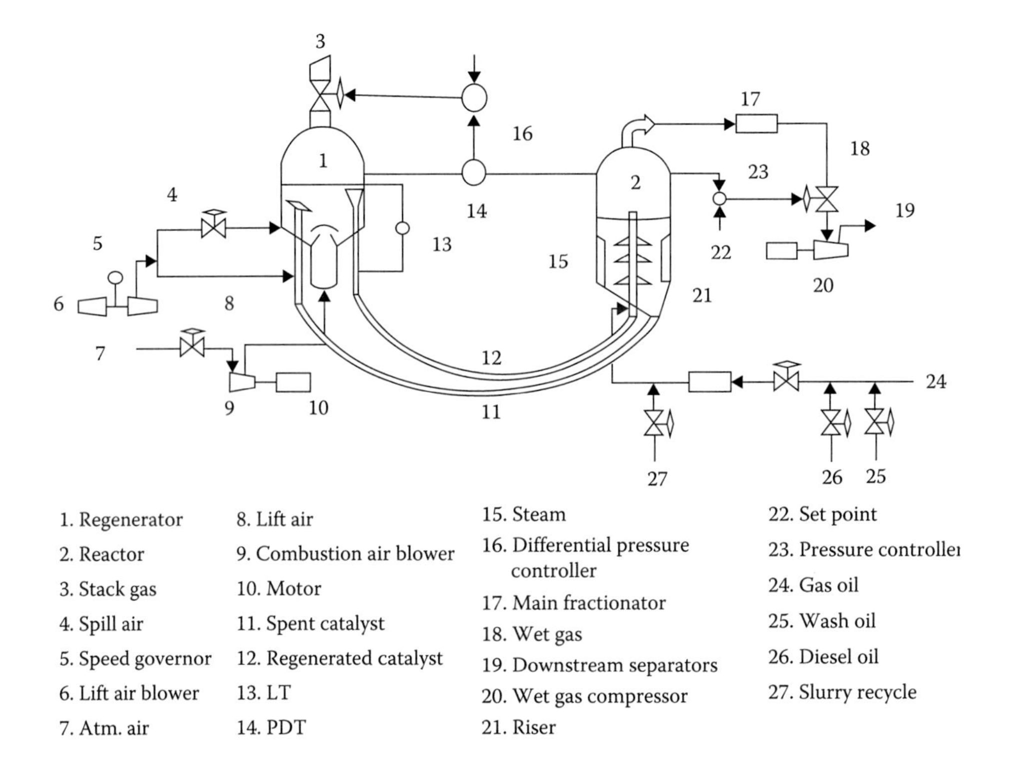

Figure 4.1 Schematic of a Model IV fluid catalytic cracking unit. (From R.C. McFarlane et al., *Comput. Chem. Eng.*, 17(3), 275, 1993.)

While most of the reactions occur in the fluidized bed, some reactions also occur in the disengaging section above the bed, where some catalyst is still present. Gas travels up the regenerator into the cyclones where entrained catalyst is removed and returned to the bed. The regenerator is maintained with respect to temperature and excess oxygen concentration to ensure complete conversion of all carbon monoxide produced in the bed to carbon dioxide before entering the cyclones. Regenerated catalyst flows over a weir into the regenerator standpipe. The heat produced by the catalyst in the standpipe provides the driving force for catalyst flow through the regenerated catalyst U-bend to the reactor riser. Head in the catalyst circulation lines and vessels, reactor/regenerator differential pressure, and lift air flow rate can be manipulated in a Model IV FCC unit to alter catalyst circulation rate. Catalyst circulation rate is an important manipulated variable affecting the severity of cracking in the riser.

4.1.3 Generation of historical database for FCC process

McFarlane et al. (1993) presented mechanistic dynamic simulator for the reactor/regenerator section of a Model IV FCCU [23]. This article [23] provides a detailed model description, along with dynamic simulation and model analysis. We have adopted that model for generating a historical database for the FCC unit. The SIMULINK® block diagram

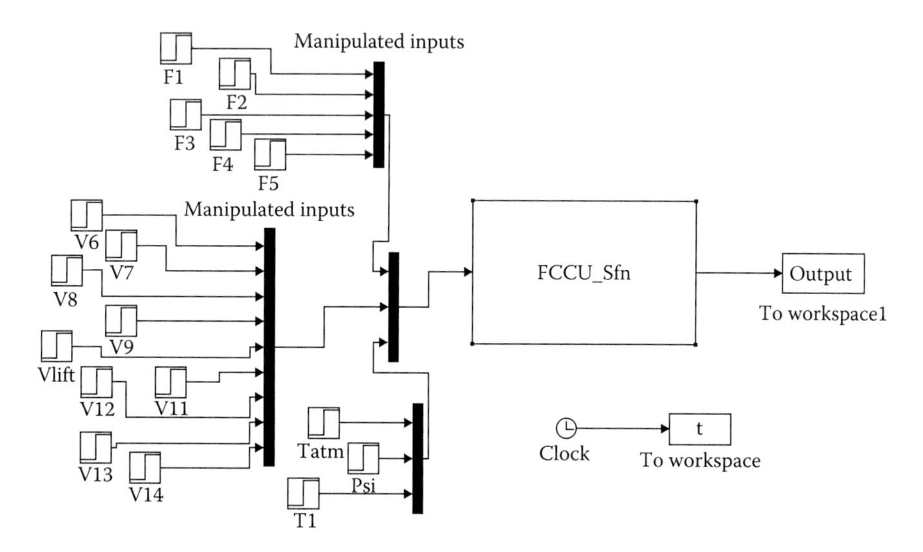

Figure 4.2 Dynamic simulator for Model IV FCC unit.

for the FCCU process is shown in Figure 4.2 [23]. Dr. Emadadeen M. Ali (Professor in Department of Chemical Engineering in College of Engineering at King Saud University) wrote the associated MATLAB® codes of the SIMULINK® modules, which are publicly available at http://faculty.ksu.edu.sa/Emad.Ali/Pages/SimulinkModule.aspx, and have been adapted here to run the SIMULINK® block shown in Figure 4.2. Seventy-nine data sets were created pertaining to 15 operating conditions as shown in Table 4.1. In Table 4.1, column 1 presents the identification number of the operating condition, the variables undergoing the change in the simulated operating condition are in column 2, descriptions of these operating conditions are in column 3, nominal values of the variables present in column 2 (around which different normal and abnormal data sets are created under the specified operating condition category) are present in column 4, and the number of data sets created under each category is in column 5 of the same table.

4.1.4 *Application of moving window-based pattern-matching algorithm on historical database of FCC process*

The generated database is used to identify different abnormal operating conditions prevailing in the FCC unit using two types of moving window-based pattern-matching algorithms, namely, combined similarity based (PCA similarity and distance similarity) and dissimilarity factor (KL expansion) based. The theoretical postulates of combined similarity, dissimilarity (based on KL expansion), and moving window-based pattern-matching with all its performance indices are provided in Chapter 2.

Table 4.2 shows the moving window-based pattern-matching efficacy when detecting the algorithm is the combined similarity based type. Column 1 of Table 4.2 presents the snapshot data such as S1 (4) representing the fourth out of seven data sets created for the S1 operating condition (around the nominal value of the responsible variable F_1 as presented

Table 4.1 Operating conditions and faults generated in FCC process

ID	Operating condition	Description	Nominal value	No. of data sets
N	Normal operation	Operation at the nominal conditions	N/A	1
S1	Setpoint change in F_1	Setpoint change for the wash oil flow	3 lb/s	7
S2	Setpoint change in F_3	Setpoint change for fresh feed flow	10 lb/s	7
S3	Setpoint change in F_4	Setpoint change for slurry recycle	3 lb/s	7
S4	Setpoint change in F_5	Setpoint change for furnace fuel flow	8 lb/s	7
F1	Change in v_6	Change in combustion air blower suction valve position (0–1)	0.6	5
F2	Change in v_7	Change in combustion air blower vent valve position (0–1)	0.6	5
F3	Change in v_8	Change in lift air blower vent valve position (0–1)	0.4	5
F4	Change in v_9	Change in spill air valve position (0–1)	0.8	5
F5	Change in v_{11}	Change in wet gas compressor suction valve position (0–1)	0.8	5
F6	Change in v_{13}	Wet gas compressor vent valve position (0–1)	0.3	5
F7	Change in v_{14}	Change in stack gas valve position (0–1)	0.8	5
F8	Change in v_{lift}	Change in manipulated variable for lift air blower steam valve (0–1.2)	0.7	5
F9	Ramp change in T_{atm}	Ambient air temperature ramps or down	ramp rate is 0.013 K/s	5
F10	Step change in T_1	Step increase in preheater outlet temperature	30 K	5

in Table 4.1). The other columns (2–8) present the performance indices of the moving window-based pattern-matching algorithm, which are as follows:

N_1 = The number of records in the candidate pool that are actually similar to the current snapshot, that is the number of correctly identified records (as per the algorithm).

N_2 = The number of records in the candidate pool that are actually not similar to the current snapshot, that is the number of incorrectly identified records (as per the algorithm). Hence, $N_P = N_1 + N_2$.

N_{DB}: The total number of historical windows that are actually similar to the current snapshot. In general, $N_{DB} \neq N_P$.

Table 4.2 Pattern-matching performance using combined similarity factor method

Snapshot	N_{DB}	N_1	N_2	N_P	$P(\%)$	$\eta(\%)$	$\xi(\%)$	Classification status
N	1	1	0	1	100	100	100	All are identified and none are misclassified
S1(4)	6	6	0	6	100	100	100	All are identified and none are misclassified
S2(4)	6	6	0	6	100	100	100	All are identified and none are misclassified
S3(4)	6	6	0	6	100	100	100	All are identified and none are misclassified
S4(4)	6	6	0	6	100	100	100	All are identified and none are misclassified
F1(3)	4	4	0	4	100	100	100	All are identified and none are misclassified
F2(3)	4	4	0	4	100	100	100	All are identified and none are misclassified
F3(3)	4	4	0	4	100	100	100	All are identified and none are misclassified
F4(3)	4	4	0	4	100	100	100	All are identified and none are misclassified
F5(3)	4	4	0	4	100	100	100	All are identified and none are misclassified
F6(3)	4	4	0	4	100	100	100	All are identified and none are misclassified
F7(3)	4	4	0	4	100	100	100	All are identified and none are misclassified
F8(3)	4	4	0	4	100	100	100	All are identified and none are misclassified
F9(3)	4	4	0	4	100	100	100	All are identified and none are misclassified
F10(3)	4	4	0	4	100	100	100	All are identified and none are misclassified

$\alpha_1 = 0.75$; $\alpha = 0.25$; $w = 245$; cutoff = 0.99.

Pool accuracy, $P = \left(\dfrac{N_1}{N_P}\right) \times 100\%$

Pattern matching efficiency, $\eta = \left(\dfrac{N_1}{N_{DB}}\right) \times 100\%$

Overall effectiveness of pattern matching, $\xi = \left(\dfrac{N_P}{N_{DB}}\right) \times 100$

The weight factors α_1 and α_2 (for PCA and distance similarity, respectively), the size of the moving window (w), and the cutoff/threshold similarity value are as follows:

$\alpha_1 = 0.75$, $\alpha_2 = 0.25$, $w = 245$, cutoff = 0.99 (as perfect similarity is indicated by 1)

Fifteen data sets (out of 79 data sets) are chosen as snapshot data sets and allowed to undergo pattern matching with all the data sets present in the database at a window-sliding rate of 245. The old data windows are discarded and replenished by new ones. Table 4.2 shows that none of the snapshot data got misclassified, and 100% performance indices reveal excellent performance of the proposed algorithm. In reality, the snapshot data symbolizes the abnormal plant data, and data windows present the process historical database. Under troubled plant conditions, one can detect the exact operating conditions (values of variables and parameters) of the snapshot condition by matching them with the template data, and hence, analyze the abnormal plant behavior followed by isolation of the abnormality/fault. The exact identification of the snapshot data is the key to successful fault analysis, isolation, and reconstruction, which seems to be promising using the proposed algorithm.

Tables 4.3 through 4.5 summarize the results of moving window-based pattern matching based on dissimilarity (KL expansion based). Column 1 of these tables present the 15 snapshot data (generated around nominal values as explained before in Table 4.2). Columns 2–8 contain the same performance indices of moving window-based pattern-matching as present in Table 4.2. The ninth column of these tables describe the classification status in terms of detectability of the data sets similar to the snapshot data. As the pattern matching algorithm in these cases is dissimilarity based, the perfect identification of a data set leads to a dissimilarity value of 0. The cutoff/threshold dissimilarity values are taken in the range of 0.1–0.5, for Tables 4.3 through 4.7 respectively, to show the algorithm performance. The cutoff dissimilarity means the data sets showing a dissimilarity value less than the threshold limit should be considered similar. The window-sliding rate in every table is 245. The snapshot data are allowed to undergo pattern matching against the template/historical data windows. Since it is a moving window-based algorithm, all the previous data windows are discarded. Table 4.3 reveals that none of the data sets similar to the snapshot could be correctly identified, but at the same time none of the data sets in the pool were misclassified. In Table 4.4, only for snapshot F1(3) could the remaining four other (pertaining to the F1 operating condition) similar data sets in the pool be correctly identified leading to 100% pool accuracy for this snapshot. In Table 4.5, snapshot N is detected correctly along with the data sets F1(1), F1(2), F1(4), F1(5). The data sets are misclassified, bringing down the pool accuracy to 20%, whereas the pattern-matching efficiency remains 100% because of the correct identification of *N*. Since the algorithm efficiency is an average of pool accuracy and pattern matching efficiency, it declines. In the same table, data sets similar to snapshot F1(3) are detected correctly, and *N* is misidentified as F1(3), resulting in reduced pool accuracy and algorithm efficiency. When the dissimilarity cutoff is raised to 0.4 in Table 4.6, there is no notable improvement in performance of the algorithm. In Table 4.7, the dissimilarity cutoff is 0.5 and data sets similar to all the snapshots are correctly identified except N, S3(4), and F1(3). The summary results based on Tables 4.3 through 4.7 reveal that the dissimilarity 0 based pattern 0 matching algorithm is not working well in comparison to the similarity one (Table 4.2). Even though the dissimilarity results with varying dissimilarity cutoffs are presented with a varying degree of performance indices, the dissimilarity cutoff of 0.5 is the borderline value of dissimilarity between two data sets. Fixing a threshold value of dissimilarity is a problem-specific parameter. This elaborate exercise manifests the robustness of the algorithm, which may work very well in other situations.

Table 4.8 documents the combined similarity of each of the 15 snapshots to each other, generated at their nominal values. Their exact similarity is revealed by the presence of 1 values all along the diagonal of the table. The objective behind producing Table 4.8 is to reveal the efficiency and robustness of the proposed algorithm in the face of widely varying operating conditions.

Table 4.3 Pattern-matching performance using dissimilarity factor (KL expansion) method

Snapshot	N_{DB}	N_1	N_2	N_P	P(%)	η(%)	ξ(%)	Classification status
N	1	0	0	0	0	0	0	Data set similar to *N* is not identified
S1(4)	6	0	0	0	0	0	0	Data sets similar to S1(4) are not identified
S2(4)	6	0	0	0	0	0	0	Data sets similar to S2(4) are not identified
S3(4)	6	0	0	0	0	0	0	Data sets similar to S3(4) are not identified
S4(4)	6	0	0	0	0	0	0	Data sets similar to S4(4) are not identified
F1(3)	4	0	0	0	0	0	0	Data sets similar to F1(3) are not identified
F2(3)	4	0	0	0	0	0	0	Data sets similar to F2(3) are not identified
F3(3)	4	0	0	0	0	0	0	Data sets similar to F3(3) are not identified
F4(3)	4	0	0	0	0	0	0	Data sets similar to F4(3) are not identified
F5(3)	4	0	0	0	0	0	0	Data sets similar to F5(3) are not identified
F6(3)	4	0	0	0	0	0	0	Data sets similar to F6(3) are not identified
F7(3)	4	0	0	0	0	0	0	Data sets similar to F7(3) are not identified
F8(3)	4	0	0	0	0	0	0	Data sets similar to F8(3) are not identified
F9(3)	4	0	0	0	0	0	0	Data sets similar to F9(3) are not identified
F10(3)	4	0	0	0	0	0	0	Data sets similar to F10(3) are not identified

w = 245; cutoff = 0.1

Table 4.4 Pattern-matching performance using dissimilarity factor (KL expansion) method

Snapshot	N_{DB}	N_1	N_2	N_P	P(%)	η(%)	ξ(%)	Classification status
N	1	0	0	0	0	0	0	Data set similar to *N* is not identified
S1(4)	6	0	0	0	0	0	0	Data sets similar to S1(4) are not identified
S2(4)	6	0	0	0	0	0	0	Data sets similar to S2(4) are not identified
S3(4)	6	0	0	0	0	0	0	Data sets similar to S3(4) are not identified
S4(4)	6	0	0	0	0	0	0	Data sets similar to S4(4) are not identified

(*Continued*)

Table 4.4 **(*Continued*)** Pattern-matching performance using dissimilarity factor (KL expansion) method

Snapshot	N_{DB}	N_1	N_2	N_P	$P(\%)$	$\eta(\%)$	$\xi(\%)$	Classification status
F1(3)	4	4	0	4	100	100	100	None were misclassified
F2(3)	4	0	0	0	0	0	0	Data sets similar to F2(3) are not identified
F3(3)	4	0	0	0	0	0	0	Data sets similar to F3(3) are not identified
F4(3)	4	0	0	0	0	0	0	Data sets similar to F4(3) are not identified
F5(3)	4	0	0	0	0	0	0	Data sets similar to F5(3) are not identified
F6(3)	4	0	0	0	0	0	0	Data sets similar to F6(3) are not identified
F7(3)	4	0	0	0	0	0	0	Data sets similar to F7(3) are not identified
F8(3)	4	0	0	0	0	0	0	Data sets similar to F8(3) are not identified
F9(3)	4	0	0	0	0	0	0	Data sets similar to F9(3) are not identified
F10(3)	4	0	0	0	0	0	0	Data sets similar to F10(3) are not identified

w = 245; cutoff = 0.2

Table 4.5 Pattern-matching performance using dissimilarity factor (KL expansion) method

Snapshot	N_{DB}	N_1	N_2	N_P	$P(\%)$	$\eta(\%)$	$\xi(\%)$	Classification status
N	1	1	4	5	20	100	60	Data set similar to *N* is detected and data sets F1(1), F1(2), F1(4), F1(5) are misclassified
S1(4)	6	0	0	0	0	0	0	Data sets similar to S1(4) are not detected
S2(4)	6	0	0	0	0	0	0	Data sets similar to S2(4) are not detected
S3(4)	6	0	0	0	0	0	0	Data sets similar to S3(4) are not detected
S4(4)	6	0	0	0	0	0	0	Data sets similar to S4 (4) are not detected
F1(3)	4	4	1	5	80	100	90	Data sets similar to F1(3) are detected and data set *N* is classified as F1(3)
F2(3)	4	0	0	0	0	0	0	Data sets similar to F2(3) are not identified
F3(3)	4	0	0	0	0	0	0	Data sets similar to F3(3) are not identified
F4(3)	4	0	0	0	0	0	0	Data sets similar to F4(3) are not identified

(*Continued*)

Table 4.5 (Continued) Pattern-matching performance using dissimilarity factor (KL expansion) method

Snapshot	N_{DB}	N_1	N_2	N_P	P(%)	η(%)	ξ(%)	Classification status
F5(3)	4	0	0	0	0	0	0	Data sets similar to F5(3) are not identified
F6(3)	4	0	0	0	0	0	0	Data sets similar to F6(3) are not identified
F7(3)	4	0	0	0	0	0	0	Data sets similar to F7(3) are not identified
F8(3)	4	0	0	0	0	0	0	Data sets similar to F8(3) are not identified
F9(3)	4	0	0	0	0	0	0	Data sets similar to F9(3) are not identified
F10(3)	4	0	0	0	0	0	0	Data sets similar to F10(3) are not identified

w = *245; cutoff* = *0.3*

Table 4.6 Pattern-matching performance using dissimilarity factor (KL expansion) method

Snapshot	N_{DB}	N_1	N_2	N_P	P(%)	η(%)	ξ(%)	Classification status
N	1	1	4	5	20	100	60	Data set similar to *N* is detected along with F1(1), F1(2), F1(4), F1(5) classified as *N*
S1(4)	6	0	0	0	0	0	0	Data sets similar to S1(4) are not detected
S2(4)	6	3	0	3	100	50	75	Data sets similar to S2(4) are detected partially and none are misclassified
S3(4)	6	0	0	0	0	0	0	Data sets similar to S3(4) are not detected
S4(4)	6	3	0	3	100	50	75	Data sets similar to S4(4) are detected partially and none are misclassified
F1(3)	4	4	1	5	80	100	90	Data sets similar to F1(3) are detected and data set *N* is misclassified as F1(3)
F2(3)	4	4	0	4	100	100	100	Data sets similar to F2(3) are detected
F3(3)	4	0	0	0	0	0	0	Data sets similar to F3(3) are not detected
F4(3)	4	4	0	4	100	100	100	Data sets similar to F4(3) are detected
F5(3)	4	2	0	2	100	50	75	Data sets similar to F5(3) are detected partially and none are misclassified
F6(3)	4	0	0	0	0	0	0	Data sets similar to F6(3) are not detected
F7(3)	4	4	0	4	100	100	100	Data sets similar to F7(3) are detected

Table 4.6 **(*Continued*)** Pattern-matching performance using dissimilarity factor (KL expansion) method

Snapshot	N_{DB}	N_1	N_2	N_P	$P(\%)$	η(%)	ξ(%)	Classification status
F8(3)	4	4	0	4	100	100	100	Data sets similar to F8(3) are detected
F9(3)	4	0	0	0	0	0	0	Data sets similar to F9(3) are not detected
F10(3)	4	4	0	4	100	100	100	Data sets similar to F10(3) are detected

w = 245; *cutoff* = 0.4

Table 4.7 Pattern-matching performance using dissimilarity factor (KL expansion) method

Snapshot	N_{DB}	N_1	N_2	N_P	$P(\%)$	η(%)	ξ(%)	Classification status
N	1	1	4	5	20	100	60	Data set similar to *N* is detected along with F1(1), F1(2), F1(4), F1(5) classified as *N*
S1(4)	6	6	0	6	100	100	100	Data sets similar to S1(4) are detected
S2(4)	6	6	0	6	100	100	100	Data sets similar to S2(4) are detected
S3(4)	6	5	1	6	80	80	80	Data sets similar to S3(4) are detected partially
S4(4)	6	6	0	6	100	100	100	Data sets similar to S4(4) are detected with no misclassification
F1(3)	4	4	1	5	80	100	90	Data sets similar to F1(3) are detected and data set *N* is classified as F1(3)
F2(3)	4	4	0	4	100	100	100	Data sets similar to F2(3) are detected with no misclassification
F3(3)	4	4	0	4	100	100	100	Data sets similar to F3(3) are detected with no misclassification
F4(3)	4	4	0	4	100	100	100	Data sets similar to F4(3) are detected with no misclassification
F5(3)	4	4	0	4	100	100	100	Data sets similar to F5(3) are detected with no misclassification
F6(3)	4	4	0	4	100	100	100	Data sets similar to F6(3) are detected with no misclassification
F7(3)	4	4	0	4	100	100	100	Data sets similar to F7(3) are detected with no misclassification

(*Continued*)

Table 4.7 (*Continued*) Pattern-matching performance using dissimilarity factor (KL expansion) method

Snapshot	N_{DB}	N_1	N_2	N_P	$P(\%)$	$\eta(\%)$	$\xi(\%)$	Classification status
F8(3)	4	4	0	4	100	100	100	Data sets similar to F8(3) are detected with no misclassification
F9(3)	4	4	0	4	100	100	100	Data sets similar to F9(3) are detected with no misclassification
F10(3)	4	4	0	4	100	100	100	Data sets similar to F10(3) are detected with no misclassification

w = 245; *cutoff* = 0.5

Table 4.9 presents the dissimilarity of each of the 15 snapshots to each other, generated at their nominal values. Their exact dissimilarity to each other should have generated a value of 0 along the diagonal of the table, whereas a departure from "zero" concludes the ineffectiveness of the dissimilarity-based pattern-matching algorithm in this FCC case study.

4.2 Detection of abnormal operating conditions in continuous stirred tank heater

4.2.1 Continuous stirred tank heater

For the continuous stirred tank heater (CSTH), the mathematical model presented in Thornhill et al. [41] has been adapted to generate a simulated historical database. A hybrid simulation is used for the CSTH using measured data captured from the original continuous stirred tank heater pilot plant and model predicted estimates. In comparison with the FCC unit, the model for the CSTH is small and does not contain any chemical reaction but it has a complete characterization of all the sensors and valves and the heat exchanger. The CSTH pilot plant was built in the Department of Chemical and Material Engineering at the University of Alberta in Alberta, Canada. The schematic of the CSTH pilot plant is shown in Figure 4.3 [41]. The pilot plant is an experimental rig in which hot and cold water are mixed and further heated using steam through a long pipe. Because the CSTH is well mixed, the temperature of liquid in the tank is the same as the temperature of liquid in the outlet pipe. The volume and height of the tank are 81 and 50 cm, respectively. The inputs (hot water, cold water and steam) to the CSTH are used by other users and thus are subjected to constant changes. The steam coming from the same central campus source is used to heat the hot water boiler. Control valves with pneumatic actuators are used in the CSTH plant. Orifice plates with differential pressure transmitters which give a nominal 4–20 mA output are used to measure flow of liquid. A differential pressure measurement is used to measure the level of liquid in the tank. The detailed mathematical model using the data collected from the pilot plant is provided in [41]. In Reference [41], the closed loop control system was developed and tested against the disturbances in cold and hot water flow and noise in the temperature of the liquid in the outlet pipe. Thornhill et al. [41] developed MATLAB® codes

Table 4.8 Values of combined similarity factor ($\alpha_1 = 0.75$, $\alpha_2 = 0.25$) for the nominal magnitudes of operating conditions in the FCC process

Op ID	N	S1(4)	S2(4)	S3(4)	S4(4)	F1(3)	F2(3)	F3(3)	F4(3)	F5(3)	F6(3)	F7(3)	F8(3)	F9(3)	F10(3)
N	1.00	0.789	0.379	0.795	0.487	0.938	0.488	0.613	0.712	0.476	0.486	0.470	0.699	0.543	0.521
S1(4)	0.789	1.00	0.560	0.963	0.601	0.767	0.612	0.598	0.640	0.579	0.504	0.579	0.636	0.618	0.583
S2(4)	0.379	0.770	1.000	0.845	0.690	0.575	0.343	0.227	0.257	0.643	0.283	0.271	0.299	0.577	0.625
S3(4)	0.795	0.963	0.635	1.000	0.635	0.767	0.564	0.576	0.635	0.584	0.508	0.540	0.649	0.596	0.599
S4(4)	0.487	0.848	0.690	0.881	1.000	0.707	0.297	0.299	0.358	0.528	0.228	0.239	0.359	0.544	0.985
F1(3)	0.938	0.767	0.365	0.767	0.461	1.00	0.489	0.601	0.695	0.477	0.487	0.460	0.666	0.548	0.489
F2(3)	0.488	0.613	0.347	0.564	0.265	0.489	0.999	0.560	0.550	0.690	0.657	0.939	0.565	0.800	0.255
F3(3)	0.613	0.599	0.267	0.576	0.303	0.601	0.560	1.000	0.552	0.416	0.413	0.495	0.917	0.687	0.311
F4(3)	0.712	0.641	0.296	0.635	0.361	0.695	0.550	0.552	1	0.389	0.468	0.570	0.553	0.590	0.379
F5(3)	0.476	0.579	0.433	0.584	0.282	0.477	0.690	0.416	0.389	1.000	0.914	0.623	0.462	0.591	0.244
F6(3)	0.486	0.504	0.322	0.508	0.231	0.487	0.657	0.413	0.468	0.914	1.000	0.641	0.443	0.594	0.214
F7(3)	0.470	0.579	0.308	0.540	0.240	0.460	0.939	0.495	0.570	0.623	0.641	0.999	0.512	0.684	0.239
F8(3)	0.699	0.636	0.338	0.649	0.362	0.666	0.565	0.917	0.553	0.462	0.443	0.512	1.000	0.657	0.357
F9(3)	0.543	0.618	0.366	0.596	0.298	0.548	0.800	0.687	0.590	0.591	0.594	0.684	0.656	1.000	0.276
F10(3)	0.521	0.832	0.625	0.848	0.985	0.739	0.290	0.311	0.379	0.493	0.214	0.240	0.357	0.525	1.000

Table 4.9 Values of dissimilarity (KL-expansion) factor for the nominal magnitudes of operating conditions in the FCC process

Op ID	N	S1(4)	S2(4)	S3(4)	S4(4)	F1(3)	F2(3)	F3(3)	F4(3)	F5(3)	F6(3)	F7(3)	F8(3)	F9(3)	F10(3)
N	0.200	0.870	0.937	0.860	0.934	0.240	0.929	0.908	0.916	0.866	0.887	0.893	0.894	0.897	0.916
S1(4)	0.870	0.200	0.911	0.843	0.924	0.899	0.896	0.921	0.886	0.921	0.890	0.883	0.926	0.876	0.919
S2(4)	0.937	0.911	0.100	0.893	0.864	0.912	0.904	0.951	0.948	0.892	0.898	0.885	0.905	0.902	0.874
S3(4)	0.860	0.843	0.893	0.2000	0.9237	0.890	0.846	0.868	0.892	0.866	0.871	0.843	0.879	0.835	0.933
S4(4)	0.934	0.924	0.864	0.923	0.100	0.911	0.925	0.937	0.924	0.940	0.907	0.921	0.934	0.958	0.677
F1(3)	0.240	0.899	0.912	0.890	0.911	0.150	0.955	0.937	0.942	0.898	0.919	0.919	0.920	0.9259	0.894
F2(3)	0.929	0.896	0.904	0.846	0.925	0.955	0.200	0.801	0.860	0.815	0.840	0.815	0.837	0.882	0.913
F3(3)	0.908	0.921	0.951	0.868	0.937	0.937	0.801	0.200	0.858	0.902	0.857	0.8699	0.797	0.914	0.936
F4(3)	0.916	0.886	0.948	0.892	0.924	0.942	0.860	0.858	0.200	0.884	0.860	0.806	0.763	0.854	0.932
F5(3)	0.866	0.921	0.892	0.866	0.940	0.898	0.815	0.902	0.884	0.200	0.883	0.828	0.902	0.931	0.926
F6(3)	0.887	0.890	0.898	0.871	0.907	0.919	0.840	0.857	0.860	0.883	0.200	0.832	0.862	0.913	0.887
F7(3)	0.893	0.883	0.885	0.843	0.921	0.919	0.815	0.869	0.806	0.828	0.832	0.200	0.873	0.885	0.912
F8(3)	0.894	0.926	0.905	0.879	0.934	0.920	0.837	0.797	0.763	0.902	0.862	0.873	0.200	0.907	0.941
F9(3)	0.897	0.876	0.902	0.835	0.958	0.925	0.882	0.914	0.854	0.931	0.913	0.885	0.907	0.199	0.937
F10(4)	0.916	0.919	0.874	0.9337	0.677	0.894	0.913	0.936	0.932	0.926	0.887	0.912	0.941	0.937	0.1000

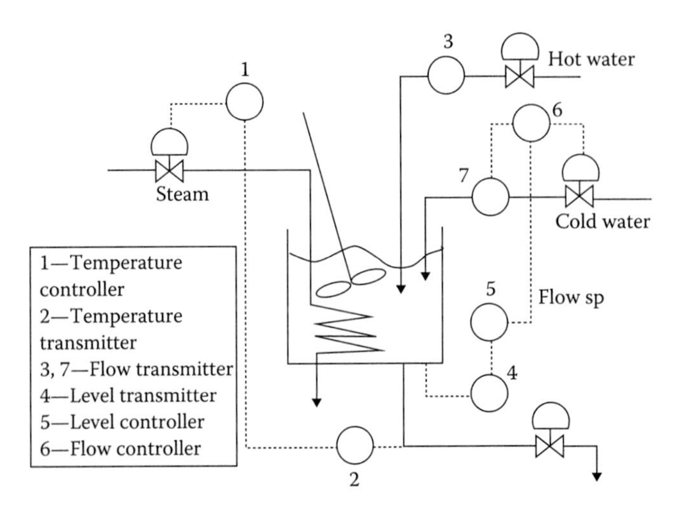

Figure 4.3 Schematic of continuous stirred tank heater pilot plant

and SIMULINK® modules and made them available at http://www.ps.ic.ac.uk/~nina/CSTHSimulation/index.htm.

4.2.2 Generation of historical database for CSTH

The pilot plant CSTH has been adopted from Thornhill et al. [41] and simulated for the generation of a historical database. Eleven different operating conditions as explained in Table 4.10 were created. Sixty-five data sets, each data set consisting of 2002 samples of level, cold water flow, and temperature, were generated by running the simulated pilot plant under each operating mode for 2000 seconds using a variable sample time. In the first and eighth operating modes, the simulated pilot plant ran only once; thus, there is only one data set generated in each mode and this is also serving as the snapshot data set for the respective operating condition. For the rest of the operating conditions, the fourth data set is acting as the snapshot data set for the corresponding operating mode. In Table 4.10, changes in process variables, such as changes in cold and hot water valve positions, changes in cold and hot water temperatures, and noise in the temperature of the outlet liquid, are considered to be faults (termed F).

4.2.3 Application of combined similarity factor and dissimilarity factor-based pattern-matching algorithm on a CSTH historical database

In this case study, both the combined similarity factor (PCA similarity factor plus distance similarity factor) and dissimilarity factor-based on KL expansion are deployed. The pattern-matching algorithm using a combined similarity factor (already explained in Chapter 2) was applied on the historical database (as generated in the previous section). Here the snapshot data set contains 2002 measurements of level, cold water flow, and temperature which were collected by operating the simulated pilot plant at the nominal value of each operating condition. The data window moved 2002 samples at a

Table 4.10 Operating conditions and faults in CSTH

ID	Operating condition	Description	Nominal value	No. of data sets
N	Nominal operation	Operation at the nominal conditions	N/A	1
F1	Change in CW valve (open loop)	Change in cold water valve position (4–20 mA)	18 mA	7
F2	Change in steam valve (open loop)	Change in steam valve position (4–20 mA)	7 mA	7
F3	Change in HW valve (open loop)	Change in hot water valve position (4–20 mA)	5 mA	7
F4	Step change in CW temperature (open loop)	Step increase in cold water temperature	5 °C	7
F5	Step change in HW temperature (open loop)	Step increase in hot water temperature	15 °C	7
F6	Ramp change in CW temperature (open loop)	Cold water temperature ramps up	0.04	7
F7	Ramp change in HW temperature (open loop)	Hot water temperature ramps up	0.002	7
F8	Disturbance in cold water flow and hot water flow, and temperature measurement noise	Simultaneous disturbance in cold and hot water flow, and noise in temperature of outlet liquid	N/A	1
S1	Setpoint change in level (closed loop)	Setpoint change in liquid level in continuous stirred tank	6 mA	7
S2	Setpoint change in outflow temperature (closed loop)	Setpoint change in temperature of liquid in outlet pipe	6 mA	7

The fourth data set in all operating modes except N and F8 is considered to be the snapshot data set. Since only one data set was created in N and F8, that data set serves as the snapshot data set also.

Table 4.11 Pattern-matching efficiency using the combined similarity factor method

Snapshot	N_{DB}	N_1	N_2	N_P	P(%)	η(%)	ξ(%)	Classification status
N	1	1	0	1	100	100	100	Data window similar to *N* is identified and none are misclassified
F1(4)	6	6	0	6	100	100	100	All are identified and none are misclassified
F2(4)	6	6	0	6	100	100	100	All are identified and none are misclassified
F3(4)	6	6	0	6	100	100	100	All are identified and none are misclassified
F4(4)	6	6	0	6	100	100	100	All are identified and none are misclassified
F5(4)	6	6	0	6	100	100	100	All are identified and none are misclassified
F6(4)	6	6	0	6	100	100	100	All are identified and none are misclassified
F7(4)	6	6	0	6	100	100	100	All are identified and none are misclassified
F8	1	1	0	1	100	100	100	Data window similar to F8 is identified and none are misclassified
S1(4)	6	6	0	6	100	100	100	All are identified and none are misclassified
S2(4)	6	6	0	6	100	100	100	All are identified and none are misclassified

$\alpha_1 = 0.75$; $\alpha_2 = 0.25$; $w = 2002$; cutoff = 0.99

Table 4.12 Pattern-matching efficiency dissimilarity factor (KL expansion) method

Snapshot	N_{DB}	N_1	N_2	N_P	P(%)	η(%)	ξ(%)	Classification Status
N	1	1	0	1	100	100	100	Data window similar to *N* is identified and none are misclassified
F1(4)	6	6	0	6	100	100	100	All are identified and none are misclassified
F2(4)	6	6	0	6	100	100	100	All are identified and none are misclassified
F3(4)	6	6	0	6	100	100	100	All are identified and none are misclassified
F4(4)	6	6	0	6	100	100	100	All are identified and none are misclassified
F5(4)	6	5	0	5	100	83.33	91.667	Out of 6 data windows which are actually similar to F5(4), only five are identified and none are misclassified

(*Continued*)

Table 4.12 (***Continued***) Pattern-matching efficiency dissimilarity factor (KL expansion) method

Snapshot	N_{DB}	N_1	N_2	N_P	$P(\%)$	η(%)	ξ(%)	Classification status
F6(4)	6	6	0	6	100	100	100	All are identified and none are misclassified
F7(4)	6	6	0	6	100	100	100	All are identified and none are misclassified
F8	1	1	0	1	100	100	100	Data window similar to F8 is identified and none are misclassified
S1(4)	6	6	0	6	100	100	100	All are identified and none are misclassified
S2(4)	6	6	0	6	100	100	100	All are identified and none are misclassified

window rate = 2002; cutoff = 10^{-4}

time, which is the size of the snapshot data set. Table 4.11 presents the detailed analysis of the results. The combined similarity factor-based pattern-matching algorithm exhibited excellent performance with 100% cluster purity and 100% algorithm efficiency. The results of the dissimilarity factor-based pattern-matching algorithm are shown in Table 4.12. The cutoff used is 10^{-4}. Out of six data windows, which are actually similar to the snapshot F5(4), only five data windows were identified by the dissimilarity-based algorithm, and none were misclassified. In this simulation case study too, the moving window-based pattern-matching algorithm using the combined similarity factor performs better for the detection of historical data windows similar to a snapshot data set. Both the pattern-matching algorithms exhibited the same performance (Tables 4.13 through 4.14) when the historical database contained only snapshot data sets.

4.3 *Detection of abnormal operating conditions in simulated industrial gas-phase polyethylene reactor*

4.3.1 *Simulated industrial gas-phase polyethylene reactor*

Polyethylene or polyethene is the most important and common plastic being produced—nearly 80 million tons annually. Polyethylene is extensively used in packaging like plastic bags, plastic films, membranes, containers (for example bottles), etc. It is generally produced in three forms: low-density polyethylene (LDPE), linear low-density polyethylene (LLDPE), and high-density polyethylene (HDPE). The first two forms (LDPE and LLDPE) are preferred for film packaging and electrical insulation, whereas the third form is used for manufacturing containers for storing household chemicals such as dish soap, etc. Polyethylene possesses a high ductility, impact strength, and low friction but is of low strength (in terms of hardness and rigidity). Under the influence of persistent mechanical stress, it tends to move slowly or deform permanently, but this property of strong creep can be reduced by the addition of short fibers. It has a smooth surface, and feels waxy when touched. The melting point for LDPE is around 105 to 115°C and for common commercial grades of medium- and high-density polyethylene, the melting point falls in the range 120 to 180°C. The

Table 4.13 Values of combined similarity factor ($\alpha_1 = 0.75$, $\alpha_2 = 0.25$) for the nominal magnitudes of operating conditions in CSTH process

Op ID	N	F1(4)	F2 (4)	F3(4)	F4(4)	F5(4)	F6(4)	F7(4)	F8	S1(4)	S2(4)
N	1	0.95849	0.98548	0.81854	0.98591	0.77376	0.73688	0.66381	0.63208	0.88206	0.84272
F1(4)	0.95849	1	0.90277	0.90277	0.98751	0.88951	0.64689	0.76139	0.67329	0.75764	0.91395
F2(4)	0.98548	0.90277	1.00000	0.75442	0.94571	0.71337	0.80187	0.63395	0.62538	0.94310	0.77707
F3(4)	0.81854	0.92623	0.75442	1	0.86525	0.99453	0.63779	0.92180	0.81579	0.64599	0.84343
F4(4)	0.98591	0.98751	0.94571	0.86525	1.0000	0.82307	0.68680	0.70035	0.63927	0.81462	0.90742
F5(4)	0.77376	0.88951	0.71337	0.99453	0.82307	1.0000	0.65920	0.95498	0.84487	0.63023	0.82461
F6(4)	0.73688	0.64689	0.80187	0.63779	0.68680	0.65920	1.0000	0.76532	0.76271	0.93223	0.62824
F7(4)	0.66381	0.76139	0.63395	0.92180	0.70035	0.95498	0.76532	0.99999	0.91041	0.64400	0.73808
F8	0.63208	0.67329	0.62538	0.81579	0.63927	0.84487	0.76271	0.91041	1	0.64689	0.63078
S1(4)	0.88206	0.75764	0.94310	0.64599	0.81462	0.63023	0.93223	0.64400	0.64689	0.99999	0.66959
S2(4)	0.84272	0.91395	0.77707	0.84343	0.90742	0.82461	0.62824	0.73808	0.63078	0.66959	1.00000

Table 4.14 Values of dissimilarity (KL expansion) factor for the nominal magnitudes of operating conditions in CSTH process

Op ID	N	F1(4)	F2(4)	F3(4)	F4(4)	F5(4)	F6(4)	F7(4)	F8	S1(4)	S2(4)
N	5.120e-25	0.17203	0.07416	0.31407	0.06646	0.38968	0.34827	0.33247	0.33634	0.22442	0.28295
F1(4)	0.17203	5.57e-26	0.30822	0.35013	0.11392	0.34607	0.38117	0.36383	0.33159	0.31965	0.30885
F2(4)	0.07416	0.30822	1.65e-25	0.30490	0.19192	0.44617	0.31687	0.34490	0.35092	0.18669	0.35480
F3(4)	0.31407	0.35013	0.30490	4.96e-25	0.43093	0.23384	0.39442	0.03992	0.60474	0.42975	0.66129
F4(4)	0.06646	0.11392	0.19192	0.43093	7.54e-27	0.37358	0.35185	0.40691	0.29434	0.24399	0.16264
F5(4)	0.38968	0.34607	0.44617	0.23384	0.37358	3.00e-24	0.34671	0.19366	0.51105	0.37041	0.56258
F6(4)	0.34827	0.38117	0.31687	0.39442	0.35185	0.34671	1.25e-27	0.39283	0.18155	0.06990	0.34785
F7(4)	0.33247	0.36383	0.34490	0.03992	0.40691	0.19366	0.39283	1.05e-23	0.60970	0.41080	0.64082
F8	0.33634	0.33159	0.35092	0.60474	0.29434	0.51105	0.18155	0.60970	1.28e-29	0.21140	0.17545
S1(4)	0.22442	0.31965	0.18669	0.42975	0.24399	0.37041	0.06990	0.41080	0.21140	6.20e-27	0.295017
S2(4)	0.28295	0.30885	0.35480	0.66129	0.16264	0.56258	0.34785	0.64082	0.17545	0.295017	3.12e-27

softening point (80°C) of polyethylene limits its commercial use. All three forms of polyethylene are good electrical insulators. As far as optical properties are concerned, LDPE possesses the largest transparency; LLDPE has slightly lower transparency; and HDPE has the least transparency. The chemical behavior of polyethylene is similar to paraffin. Polyethylene can be produced by addition polymerization of ethene. The industrial gas-phase polyethylene reactor is schematically shown in Figure 4.4. The gas-phase polyethylene production has two advantages. The first one is that the reactor does not contain any solvent to be recovered or processed, and the second one is that, in comparison to liquid-phase polymerization process, the gas-phase processes commonly operate at much more moderate temperatures and pressures. To circumvent condensation, particle melting, and agglomeration, the temperature in the reaction zone must be maintained above the dew point of the reactants; this can be considered the potential disadvantage of the gas-phase polymerization process. More information on gas-phase polyethylene production technology can be found in Choi and Ray (1985) [42] and Xie et al. (1994) [43]. Operating the commercial gas-phase fluidized-bed polyethylene reactors in the temperature range 75°C–110°C ensures prevention of the temperature dropping below the polymer melting point and makes sure that the polymerization takes place at sufficient rates [43]. In the absence of an appropriate controller, the reaction temperature may change, hence, polyethylene reactors may behave abnormally, leading to serious runaway. Even in the event of operating the gas-phase fluidized-bed polyethylene reactors in the above-mentioned temperature range with a robust controller, changes may take place in the product properties because of high nonlinearity and strong interaction between the process variables.

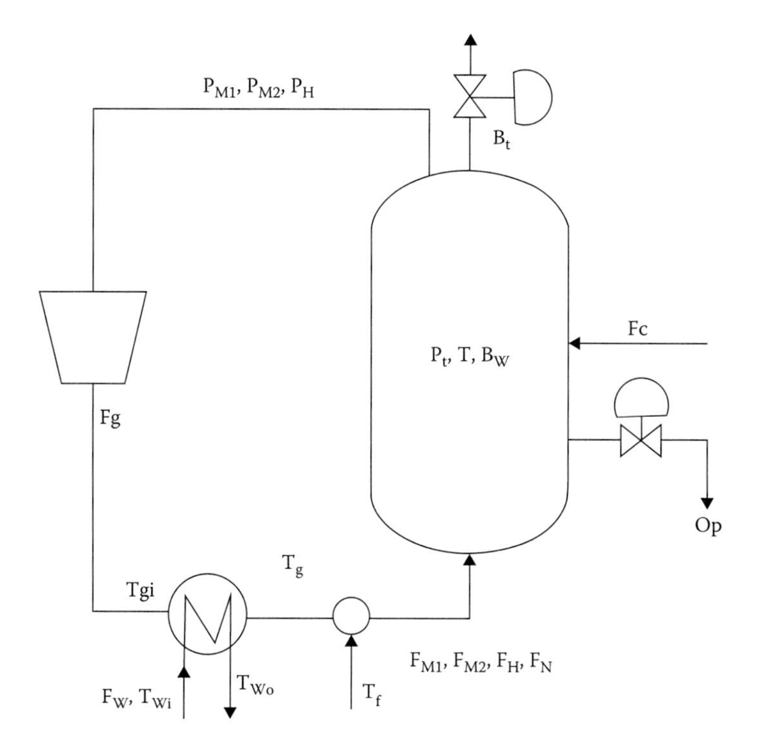

Figure 4.4 Polyethylene reactor.

Therefore, the industrial gas-phase fluidized-bed polyethylene reactors need to be monitored continuously. The simulated industrial gas-phase fluidized-bed polyethylene reactor presented in Ali et al. [44] has been adopted here to generate a historical database. The SIMULINK® block diagram for the polyethylene reactor is shown in Figure 4.5. The SIMULINK® modules with the associated MATLAB® codes are available at http://faculty.ksu.edu.sa/Emad.Ali/Pages/SimulinkModule.aspx.

4.3.2 *Generation of historical database for simulated industrial polyethylene reactor*

The industrial fluidized-bed polyethylene reactor was simulated over a period of 4000 minutes. For every 600 seconds, the readings of outputs (partial pressure of monomer, partial pressure of comonomer, partial pressure of hydrogen, total pressure, and bed temperature) were noted. By changing inputs (manipulated variables and disturbances) in step and sinusoidal fashion, 17 operating conditions were created (Table 4.15). Under each operating condition (except the first operating condition, termed as *N*), seven data sets were generated, out of which the fourth one is treated as the snapshot data set. Each data set consists of 401 samples of five outputs. For the first operating condition, only one data set was created, and this is the snapshot for the first operating condition.

4.3.3 *Application of combined similarity factor and dissimilarity factor-based pattern-matching algorithm on polyethylene historical database*

Similar to case studies 1 and 2, the two developed pattern-matching algorithms (combined similarity factor based and dissimilarity factor based, respectively) were implemented on the historical database generated for the fluidized-bed polyethylene reactor. Table 4.16 presents the detailed analysis of the results of the combined similarity factor-based pattern-matching algorithm. As observed in Table 4.16, the combined

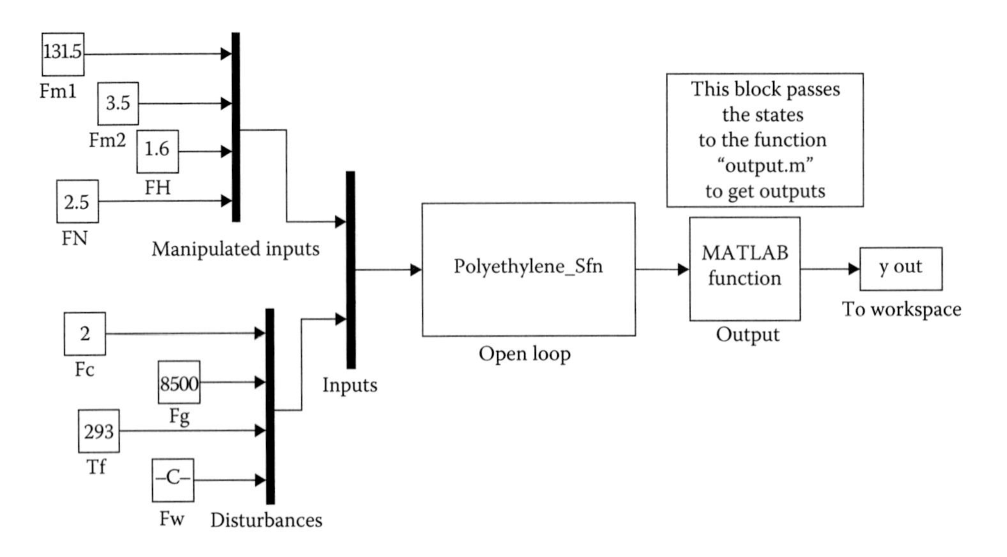

Figure 4.5 Dynamic simulator for polyethylene reactor.

Table 4.15 Operating conditions and faults generated in a polyethylene reactor

ID	Operating condition	Description	Nominal value	No. of data sets
N	Normal operation	Operation at the nominal conditions	N/A	1
F1	Step change in F_{m1}	Step increase in monomer flow rate	30 mole/s	7
F2	Step change in F_{m2}	Step increase in co-monomer flow rate	2.5 mole/s	7
F3	Step change in F_H	Step increase in hydrogen flow rate	1 mole/s	7
F4	Step change in F_N	Step increase in nitrogen flow rate	2 mole/s	7
F5	Step change in F_C	Step increase in catalyst flow rate	1.5 mole/s	7
F6	Step change in F_g	Step increase in recycle flow rate	1000 mole/s	7
F7	Step change in T_f	Step increase in feed temperature	50 °C	7
F8	Step change in F_W	Step increase in cooling water flow rate	10^3 mole/s	7
F9	High frequency oscillations in F_{m1}	Sustained oscillations of frequency of 10 cycles/min in monomer flow rate	40 mole/s	7
F10	High frequency oscillations in F_{m2}	Sustained oscillations of frequency of 10 cycles/min in co-monomer flow rate	3 mole/s	7
F11	High frequency oscillations in F_H	Sustained oscillations of frequency of 10 cycles/min in hydrogen flow rate	1.2 mole/s	7
F12	High frequency oscillations in F_N	Sustained oscillations of frequency of 10 cycles/min in nitrogen flow rate	1.5 mole/s	7
F13	High frequency oscillations in F_C	Sustained oscillations of frequency of 10 cycles/min in catalyst flow rate	1 mole/s	7
F14	High frequency oscillations in F_g	Sustained oscillations of frequency of 5 cycles/min in recycle flow rate	2000 mole/s	7
F15	High frequency oscillations in T_f	Sustained oscillations of frequency of 10 cycles/min in feed temperature	100 °C	7
F16	High frequency oscillations in F_W	Sustained oscillations of frequency of 10 cycles/min cooling water flow rate	2×10^3 mole/s	7

similarity factor-based pattern-matching algorithm maintained consistency in successfully identifying the data windows that are similar to a snapshot data set. While using the dissimilarity factor-based pattern-matching algorithm, the data windows, which are exactly similar to snapshot data sets F3(4), F4(4), F11(4), and F12(4), were only identified, and no data window was misclassified (Table 4.17). Tables 4.18 and

Table 4.16 Pattern-matching performance using the combined similarity factor method

Snapshot	N_{DB}	N_1	N_2	N_P	$P(\%)$	$\eta(\%)$	$\xi(\%)$	Classification status
N	1	1	0	1	100	100	100	All are identified and none are misclassified
F1(4)	6	6	0	6	100	100	100	All are identified and none are misclassified
F2(4)	6	6	0	6	100	100	100	All are identified and none are misclassified
F3(4)	6	6	0	6	100	100	100	All are identified and none are misclassified
F4(4)	6	6	0	6	100	100	100	All are identified and none are misclassified
F5(4)	6	6	0	6	100	100	100	All are identified and none are misclassified
F6(4)	6	6	0	6	100	100	100	All are identified and none are misclassified
F7(4)	6	6	0	6	100	100	100	All are identified and none are misclassified
F8(4)	6	6	0	6	100	100	100	All are identified and none are misclassified
F9(4)	6	6	0	6	100	100	100	All are identified and none are misclassified
F10(4)	6	6	0	6	100	100	100	All are identified and none are misclassified
F11(4)	6	6	0	6	100	100	100	All are identified and none are misclassified
F12(4)	6	6	0	6	100	100	100	All are identified and none are misclassified
F13(4)	6	6	0	6	100	100	100	All are identified and none are misclassified
F14(4)	6	6	0	6	100	100	100	All are identified and none are misclassified
F15(4)	6	6	0	6	100	100	100	All are identified and none are misclassified
F16(4)	6	6	0	6	100	100	100	All are identified and none are misclassified

$\alpha_1 = 0.75$; $\alpha_2 = 0.25$; $w = 401$; $cutoff = 0.99$

Table 4.17 Pattern-matching performance using the dissimilarity factor method

Snapshot	N_{DB}	N_1	N_2	N_P	P(%)	η(%)	ξ(%)	Classification status
N	1	0	0	0	NA	0	NA	Data window similar to *N* is not identified
F1(4)	6	0	0	0	NA	0	NA	Data windows similar to F1(4) are not identified
F2(4)	6	0	0	0	NA	0	NA	Data windows similar to F2(4) are not identified
F3(4)	6	6	0	6	100	100	100	All is identified and none is misclassified
F4(4)	6	6	0	6	100	100	100	All is identified and none is misclassified
F5(4)	6	0	0	0	NA	0	NA	Data windows similar to F5(4) are not identified
F6(4)	6	0	0	0	NA	0	NA	Data windows similar to F6(4) are not identified
F7(4)	6	0	0	0	NA	0	NA	Data windows similar to F6(4) are not identified
F8(4)	6	0	0	0	NA	0	NA	Data windows similar to F6(4) are not identified
F9(4)	6	0	0	0	NA	0	NA	Data windows similar to F6(4) are not identified
F10(4)	6	0	0	0	NA	0	NA	Data windows similar to F6(4) are not identified
F11(4)	6	6	0	6	100	100	100	All is identified and none are misclassified
F12(4)	6	6	0	6	100	100	100	All is identified and none are misclassified
F13(4)	6	0	0	0	NA	0	NA	Data windows similar to F13(4) are not identified
F14(4)	6	0	0	0	NA	0	NA	Data windows similar to F14(4) are not identified
F15(4)	6	0	0	0	NA	0	NA	Data windows similar to F15(4) are not identified
F16(4)	6	0	0	0	NA	0	NA	Data windows similar to F16(4) are not identified

w = 401; cutoff = 0.001

4.19 present the similarity and dissimilarity values, respectively, obtained when the data window moved 401 samples at a time looking for exactly the same data set, in which the historical databases were similar to snapshot data sets. The performance of the dissimilarity factor-based pattern-matching algorithm remained far from satisfactory.

Table 4.18 Values of combined similarity factor ($\alpha_1 = 0.75$, $\alpha_2 = 0.25$) for the nominal magnitudes of operating conditions for a polyethylene reactor

Op ID	N	F1(4)	F2(4)	F3(4)	F4(4)	F5(4)	F6(4)	F7(4)	F8(4)	F9(4)	F10(4)	F11(4)	F12(4)	F13(4)	F14(4)	F15(4)	F16(4)
N	1	0.884	0.643	0.794	0.475	0.656	0.708	0.804	0.677	0.579	0.602	0.630	0.399	0.477	0.482	0.454	0.572
F1(4)	0.884	1	0.585	0.722	0.530	0.529	0.864	0.933	0.589	0.741	0.624	0.527	0.378	0.618	0.631	0.500	0.614
F2(4)	0.643	0.585	1	0.897	0.601	0.749	0.665	0.663	0.655	0.391	0.465	0.707	0.497	0.561	0.413	0.387	0.705
F3(4)	0.794	0.722	0.897	1	0.674	0.802	0.690	0.754	0.747	0.401	0.413	0.724	0.512	0.539	0.440	0.411	0.670
F4(4)	0.475	0.530	0.601	0.674	1	0.535	0.628	0.612	0.554	0.391	0.527	0.624	0.750	0.668	0.629	0.643	0.552
F5(4)	0.656	0.529	0.749	0.802	0.535	1	0.571	0.594	0.961	0.310	0.292	0.901	0.700	0.581	0.390	0.474	0.619
F6(4)	0.708	0.864	0.665	0.690	0.628	0.571	1	0.975	0.611	0.697	0.622	0.609	0.579	0.730	0.659	0.537	0.602
F7(4)	0.804	0.933	0.663	0.754	0.612	0.594	0.975	0.999	0.624	0.673	0.601	0.580	0.513	0.663	0.598	0.481	0.611
8(4)	0.677	0.589	0.655	0.747	0.554	0.961	0.611	0.624	0.999	0.405	0.338	0.946	0.711	0.708	0.536	0.606	0.628
F9(4)	0.579	0.741	0.391	0.401	0.391	0.310	0.697	0.673	0.405	1	0.684	0.416	0.469	0.709	0.873	0.743	0.599
F10(4)	0.602	0.624	0.465	0.413	0.527	0.292	0.622	0.601	0.338	0.684	1	0.418	0.494	0.479	0.592	0.603	0.334
F11(4)	0.630	0.527	0.707	0.724	0.624	0.901	0.609	0.580	0.946	0.416	0.418	1	0.789	0.749	0.593	0.671	0.577
F12(4)	0.399	0.378	0.497	0.512	0.750	0.700	0.579	0.513	0.711	0.469	0.494	0.789	1	0.654	0.616	0.711	0.418
F13(4)	0.477	0.618	0.561	0.539	0.668	0.581	0.730	0.663	0.708	0.709	0.479	0.749	0.654	1	0.893	0.843	0.788
F14(4)	0.482	0.631	0.413	0.440	0.629	0.390	0.659	0.598	0.536	0.873	0.592	0.593	0.616	0.893	1	0.909	0.678
F15(4)	0.454	0.500	0.387	0.411	0.643	0.474	0.537	0.481	0.606	0.743	0.603	0.671	0.711	0.843	0.909	1	0.626
F16(4)	0.572	0.614	0.705	0.670	0.552	0.619	0.602	0.611	0.628	0.599	0.334	0.577	0.418	0.788	0.678	0.626	1

Table 4.19 Values of dissimilarity factor for the nominal magnitudes of operating conditions for a polyethylene reactor

Op ID	N	F1(4)	F2(4)	F3(4)	F4(4)	F5(4)	F6(4)	F7(4)	F8(4)	F9(4)	F10(4)	F11(4)	F12(4)	F13(4)	F14(4)	F15(4)	F16(4)
N	0.199	0.463	0.582	0.477	0.609	0.552	0.530	0.483	0.673	0.837	0.739	0.718	0.726	0.717	0.713	0.847	0.543
F1(4)	0.463	0.199	0.698	0.580	0.607	0.596	0.437	0.435	0.742	0.776	0.734	0.742	0.737	0.644	0.677	0.755	0.564
F2(4)	0.582	0.698	0.199	0.660	0.711	0.760	0.619	0.655	0.710	0.967	0.756	0.778	0.800	0.719	0.769	0.859	0.611
F3(4)	0.477	0.580	0.660	3.0e-26	0.403	0.630	0.680	0.612	0.653	0.913	0.850	0.838	0.802	0.740	0.871	0.764	0.710
F4(4)	0.609	0.607	0.711	0.403	4.1e-25	0.668	0.568	0.584	0.617	0.834	0.765	0.744	0.562	0.541	0.559	0.777	0.708
F5(4)	0.552	0.596	0.760	0.630	0.668	0.199	0.648	0.603	0.566	0.881	0.930	0.708	0.705	0.694	0.864	0.713	0.694
F6(4)	0.530	0.437	0.619	0.680	0.568	0.648	0.199	0.426	0.567	0.781	0.819	0.680	0.712	0.699	0.729	0.797	0.578
F7(4)	0.483	0.435	0.655	0.612	0.584	0.603	0.426	0.199	0.551	0.818	0.805	0.688	0.753	0.755	0.739	0.847	0.558
8(4)	0.673	0.742	0.710	0.653	0.617	0.566	0.567	0.551	0.199	0.808	0.914	0.580	0.584	0.591	0.724	0.793	0.569
F9(4)	0.837	0.776	0.967	0.913	0.834	0.881	0.781	0.818	0.808	0.199	0.831	0.979	0.910	0.946	0.742	0.626	0.890
F10(4)	0.739	0.734	0.756	0.850	0.765	0.930	0.819	0.805	0.914	0.831	0.199	0.774	0.903	0.789	0.859	0.652	0.814
F11(4)	0.718	0.742	0.778	0.838	0.744	0.708	0.680	0.688	0.580	0.979	0.774	8.e-27	0.402	0.705	0.777	0.861	0.637
F12(4)	0.726	0.737	0.800	0.802	0.562	0.705	0.712	0.753	0.584	0.910	0.903	0.402	3.5e-26	0.531	0.617	0.664	0.683
F13(4)	0.717	0.644	0.719	0.740	0.541	0.694	0.699	0.755	0.591	0.946	0.789	0.705	0.531	0.199	0.681	0.725	0.677
F14(4)	0.713	0.677	0.769	0.871	0.559	0.864	0.729	0.739	0.724	0.742	0.859	0.777	0.617	0.681	0.199	0.710	0.589
F15(4)	0.847	0.755	0.859	0.764	0.777	0.713	0.797	0.847	0.793	0.626	0.652	0.861	0.664	0.725	0.710	0.199	0.712
F16(4)	0.543	0.564	0.611	0.710	0.708	0.694	0.578	0.558	0.569	0.890	0.814	0.637	0.683	0.677	0.589	0.712	0.199

References

1. V. Venkatasubramanian, R. Rengaswamy, K. Yin, S.N. Kavuri (2003a). A review of process fault detection and diagnosis. Part I: Quantitative model based methods. *Computers and Chemical Engineering*, vol. 27, 293–311.
2. V. Venkatasubramanian, R. Rengaswamy, K. Yin, S.N. Kavuri (2003b). A review of process fault detection and diagnosis. Part II: Qualitative model and search strategies. *Computers and Chemical Engineering*, vol. 27, 313–326.
3. V. Venkatasubramanian, R. Rengaswamy, S.N. Kavuri, K. Yin (2003c). A review of process fault detection and diagnosis. Part III: Process history based methods. *Computers and Chemical Engineering*, vol. 27, 327–346.
4. E.T.C. Vogt, B.M. Weckhuysen (2015). Fluid catalytic cracking: Recent developments on the grand old lady of zeolite catalysis. *Chemical Society Reviews*, vol. 44, 7342–7370.
5. C. Ye-Mon (2011). Evolution of FCC—Past, present and future and the challenges of operating a high temperature CFB system. *10th International Conference on Circulating Fluidized Beds and Fluidization Technology-CFB-10* (5 April 2011).
6. A.A. Avidan, R. Shinnar (1990). Development of catalytic cracking technology. A lesson in chemical reactor design. *Industrial and Engineering Chemistry Research*, vol. 29, 931–942.
7. A.A. Avidan, M. Edwards, H. Owen (1990). Innovative improvements highlight FCC's past and future. *Oil and Gas Journal*, vol. 88(11), 33–58.
8. R.B. Ewell, G. Gadmer (1978). Design cat crackers by computer. *Hydrocarbon Processing*, vol. 4, 125–134.
9. G.W.G. McDonald, B.L. Harkins (1987) *Maximizing FCC Profits by Process Optimization*. San Antonio, TX: Presented at the NPRA Annual Meeting.
10. W.L. Luyben, D.E. Lamb (1963). Feed forward control of a fluidized catalytic reactor-regenerator system. *Chemical Engineering Progress Symposium Series*, vol. 46, 165–171.
11. H. Kurihara (1967). Optimal control of fluid catalytic cracking processes. PhD Dissertation, MIT.
12. L. Iscol (1970). The dynamics and stability of a fluid catalytic cracker. In *Proceedings of the American Control Conference*, Atlanta, GA.
13. W. Lee, A.M. Kugelman (1973). Number of steady-state operating points and local stability of open-loop fluid catalytic cracker. *Industrial and Engineering Chemistry Process Design and Development*, vol. 12(21), 197–204.
14. H. Seko, S. Tone, T. Otake (1978). Operation and control of a fluid catalytic cracker. *Journal of Chemical Engineering of Japan*, vol. 11(21), 130–135.
15. E. Lee, F.R. Jr. Groves (1985). Mathematical model of the fluidized bed catalytic cracking plant. *Transactions of the Society for Computer Simulation*, vol. 2, 219–236.
16. J.M. Arandes, H.I. de Lasa (1992). Simulation and multiplicity of steady states in fluidized FCCUs. *Chemical Engineering Science*, vol. 47, 2535–2540.
17. S.S.E.H. Elnashaie, S.S. Elshishini (1993a). Comparison between different mathematical models for the simulation of industrial fluid catalytic cracking (FCC) units. *Mathematical and Computer Modelling*, vol. 18, 91–110.
18. S.S.E.H. Elnashaie, S.S. Elshishini (1993b). Digital simulation of industrial fluid catalytic cracking units-IV. Dynamic behavior. *Chemical Engineering Science*, vol. 48, 567–583.
19. S.S.E.H. Elnashaie, A.E. Abasaeed, S.S. Elshishini (1995). Digital simulation of industrial fluid catalytic cracking units-V. Static and dynamic bifurcation. *Chemical Engineering Science*, vol. 50, 1635–1643.
20. S.S. Elshishini, S.E.E.H. Elnashaie (1990a). Digital simulation of industrial fluid catalytic cracking units-I: Bifurcation and its implications. *Chemical Engineering Science*, vol. 45, 553–559.
21. S.S. Elshishini, S.E.E.H. Elnashaie (1990b). Digital simulation of industrial fluid catalytic cracking units-II: Effect of charge stock composition on bifurcation and gasoline yield. *Chemical Engineering Science*, vol. 45, 2959–2964.
22. S.S.E.H. Elnashaie, I.M. El-Hennawi (1979). Multiplicity of the steady state in fluidized bed reactors-IV. Fluid catalytic cracking (FCC). *Chemical Engineering Science*, vol. 34, 1113–1121.
23. R.C. McFarlane, R.C. Reinman, J.F. Bartee, C. Georgakis (1993). Dynamic simulation for model VI fluid catalytic cracking unit. *Computers and Chemical Engineering*, vol. 17(3), 275.

24. Y. Zheng (1994). Dynamic modeling and simulation of a catalytic cracking unit. *Computers and Chemical Engineering*, vol. 18(1), 39–44.
25. W.D. Ford, R.C. Reineman, I.A. Vasalos, R.J. Fahrig (1976). Modeling catalytic cracking regenerators. Presented at the NPRA Annual Meeting, San Antonio, TX.
26. H.I. de Lasa, A. Errazu, E. Barreiro, S. Solioz (1981). Analysis of fluidized bed catalkic cracking regenerator models' in an industrial scale unit. *The Canadian Journal of Chemical Engineering*, vol. 59, 549–553.
27. P. Guigon, J.F. Large (1984). Application of the Kunii-Levenspiel model to a multistage baffled catalytic cracking regenerator. *The Chemical Engineering Journal*, vol. 28, 131–138.
28. A.S. Krishna, E.S. Parkin (1985). Modeling the regenerator in commercial fluid catalytic cracking units. *Chemical Engineering Progress*, vol. 81(41), 57–62.
29. L. Lee, S. Yu, C. Cheng (1989). Fluidized-bed catalyst cracking regenerator modelling and analysis. *The Chemical Engineering Journal*, vol. 40, 71–82.
30. V.W. Jr. Weekman, D.M. Nace (1970). Kinetics of catalytic cracking selectivity in fixed, moving, and fluid bed reactors. *AIChE Journal*, vol. 16, 397–404.
31. L. Lee, Y. Chen, T. Huang, W. Pan (1989). Four-lump kinetic model for fluid catalytic cracking process. *The Canadian Journal of Chemical Engineering*, vol. 67, 615.
32. M. Larocca, S. Ng, H. Lasa (1990). Fast catalytic cracking of heavy gas oils: Modeling coke deactivation. *Industrial and Engineering Chemistry Research*, vol. 29, 171–180.
33. G.S. Shnaider, A.G. Shnaider (1990). Kinetic models of catalytic cracking of oil fractions in single-stage and multistage reactors with fluidized beds and interpretation of experimental data of catalytic cracking performed in these reactors. *The Chemical Engineering Journal*, vol. 44, 53–72.
34. S.H. Yang, B.H. Chen, X.Z. Wang (2000). Neural network based fault diagnosis using unmeasurable inputs. *Engineering Applications of Artificial Intelligence*, vol. 13(3), 345–356.
35. B. Heim, S. Gentil, B. Celse, S. Cauvin, L. TraveMassuyes (2003). FCC diagnosis using several causal and knowledge based models. *In 5th IFAC Symposium on Fault Detection, Supervision and Safety of Technical Processes, SAFEPROCESS.2003*, Washington, DC.
36. Y. Huang, G.V. Reklaitis, V. Venkatasubramanian (2003). A heuristic extended Kalman filter based estimator for fault identification in a fluid catalytic cracking unit. *Industrial and Engineering Chemistry Research*, vol. 42(14), 3361–3371.
37. A. Sundarraman, R. Srinivasan (2003). Monitoring transitions in chemical plants using enhanced trend analysis. *Computers and Chemical Engineering*, vol. 27(10), 1455–1472.
38. X. Wang, U. Kruger, B. Lennox (2003). Recursive partial least squares algorithms for monitoring complex industrial processes. *Control Engineering Practice*, vol. 11(6), 613–632.
39. T.N. Pranatyasto, S.J. Qin (2001). Sensor validation and process fault diagnosis for FCC units under MPC feedback. *Control Engineering Practice*, vol. 9(8), 877–888.
40. O.A.Z. Sotomayor, D. Odloak, E. Alcorta-Garcia, P. de Leon-Canton (2004). *Observer-based supervision and fault detection of a FCC Unit Model Predictive Control*. IFAC.
41. N.F. Thornhill, S.C. Patwardhan, S.L. Shah (2008). A continuous stirred tank heater simulation model with applications. *Journal of Process Control*, vol. 18, 347–360.
42. K.Y. Choi, W.H. Ray (1985). Recent developments in transition metal catalysed olefin polymerization—A survey. I. Ethylene polymerization. *Journal of Macromolecular Science Reviews Macromolecular Chemistry Physics*, vol. 25(1), 1–55.
43. T. Xie, K.B. McAuley, C.C. Hsu, D.W. Bacon (1994). Gas phase ethylene polymerization: Production processes, polymer properties and reactor modelling. *Industrial and Engineering Chemistry Research*, vol. 33(3), 449–479.
44. E. Ali, K. Al-Humaizi, A. Ajbar (2003). Multivariable control of a simulated industrial gas-phase polyethylene reactor. *Industrial and Engineering Chemistry Research*, vol. 42, 2349–2364.

chapter five

Design of an automated tea grader

This chapter is dedicated towards the design of a commercial tea grader using BIS (Bureau of Indian Standards) certified tea brands available in the Indian marketplace. Tea is the most consumed beverage (not counting water) in the world because of its health, dietetic, and therapeutic benefits. The taste of tea is one of the most important factors in its quality grading, and generally has been assessed by professional tea tasters. However, the judgment conjectured with human perception may be subjective and prone to suffer from inconsistency and unpredictability due to various mundane factors. An electronic tongue (e-tongue or ET) has been used alternatively as a tea-taste recognizer and classifier [1–3]. Five grades of Xihulongjing tea (grade: AAA, AA, A, B and C, from the same region, processed with the same processing method) were discriminated using an Alpha MOS ASTREE II electronic tongue™ coupled with pattern recognition methods including principal component analysis (PCA), canonical discriminant analysis (CDA), and back-propagation neural networks (BPNN) [1]. In the reference Palit et al. sampled data that was compressed using discrete wavelet transform (DWT) and was then processed using a principal component analysis (PCA) and linear discriminant analysis (LDA) for visualization of underlying clusters. Finally, different pattern recognition models based on neural networks (back-propagation multilayer perceptron (BP-MLP), radial basis function (RBF) model, and probabilistic neural network (PNN)) were investigated to carry out a correlation study with the tea tasters' score of five different grades of black tea samples obtained from a tea garden in India [3]. The process of classifier design consists of steps including electrical signature generation and capturing of the selected tea brands, using a voltammetric electronic tongue (ET), preprocessing of signals, and development of automated machine-learning algorithms. The algorithms, which include moving window-based dissimilarity, moving window-based recursive PCA, and Fisher discriminant analysis (FDA) enhanced by the decision-directed acyclic graph (DDAG) method have been used as machine-learning components of the proposed tea grader. An introduction to the electronic tongue with a brief review on different kinds of ETs referring to their applications will provide the proper perspective and illustrate the significance of this chapter.

5.1 Electronic tongue: A biomimetic device

Biomimetics is the study of biologically inspired design, adaptation, or derivation from nature for development of new materials or products. Rapidly developing sensor technology and *chemometric* techniques have orchestrated the phenomenal concept of bionic devices like the electronic nose (e-nose), electronic tongue (e-tongue), etc. Electrochemistry plays a major role in e-tongue device operation. Analytical techniques that use a measurement of potential, charge, or current to determine an analyte's concentration to characterize the chemical reactivity of the analyte are collectively called electrochemistry because such techniques originated from the study of the movement of electrons in an electrolytic solution. Conventional analytical instruments like gas chromatography (GC), high-performance liquid chromatography (HPLC), atomic absorption and atomic emission spectroscopy (AAS and AES), capillary electrophoresis (CE), and colorimetric and elemental analysis have been in use for analysis of complex multi-component liquids. However, all these methods are time consuming, laborious, expensive, and not suitable for *in situ* analysis of bulk samples. In this context, the e-tongue, based on an electrochemical method of analysis, emerged as an alternative method. The first e-tongue system (called a "taste sensor") was introduced by Toko from Kyushu University, Japan, and consisted of eight potentiometric electrodes with lipid-polymeric membranes (PVC membranes with lipid derivatives) [4]. Legin et al. [5] from St. Petersburg University, Russia reported a solid-state crystalline ion-selective electrode (ISE) based on chalcogenide glass. As a result of the joint work conducted by St. Petersburg University, Russia with the Chemical Sensors Group of "Tor Vergata" University, Rome, Italy, the term "electronic tongue" was coined for a multisensory system composed of an array of chemical sensors and an appropriate data-processing tool [6]. Legin et al. [7] pointed out that the electronic tongue can be thought of as analogous to both olfaction and taste, and it can be used for the detection of all types of dissolved compounds, including volatile compounds, which create odors after evaporation. Various possible architectures of ET, such as potentiometric, impedentiometric, and voltammetric, as well as an e-tongue based on optical and mass sensors, have been proposed so far.

The research and development activities in the electronic tongue field have been carried out along the following avenues:

- On the development of new measurement methods/techniques
- Application of new data analysis tools
- Search for new chemosensitive materials and membrane components dedicated to multisensing purposes along with new techniques for the preparation of chemosensitive layers for ET applications (like pulsed laser deposition, the Langmuir–Blodgett technique, screen printing, electrodeposition, and electropolymerization)

Most of the e-tongue systems developed initially were based on potentiometric sensors. They were basically potentiometric electrodes with lipid-polymeric membranes (PVC membranes with lipid derivatives). Eventually, solid-state crystalline ion-selective electrodes were developed based on chalcogenide glass followed by coupled chalcogenide glass plus PVC membrane-based ion-selective electrodes. The principle of operation of the ion-selective electrodes is based on the measurement of potential changes of working electrodes against a reference electrode in zero-current conditions. The main disadvantages of potentiometric measurements are their temperature dependence, the influence of solution change, and adsorption of solution components that affect the nature of charge transfer. The advantages of ion-selective electrodes include a well-known principle of operation, low cost, easy fabrication, simple setup, and the possibility of obtaining sensors selective to various species. The commercialized version of lipid membrane-based e-tongue systems (i.e., taste sensing system SA401 and SA402B) is applied mainly for quality control in the foodstuffs and pharmaceuticals industry [8–11]. Potentiometric ion and chemical sensors based on field-effect devices form another group of transducers that can be easily miniaturized and are fabricated by means of a microelectronic technology/ screen-printing technique. There is no true "microreference electrode" available, hence, in most applications a conventional macroscopic reference electrode is used.

In voltammetric measurements, we get the current–potential relationship. The current is a function of the rate of electrolysis, which in turn is governed by the diffusion coefficients and concentrations of electroactive species present in the solution. Voltammetric sensors are advantageous devices for multicomponent measurements because of their high selectivity and sensitivity, high signal-to-noise ratio, low detection limits, and various modes of measurement (square wave, large pulse voltammetry, cyclic voltammetry, anodic stripping voltammetry, etc.). In voltammetry, the response is fast (less than 30 s), while potentiometry requires several minutes to complete a measurement sequence. The time aspect is important in a quality monitoring system since quality change can occur quickly, and a quick response is, therefore, required. Moreover, there are many possible modifications of electrodes in voltammetry that ensure the availability of sensors of various sensitivity and selectivity towards various species. However, the principle of operation limits the application of voltammetric sensors only to redox-active substances. Voltammetric ETs have three electrode configurations. The type-I voltammetric ET is based on a sensor array that consists of a few working electrodes made of various metals (copper, nickel, palladium, silver, tin, titanium, zirconium, gold, platinum, and rhodium), a reference electrode, and an auxiliary electrode. Type-II: involes metallic electrodes or carbon paste electrodes (CPE) modified with chemosensitive materials resulting in a chemically modified electrode (CME). And type-III involves modification of voltammetric electrodes with the application of enzymatic layers, providing the possibility of monitoring analytes such as glucose.

A new type of electronic tongue, an impedentiometric system, was presented in [12–16]. It consists of a bare interdigitated electrode, and interdigitated electrodes coated with various chemosensitive materials (polyaniline, polypyrrole, lipids, chitosan) deposited by the Langmuir–Blodgett technique. Because of the AC measurements, there is a reduction in the number of sensing units used and consequently a reduction in the whole size of the impedentiometric device. Further distinct differences of such types of sensors based on AC measurements are: (i) they do not have limited sensitivity for non-electrolyte substances like the taste sensor based on potentiometry; (ii) they do not require the use of redox active compounds to be either reduced or oxidized at the working electrode; (iii) voltammetry in complex liquid media provides complicated spectra; whose interpretation requires sound machine-learning algorithms unlike that of impedance spectra; and (iv) there is no requirement for active species in the impedentiometric system and no need for a standard reference electrode (a reliable reference is a critical issue in miniaturized sensor array-based devices).

The fusion of various measurement techniques has been proposed to improve the recognition capabilities of the electronic tongue systems. One of these approaches involved an array of coupled electrochemical sensors (potentiometric, amperometric, and conductometric) applied to the classification of various dairy products (milk, kefir, yogurts) [17]. In [18–21], a coupling of an electronic nose and tongue was used to combine liquid and gas sensors. However, a closer simulation of a natural sense of taste demands the coupling of mature sensing technologies, identification of signal attributes to the corresponding taste ingredients, and an exhaustive machine-learning algorithm.

So far, not much commercialization of ET has taken place. Results of apple juice measurements performed by two commercialized instruments—the ASTREE electronic tongue and Prometheus electronic nose (Alpha MOS, France)—were compared with sensory characteristics prepared by customers [22]. The same ET, with a FOX4000 electronic nose (Alpha MOS, France), was applied to the analysis of nutritive drinks [23]. The taste Sensing System from Anritsu/Atsugi Corp. (Japan), which provided a sensor array of seven transducers with different lipid/polymer membranes [24] along with the MACS-multi-array chemical sensor with seven solid-state ion selective electrodes (ISEs) produced by Mc. Science (S. Korea), are the other two commercialized versions of ET [25].

The development of miniaturized sensors and miniaturized sensor arrays faces several challenges including construction difficulties (the cost of preparation, technological problems); sample handling; sensor durability; and in some cases, the deterioration of working parameters. Descending down to lower scales brings unwarranted limitations, such as the evaporation of small volumes of samples, manufacturing constraints (e.g, spotting spacing), and problems with the receiving, amplification, and transduction of signals. For widespread acceptance, the newly proposed technologies must be significantly better and/or cheaper than the existing ones. As an illustration, despite the fact that, since 1976, more than 400 patents has been issued on ISFET (ion-sensitive field-effect transistor) technology in the U.S., no U.S. commercial products for medical applications resulted from these efforts [26].

Calibration is the most important phase of a multicomponent analysis using sensor arrays. A quantitative restriction prevails in preparing mixed calibration solutions. For the design of experiments (DOE), fractional factorial designs are often adapted. In fractional design, the main effects are computed correctly ignoring interaction effects. Hence, experimental plans require fewer runs at an affordable accuracy of calibration. For quantitative estimation using ET, additional methods like standard addition are required for sensor/ ensor array calibration. Reproducibility of parameters of the sensor forming the array is a matter of concern. This requirement becomes more important in the case of large arrays, because poor reproducibility of sensor, characteristics from array to array will require additional calibration of sensors, which will result in a longer and more expensive analysis. In other words, it is crucial to assess how the individual failure of an element sensor may affect the response of the array as a whole.

A comprehensive review on food and beverage analysis, monitoring, etc., using e-tongue has been published by Gilabert and Peris [27]. In particular, e-tongues have been used in wine analysis [28–32], honey classification [33,34], soy sauce analysis [35], water analysis [36–38], soft drink, beer, tea, coffee, and milk analysis [39–49], heavy metal detection [50–52], and rare earth metal ion detection [53]. Moreover, the e-tongue has been applied for the detection of microbial activity [54–56], which is of special importance in the food industry. Studies have shown that the voltammetric e-tongue can assess the growth of mold and bacteria as well as distinguish between different strains of molds. Similarly, the deterioration of milk due to microbial growth has also been followed and correlated with colony-forming units using e-tongue methods [56]. Riul et al. [57] made an exhaustive documentation on the advances in the electronic tongue encompassing the principles of detection, materials used in the sensing units, and data analysis with prospective applications. Ciosek and Wroblewski [58] documented almost all possible architectures of electronic tongue systems with various pattern-recognition techniques adapted to analyze the sensor array data, commercialization of e-tongues, and issues regarding miniaturization.

The ET device is important for cognitive purposes as well. The structural and functional aspects of chemical information being processed in living organisms giving rise to the human perception of taste and smell is an implicit function of some ET-monitored parameters/features. This, in turn, might lead to digitalization and quantization of sensations such as taste and smell. ET-based research is succeeding as an interdisciplinary field, which has necessitated the participation and collaboration of electrical, electronics, and chemical engineers as well as chemists across the globe.

5.2 Experimentation

5.2.1 E-tongue-based instrumentation and principles

Figure 5.1 shows the functional blocks of ET-based experimentation using the pulse voltammetric method. A typical e-tongue experimentation system consists of three major components: an electrochemical cell with sensor array, potentiostat, and data acquisition software (DAS). The electrochemical cell holds the analyte solution along with the set of electrodes (sensory array). The most popular configuration is the three-electrode setup with a working electrode, a reference electrode, and an auxiliary electrode. The potentiostat system is generally connected to the personal computer through an RS232 (serial data) or USB (universal serial bus) interface. DAS is required to generate a voltammetry signal to be applied between the working and reference electrodes and to capture current response from the liquid under test through the counter electrode. A potentiostat is used for electrochemical characterization of redox active species and in evaluating thermodynamic and kinetic parameters of electron transfer events. A potentiostat is also capable of imposing electrical-potential waveforms across a working electrode relative to a reference electrode. It also measures the resultant current through the electrochemical cell at the third electrode. In general, working electrode potential is used to derive an electron-transfer reaction:

$$O + n \times e \leftrightarrows \mathrm{R} \tag{5.1}$$

where O and R are oxidized and reduced forms of the redox couple, and the resulting current is measured with the help of an auxiliary (counter) electrode. Minimal current is drawn through the reference electrode because its current signal is made to input to a very high impedance electrometer, thus ensuring the constant potential condition. Popular

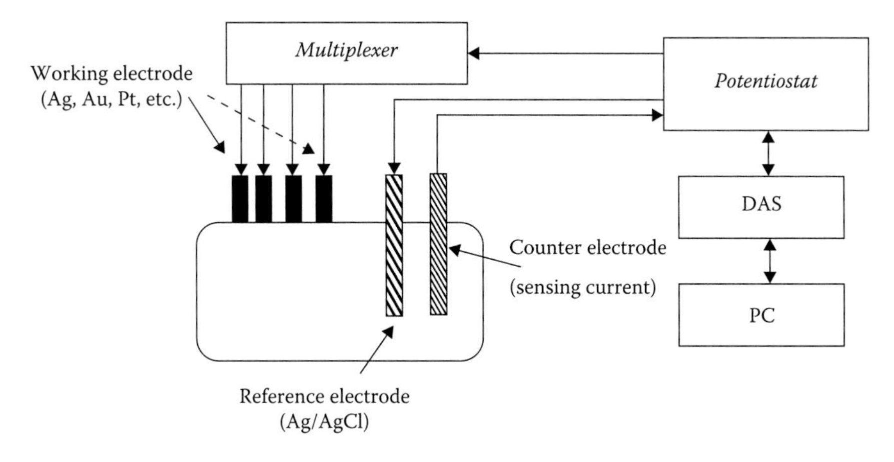

Figure 5.1 Functional blocks of e-tongue.

choices for the working electrode material include platinum, gold, silver, and iridium. Current arising from the electron transfer events at the working electrode is measured at the counter electrode, and therefore, this electrode must be geometrically larger than the working electrode so that it does not limit the current density at the working electrode. When a surface is immersed in an aqueous solution, a discontinuity is developed at the interface and the physicochemical variables such as electrical potential and electrolyte concentration vary significantly from the aqueous phase to interphase. Because of the different chemical potentials between the two phases, charge separation often occurs at the interfacial region and an electrical double layer is formed. This layer extends as far as 100 nm in a very dilute solution, to only a few angstroms in a concentrated solution. The electrical double layer exists on the interface between an electrode and its surrounding electrolyte in the electrochemical cell. Charges are separated by an insulator capacitor. On a bare metal immersed in an electrolyte, one can estimate that there will be 20–60 μF of capacitance for every 1 cm^2 of electrode area. The most commonly used pulse voltammetry methods are large amplitude pulse voltammetry (LAPV) and small amplitude pulse voltammetry (SAPV) and differential pulse voltammetry (DPV). In the experimentation referred to here, large amplitude pulse voltammetry was used. In LAPV, the working electrode is held at a base potential at which negligible electrode reactions occur. At the onset of the pulse, the current flows to the electrode surface, initially sharp when the Helmholtz double layer is formed. The current, therein, decays until the diffusion-limited faradic current remains, as described by Equation 5.2. The size and shape of the transient response reflect the amount and diffusion coefficients of both electroactive and charged compounds in the solution. At the onset of zero electrode potential, similar but opposite reactions occur, omitting the faradic current. In normal LAPV, there are successive potential pulses of gradually changing amplitude, between which the base potentials are applied to the working electrode.

$$i_r = E * e(-t/Rs * B)/Rs \tag{5.2}$$

where Rs is the resistance of the circuit (= solution), E is the applied potential, t is the time, and B represents electrode equivalent capacitance representing the double layer capacitance. i_r is redox current, or the Faradaic current (it behaves in a similar way, initially a large current pulse is generated when compounds close to the electrode surface are oxidized or reduced, but decays with time). In addition to the Faradaic current, the current in an electrochemical cell includes other, non-Faradaic sources, as mentioned earlier. Suppose the charge on an electrode is zero, and suddenly, there is a change in its potential so that the electrode's surface acquires a positive charge. Cations near the electrode's surface respond to this positive charge by migrating away from the electrode; anions, on the other hand, migrate toward the electrode. This migration of ions occurs until the electrode's positive surface charge and the negative charge of the solution near the electrode are equal. Because the movement of ions and the movement of electrons are indistinguishable, the result is a small, short-lived non-Faradaic current that is called the charging current. Every time the electrode's potential is changed, a transient charging current flows. The current response can be modeled as the sum of two exponentially decaying time functions (because of Faradaic and capacitive components) as given by:

$i(t) = i_F e^{-t/\tau_F} + i_C e^{-t/\tau_C}$, where i_F, i_C, $-1/\tau_F$ and $-1/\tau_C$ are the model parameters for a given electroactive compound.

A single potential step may not be sufficient to provide the information needed for the feature extraction, hence, classification of samples. For this reason, a voltage waveform comprising a sequence of potential pulses of varying amplitude is imposed onto the working electrode and the resulting responses; the voltammogram is recorded.

It is almost impossible to discriminate just by performing a visual check of the output waveforms from an ET-based instrumentation system. Cross selectivity (partial overlapping selectivity) of sensors used, nonlinearity, multiscale behavior, and nonstationarities in signals are the reasons that necessitated the development of automated tools to devise computer-based authentication systems. In one possible methodology, the development of such an authentication system can be configured as a two-stage problem.

Stage 1: Determination/extraction of suitable features.
Stage 2: Development of a robust authenticator based on these extracted features.

5.2.2 *E-tongue signature generation using various commercial brands of tea*

The experiments were carried out in a 150-mL electrochemical cell equipped with a magnetic stirrer, with a standard three-electrode configuration each time. A working electrode (which is made of either silver/platinum/gold) of 5 mm length and a diameter of 1 mm, counter electrode (made of stainless steel) with dimensions of 20×50 mm^2 and Ag/AgCl, 3 M KCl as a reference electrode make up the electrode assembly. A pulse voltammetry analysis was carried out using a potentiostat (Gamry Instruments, USA; Reference 600 Potentiostat), which was connected via DAS to a PC with a Pentinum-IV processor. PHE200 physical electrochemistry and PV220 pulse voltammetry software packages were used to record the data and corresponding waveform (voltammogram).

A stock of 500-mL tea liquor was prepared for each of the six categories. The mass of tea by 2% corresponding to 2 g of tea per 100 mL of liquor were used as per British standard 6008:1980. This amount was poured into a white porcelain earthenware pot and was mixed with freshly boiling water to within 4–6 mm of the brim. The mixture was allowed to brew for at least 6 minutes, after which the prepared tea liquor was poured into a 500 mL flask. From this stock of 500 mL, three liquor samples of 100 mL each by volume were prepared. The foregoing method was followed to prepare the liquor samples of six different categories. The pH values of all samples were measured with a pH meter and were maintained at 6 by adding buffer solution. During the experiment, the temperature of all samples was maintained at 20°C. Data acquisition was performed with an initial pulse magnitude of 600mV, duration of 1s and step size of –100 mV and a measuring cycle of 11s. Thus, 10 pulses with decreasing magnitudes were generated and the sequence terminated at –400 mV. The input waveform formed by potential pulses of varying amplitudes with a base potential in between is shown in Figure 5.2. Different amplitudes were used to increase the sensitivity and improve the discrimination between samples. Applying the same input waveform, the current transients were measured for all branded tea samples with different electrodes (Ag, Au, and Pt) with the help of the DAS using a sampling rate of 400 samples/s. The effect of a particular working electrode in classifier design was investigated by conducting experiments using

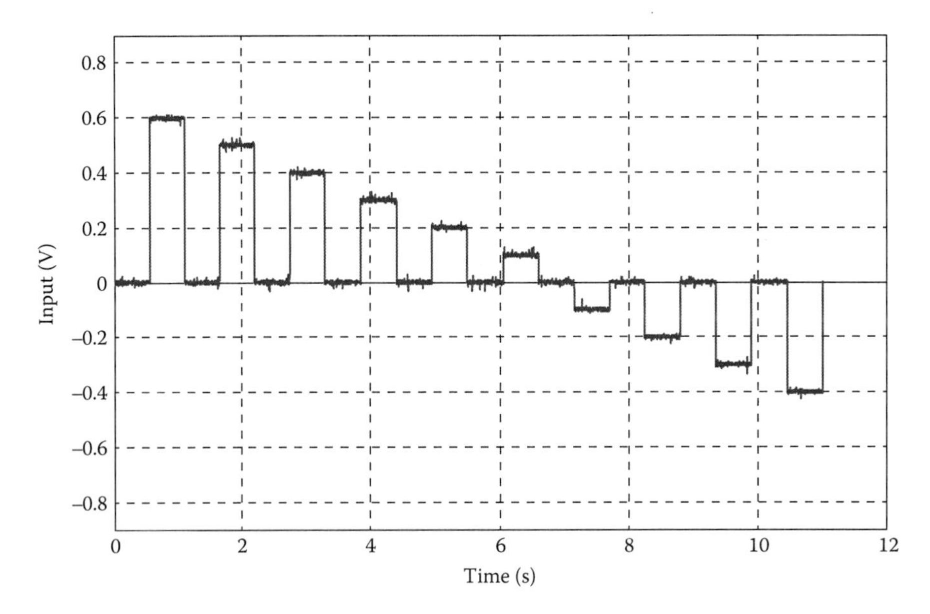

Figure 5.2 Input waveform applied to e-tongue.

silver, gold, and platinum, separately, as working electrodes to generate different sets of data for all of the six brands of tea. Figures 5.3 through 5.5 show some representative output current waveforms using platinum, glassy carbon, and gold as the working electrodes. In order to take care of uncertainties in the measurements and ensure reproducibility, the experiments were repeated for three samples of each category of tea brands. The expected uncertainties are $\pm 10^{-3}$, depending on the drift in the electronic circuitry of a data acquisition system. Temperature dependence and large surface alteration cause drifts in sensor response. To overcome this problem, electrochemical cleaning steps, or mechanical polishing, were carried out.

5.3 *Tea data preprocessing*

Time series data are statistical data collected and recorded over successive increments of time. Time series such as electrical signals that are continuous in time are treated herein as discrete time series because only digitized values in discrete time intervals are used for the computation. GARCH (generalized autoregressive conditional heteroskedasticity) is a time series modeling technique that uses past variances, and past variance forecasts, to forecast future variances. Whenever a time series is said to have GARCH effects, the series is heteroskedastic; that is, its variances vary with time. Using the GARCH toolbox, the auto-correlation function (ACF) and partial auto-correlation function (PACF) are derived for time series data of tea. Figure 5.6 presents the ACFs of representative tea samples for each electrode system, namely, silver (Ag), platinum (Pt), gold (Au), and glassy carbon along with the upper and lower standard deviation confidence bounds. Figure 5.7 presents the PACFs of representative

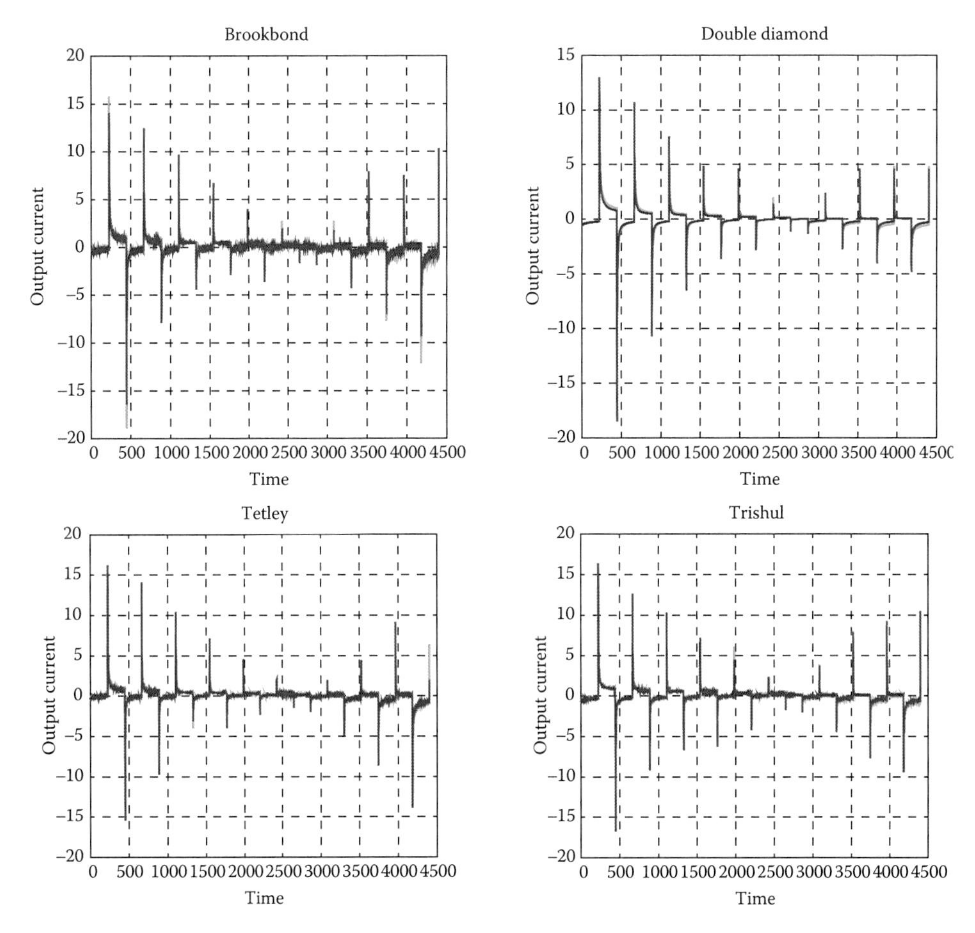

Figure 5.3 Output current signatures for various tea brands using platinum as the working electrode.

tea samples for each electrode system considered along with the upper and lower standard deviation confidence bounds. From the above correlograms, the presence of correlation among the successive observations in the time series of tea samples (collected over the Ag, Pt, and Au electrodes) is prevalent. The observations indicate that a significant amount of nonstationarity is present in the time series corresponding to various tea brands. In this regard, the glassy-carbon electrode manifested different behavior; the time series data collected using it did not show any significant correlation. The Ljung-Box Q-statistic and Engle's ARCH (autoregressive conditional heteroskedastic effects) tests (lack-of-fit hypothesis tests) are performed on mean-centered time series data. The presence of an ARCH effect and significant Q statistics in the data of various tea brands indicates the presence of nonstationarity in the data. The nonstationary behavior may be due to the change of covariance structure as well as the time-varying variance.

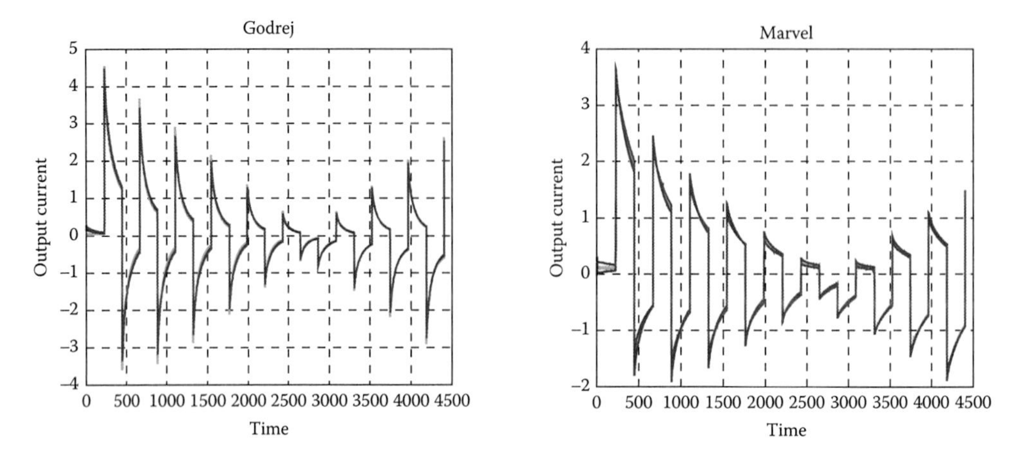

Figure 5.4 Output current signatures for various tea brands using glassy carbon as the working electrode.

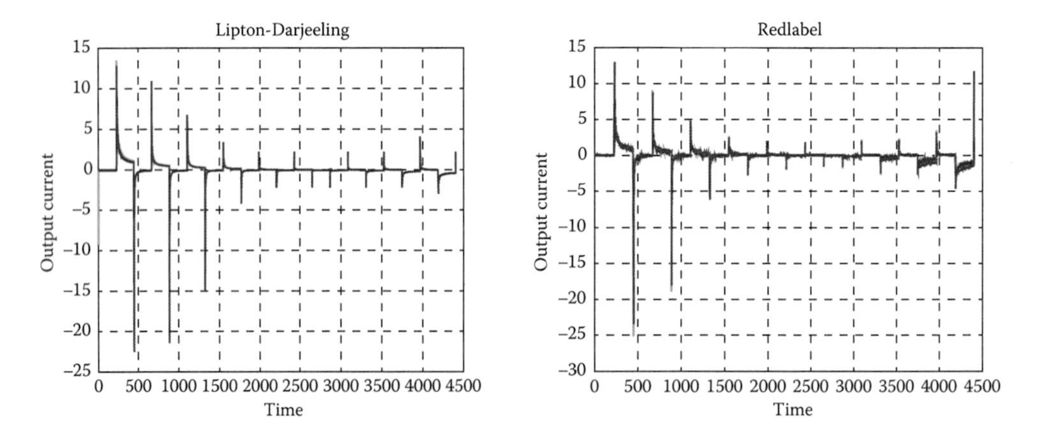

Figure 5.5 Output current signatures for various tea brands using gold as the working electrode.

5.4 Dissimilarity-based tea grader

E-tongue signatures of commercial tea brands (available in the Indian marketplace) including Brookebond (BR), Double Diamond (DD), Godrej (GD), Lipton (LP), Lipton-Darjeeling (LD), Marvel (MV), Maryada (MR), Redlabel (RD), Tajmahal (TJ), Tatagold (TG), Tetley (TT), and Trishul (TR) for each of the working electrodes (*viz.*, silver, gold , and platinum) have been used for the design of automated tea grading. The basis of authentication used is moving window-based pattern-matching using a dissimilarity index.

5.4.1 Authentication/classification algorithm

A historical database is created of the size $\{(4402 \times 3) \times 12 = 13206 \times 12\}$, that is 4402 observations of three replicate runs for each of the 12 categories of commercial tea brands for

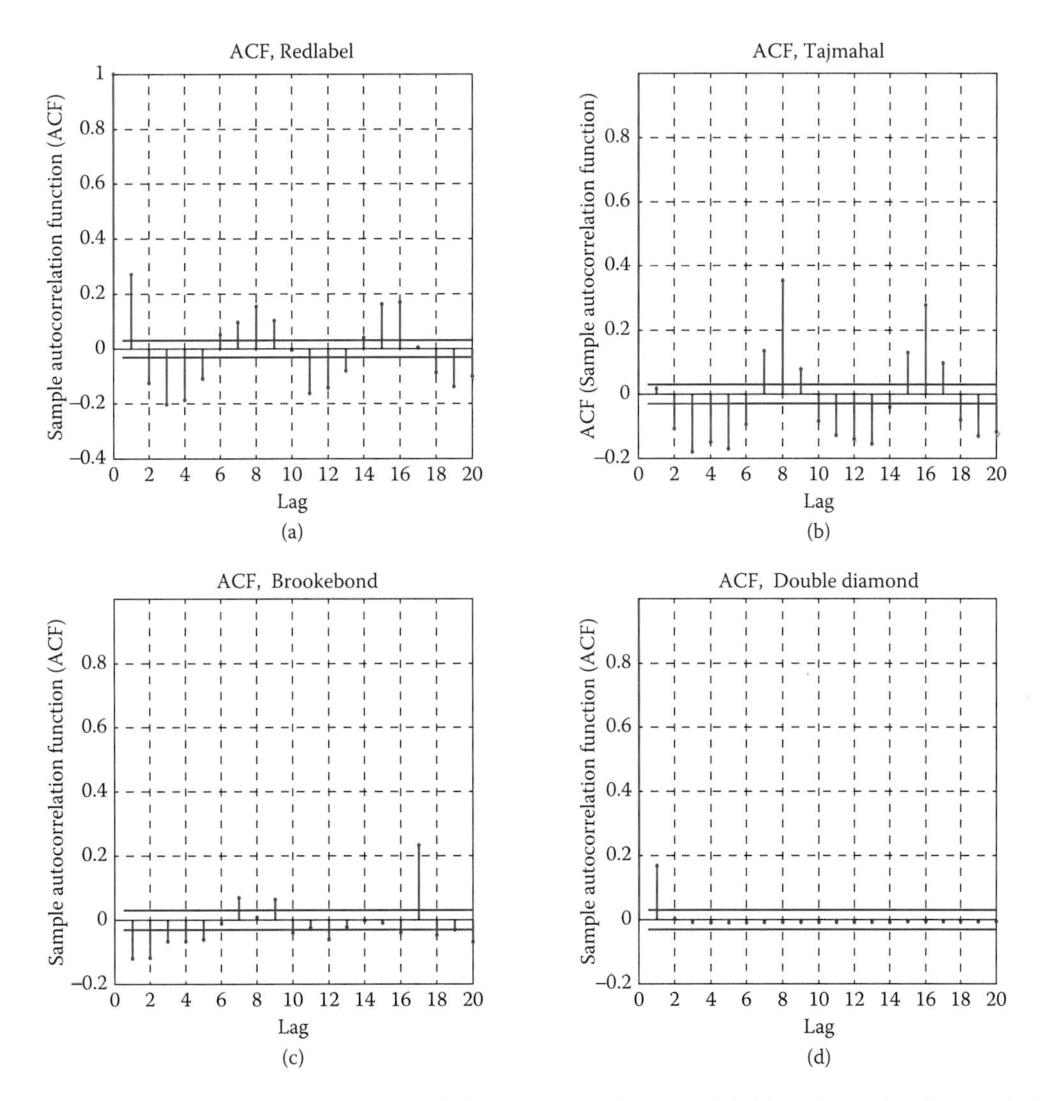

Figure 5.6 Auto correlation functions for different tea brands: (a) Redlabel (Ag electrode), (b) Tajmahal (Pt electrode), (c) Brookebond (Au electrode), (d) Double Diamond (glassy-carbon electrode).

each of the working electrodes (*viz.*, silver, gold, and platinum). Moving window-based pattern-matching using a dissimilarity factor (Karhunen-Loeve [KL] expansion) has been considered for unknown sample authentication and classification. The proposed method is completely data driven and unsupervised; no process models or training data are required. The template or reference database (historical database) is normalized to zero mean and unit variance. The historical database for one working electrode is divided into 12 data windows; each one is of 4402 × 3 size. Snapshot/test data is simulated by adding noise to e-tongue signals of all 12 tea brands present in triplicate with signal to noise ratios (SNR) of 100, 70, 50 and 40 (for each of the working electrodes). Tables 5.1 through 5.12 present the dissimilarity-based authentication performances of 12 simulated unknown tea samples (at SNR of 100, 70, 50, and 40 for the silver, gold, and

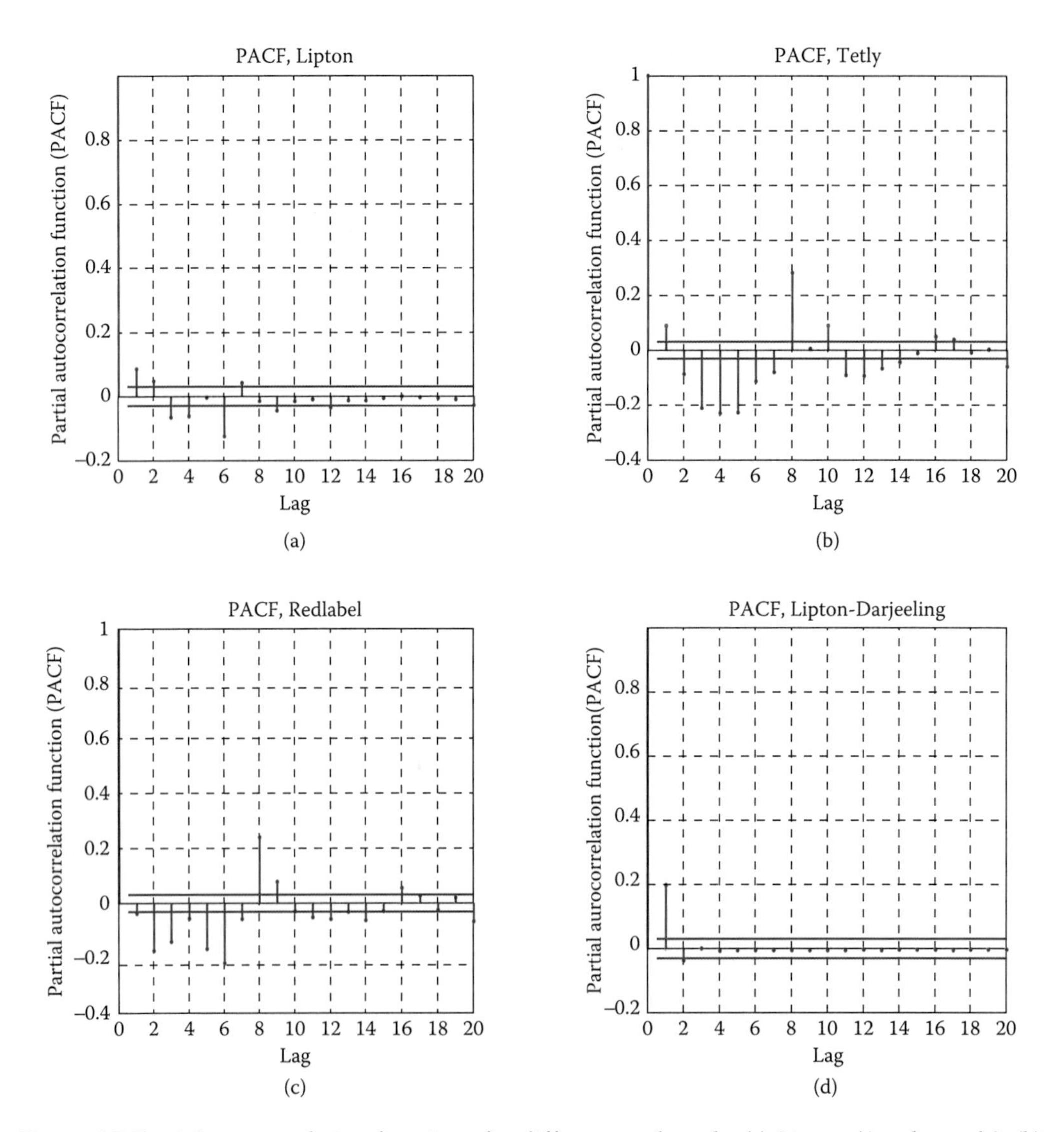

Figure 5.7 Partial autocorrelation functions for different tea brands: (a) Lipton (Ag electrode), (b) Tetley (Pt electrode), (c) Redlabel (Au electrode), (d) Lipton-Darjeeling (glassy-carbon electrode).

platinum electrodes). For each of the working electrodes, each of the four resulting dissimilarity tables is of dimension 12 by 12. In order to authenticate 12 test tea brands of 12 unknown (snapshot) categories, 12 template data windows undergo pattern matching with the 12 categories of test snapshot windows for each electrode system. Each time, the template window composition remains unchanged, but the composition of the snapshot for each table keeps on changing with respect to its infiltrated level of noise. In each table, against the template composition, 12 categories of snapshots are tested, taking one at a time, resulting in a dissimilarity factor of zero or near zero at least once in each column. Dissimilarity is calculated as described previously in Chapter 2. The proposed classification algorithm flow chart is as follows:

For each electrode system (e = 1–3) (e = 1: silver; 2: gold; 3: platinum)

Step 1: Generation of the 12 autoscaled historical data windows (i = 1–12) from the experimental measurements.
Step 2: Creation of 12 simulated tea brands/snapshots (j = 1–12) in triplicate (each containing three samples) resembling any of the particular tea brands considered for the classifier development. SNR considered are 100, 70, 50, and 40.
Step 3: For any test brand j, the first snapshot window considered (4402 × 3) samples, and then the snapshot window is updated batchwise ignoring the previous batch data.
Step 4: Creation of dissimilarity table:

```
For j = 1-12
For i = 1-12
Calculation of dissimilarity between jth snapshot and ith template
data window
End (i)
End (j)
```

Step 5: Generation of four numbers (for SNR of 100, 70, 50, and 40) of dissimilarity values table of dimension (12 × 12) according to step 4.
Step 6: Unknown sample authentication: Along each column of the generated dissimilarity values table, there remains at least 1 minimum entry value within a range of ($0–10^{-4}$). The minimum entry value authenticates a specific brand.

End (e)

Table 5.1 Authentication performance of dissimilarity-based unknown tea brand classifier with SNR = 100 (silver working electrode)

	BR	DD	GD	LP	LD	MV	MR	RD	TJ	TG	TT	TR
BR	**0.0000**	0.3518	0.2084	0.1153	0.1546	0.0296	0.2079	0.1393	0.1061	0.5434	0.1069	0.1664
DD	0.3518	**0.0000**	0.1605	0.3113	0.2637	0.2890	0.0928	0.4841	0.3624	0.3475	0.1699	0.2911
GD	0.2084	0.1605	**0.0000**	0.1351	0.1563	0.2043	0.0626	0.3081	0.2272	0.3922	0.1666	0.1518
LP	0.1153	0.3113	0.1351	**0.0000**	0.0559	0.1269	0.1570	0.2480	0.1353	0.4623	0.1254	0.0317
LD	0.1546	0.2637	0.1563	0.0559	**0.0000**	0.1206	0.1101	0.2538	0.2394	0.3797	0.0910	0.0112
MV	0.0296	0.2890	0.2044	0.1269	0.1206	**0.0000**	0.1617	0.1893	0.1793	0.4960	0.0432	0.1485
MR	0.2079	0.0928	0.0626	0.1570	0.1101	0.1617	**0.0000**	0.3375	0.2507	0.3466	0.0809	0.1385
RD	0.1393	0.4841	0.3081	0.2480	0.2538	0.1893	0.3375	**0.0000**	0.2468	0.6205	0.2925	0.2646
TJ	0.1061	0.3624	0.2272	0.1353	0.2394	0.1793	0.2507	0.2468	**0.0000**	0.5909	0.2089	0.2282
TG	0.5434	0.3475	0.3922	0.4623	0.3797	0.4960	0.3466	0.6205	0.5909	**0.0000**	0.4120	0.3973
TT	0.1069	0.1699	0.1666	0.1254	0.0910	0.0432	0.0809	0.2925	0.2089	0.4120	**0.0000**	0.1250
TR	0.1664	0.2911	0.1518	0.0317	0.0112	0.1485	0.1385	0.2646	0.2282	0.3973	0.1250	**0.0000**

Table 5.2 Authentication performance of dissimilarity-based unknown tea brand classifier with SNR = 70 (silver working electrode)

	BR	DD	GD	LP	LD	MV	MR	RD	TJ	TG	TT	TR
BR	**0.0000**	0.3518	0.2084	0.1153	0.1546	0.0296	0.2079	0.1393	0.1061	0.5434	0.1069	0.1664
DD	0.3517	**0.0000**	0.1605	0.3113	0.2636	0.2890	0.0928	0.4841	0.3624	0.3475	0.1698	0.2910
GD	0.2084	0.1605	**0.0000**	0.1351	0.1563	0.2044	0.0626	0.3081	0.2271	0.3922	0.1667	0.1518
LP	0.1153	0.3114	0.1351	**0.0000**	0.0559	0.1268	0.1571	0.2479	0.1352	0.4623	0.1254	0.0317
LD	0.1546	0.2637	0.1563	0.0559	**0.0000**	0.1205	0.1102	0.2538	0.2394	0.3797	0.0909	0.0112
MV	0.0296	0.2890	0.2044	0.1269	0.1206	**0.0000**	0.1617	0.1893	0.1794	0.4959	0.0432	0.1484
MR	0.2079	0.0928	0.0626	0.1570	0.1101	0.1617	**0.0000**	0.3375	0.2507	0.3466	0.0809	0.1385
RD	0.1393	0.4841	0.3081	0.2480	0.2538	0.1893	0.3376	**0.0000**	0.2468	0.6205	0.2925	0.2646
TJ	0.1061	0.3624	0.2272	0.1352	0.2393	0.1793	0.2507	0.2468	**0.0000**	0.5909	0.2089	0.2281
TG	0.5434	0.3475	0.3922	0.4623	0.3797	0.4959	0.3466	0.6205	0.5909	**0.0000**	0.4120	0.3973
TT	0.1068	0.1700	0.1667	0.1254	0.0910	0.0432	0.0810	0.2924	0.2089	0.4121	**0.0000**	**0.1250**
TR	0.1664	0.2912	0.1519	0.0317	0.0112	0.1484	0.1386	0.2646	0.2282	0.3973	0.1250	0.0000

Table 5.3 Authentication performance of dissimilarity-based unknown tea brand classifier with SNR = 50 (silver working electrode)

	BR	DD	GD	LP	LD	MV	MR	RD	TJ	TG	TT	TR
BR	**0.0000**	0.3531	0.2090	0.1151	0.1542	0.0296	0.2087	0.1383	0.1065	0.5436	0.1076	0.1659
DD	0.3455	**0.0002**	0.1589	0.3053	0.2579	0.2825	0.0885	0.4801	0.3567	0.3467	0.1631	0.2855
GD	0.2045	0.1605	**0.0002**	0.1291	0.1512	0.2004	0.0598	0.3063	0.2229	0.3929	0.1619	0.1466
LP	0.1148	0.3124	0.1361	**0.0000**	0.0558	0.1264	0.1580	0.2476	0.1351	0.4627	0.1255	0.0317
LD	0.1537	0.2648	0.1566	0.0554	**0.0000**	0.1197	0.1111	0.2532	0.2389	0.3804	0.0908	0.0111
MV	0.0289	0.2906	0.2041	0.1259	0.1199	**0.0000**	0.1623	0.1880	0.1787	0.4963	0.0440	0.1476
MR	0.2051	0.0935	0.0633	0.1541	0.1076	0.1584	**0.0001**	0.3366	0.2488	0.3475	0.0774	0.1359
RD	0.1396	0.4846	0.3083	0.2477	0.2537	0.1897	0.3380	**0.0000**	0.2465	0.6205	0.2930	0.2643
TJ	0.1049	0.3648	0.2288	0.1343	0.2378	0.1780	0.2519	0.2453	**0.0000**	0.5910	0.2088	0.2265
TG	0.5404	0.3483	0.3889	0.4584	0.3765	0.4923	0.3455	0.6192	0.5888	**0.0001**	0.4083	0.3937
TT	0.1047	0.1733	0.1672	0.1241	0.0896	0.0414	0.0821	0.2896	0.2082	0.4134	**0.0000**	0.1237
TR	0.1648	0.2926	0.1523	0.0309	0.0114	0.1472	0.1398	0.2638	0.2268	0.3986	0.1246	**0.0000**

Table 5.4 Authentication performance of dissimilarity-based unknown tea brand classifier with SNR = 40 (silver working electrode).

	BR	DD	GD	LP	LD	MV	MR	RD	TJ	TG	TT	TR
BR	**0.0006**	0.3647	0.2136	0.1131	0.1510	0.0324	0.2164	0.1301	0.1073	0.5458	0.1151	0.1618
DD	0.2895	**#0.0129**	0.1469	0.2527	0.2053	0.2232	0.0581	0.4423	0.3119	0.3453	0.1057	0.2357
GD	0.1747	0.1697	**#0.0100**	0.0882	0.1165	0.1700	0.0498	0.2926	0.1932	0.4014	0.1292	0.1110
LP	0.1065	0.3278	0.1515	**0.0011**	0.0554	0.1189	0.1730	0.2406	0.1322	0.4688	0.1265	0.0318
LD	0.1453	0.2810	0.1646	0.0538	**0.0013**	0.1114	0.1261	0.2463	0.2369	0.3898	0.0910	0.0115
MV	0.0270	0.3053	0.2060	0.1221	0.1146	**0.0010**	0.1693	0.1761	0.1796	0.4995	0.0518	0.1416
MR	0.1800	0.1174	0.0739	0.1229	0.0774	0.1339	**0.0053**	0.3170	0.2315	0.3535	0.0586	0.1046
RD	0.1433	0.4897	0.3110	0.2462	0.2535	0.1954	0.3432	**0.0002**	0.2442	0.6206	0.2992	0.2624
TJ	0.0983	0.3801	0.2398	0.1268	0.2264	0.1706	0.2604	0.2365	**0.0016**	0.5910	0.2085	0.2145
TG	0.5238	0.3538	0.3724	0.4380	0.3612	0.4727	0.3416	0.6121	0.5773	**0.0030**	0.3898	0.3760
TT	0.0905	0.1969	0.1715	0.1163	0.0818	0.0303	0.0918	0.2700	0.2041	0.4224	**0.0021**	0.1158
TR	0.1584	0.3083	0.1644	0.0322	0.0127	0.1401	0.1560	0.2587	0.2269	0.4054	0.1253	**0.0014**

Misclassification = (2/144) × 100 = 1.38%

Table 5.5 Authentication performance of dissimilarity-based unknown tea brand classifier with SNR = 100 (gold working electrode)

	BR	DD	GD	LP	LD	MV	MR	RD	TJ	TG	TT	TR
BR	**0.0000**	0.3102	0.2969	0.0086	0.0436	0.0697	0.3091	0.1841	0.1288	0.2184	0.1924	0.0204
DD	0.3102	**0.0000**	0.1072	0.3237	0.3307	0.3414	0.3542	0.3710	0.3722	0.2394	0.4241	0.3082
GD	0.2969	0.1072	**0.0000**	0.3289	0.3316	0.2534	0.1702	0.3113	0.3734	0.1689	0.3008	0.2654
LP	0.0086	0.3237	0.3289	**0.0000**	0.0513	0.1120	0.3591	0.1896	0.1345	0.2291	0.2430	0.0464
LD	0.0436	0.3307	0.3316	0.0513	**0.0000**	0.1252	0.3252	0.2305	0.0307	0.1389	0.2237	0.0583
MV	0.0697	0.3414	0.2534	0.1120	0.1252	**0.0000**	0.1631	0.1442	0.1994	0.2753	0.0587	0.0291
MR	0.3091	0.3542	0.1702	0.3591	0.3252	0.1631	**0.0000**	0.2877	0.3344	0.2167	0.0895	0.2514
RD	0.1841	0.3710	0.3113	0.1896	0.2305	0.1442	0.2877	**0.0000**	0.3166	0.3846	0.2137	0.1269
TJ	0.1288	0.3722	0.3734	0.1345	0.0307	0.1994	0.3344	0.3166	**0.0000**	0.1210	0.2609	0.1413
TG	0.2184	0.2394	0.1689	0.2291	0.1389	0.2753	0.2167	0.3846	0.1210	**0.0000**	0.3239	0.2253
TT	0.1924	0.4241	0.3008	0.2430	0.2237	0.0587	0.0895	0.2137	0.2609	0.3239	**0.0000**	0.1404
TR	0.0204	0.3082	0.2654	0.0464	0.0583	0.0291	0.2514	0.1269	0.1413	0.2253	0.1404	**0.0000**

Table 5.6 Authentication performance of dissimilarity-based unknown tea brand classifier with SNR = 70 (gold working electrode)

	BR	DD	GD	LP	LD	MV	MR	RD	TJ	TG	TT	TR
BR	**0.0000**	0.3102	0.2969	0.0086	0.0437	0.0697	0.3091	0.1841	0.1288	0.2184	0.1924	0.0204
DD	0.3102	**0.0000**	0.1071	0.3237	0.3307	0.3413	0.3541	0.3710	0.3722	0.2394	0.4241	0.3082
GD	0.2968	0.1073	**0.0000**	0.3289	0.3315	0.2533	0.1702	0.3113	0.3733	0.1688	0.3008	0.2653
LP	0.0086	0.3237	0.3289	**0.0000**	0.0513	0.1120	0.3591	0.1896	0.1345	0.2291	0.2430	0.0464
LD	0.0436	0.3307	0.3316	0.0513	**0.0000**	0.1252	0.3252	0.2305	0.0307	0.1389	0.2237	0.0583
MV	0.0697	0.3414	0.2534	0.1120	0.1252	**0.0000**	0.1630	0.1442	0.1994	0.2753	0.0587	0.0291
MR	0.3091	0.3542	0.1703	0.3591	0.3252	0.1630	**0.0000**	0.2876	0.3344	0.2167	0.0895	0.2513
RD	0.1841	0.3710	0.3113	0.1896	0.2305	0.1442	0.2877	**0.0000**	0.3166	0.3846	0.2137	0.1269
TJ	0.1288	0.3722	0.3734	0.1345	0.0306	0.1994	0.3344	0.3165	**0.0000**	0.1211	0.2609	0.1413
TG	0.2184	0.2394	0.1690	0.2291	0.1389	0.2753	0.2167	0.3846	0.1210	**0.0000**	0.3239	0.2253
TT	0.1924	0.4241	0.3009	0.2430	0.2237	0.0588	0.0895	0.2137	0.2609	0.3239	**0.0000**	0.1404
TR	0.0204	0.3082	0.2654	0.0464	0.0583	0.0291	0.2514	0.1269	0.1413	0.2253	0.1404	**0.0000**

Table 5.7 Authentication performance of dissimilarity-based unknown tea brand classifier with SNR = 50 (gold working electrode)

	BR	DD	GD	LP	LD	MV	MR	RD	TJ	TG	TT	TR
BR	**0.0000**	0.3102	0.2970	0.0086	0.0436	0.0697	0.3092	0.1841	0.1288	0.2184	0.1925	0.0204
DD	0.3097	**0.0000**	0.1049	0.3232	0.3299	0.3407	0.3522	0.3709	0.3712	0.2374	0.4232	0.3076
GD	0.2964	0.1086	**0.0000**	0.3286	0.3310	0.2525	0.1686	0.3109	0.3726	0.1678	0.2996	0.2647
LP	0.0087	0.3237	0.3291	**0.0000**	0.0513	0.1122	0.3593	0.1898	0.1344	0.2291	0.2432	0.0465
LD	0.0438	0.3307	0.3315	0.0515	**0.0000**	0.1253	0.3251	0.2308	0.0306	0.1387	0.2237	0.0584
MV	0.0696	0.3413	0.2534	0.1119	0.1251	**0.0000**	0.1632	0.1441	0.1993	0.2752	0.0589	0.0290
MR	0.3088	0.3539	0.1700	0.3588	0.3249	0.1627	**0.0000**	0.2874	0.3342	0.2165	0.0894	0.2510
RD	0.1837	0.3707	0.3112	0.1893	0.2301	0.1441	0.2879	**0.0000**	0.3162	0.3842	0.2137	0.1266
TJ	0.1287	0.3723	0.3736	0.1343	0.0306	0.1993	0.3346	0.3163	**0.0000**	0.1213	0.2609	0.1412
TG	0.2178	0.2400	0.1698	0.2286	0.1382	0.2746	0.2166	0.3843	0.1202	**0.0000**	0.3232	0.2246
TT	0.1919	0.4239	0.3008	0.2425	0.2234	0.0584	0.0899	0.2135	0.2608	0.3240	**0.0000**	0.1400
TR	0.0203	0.3082	0.2655	0.0463	0.0583	0.0291	0.2515	0.1269	0.1413	0.2253	0.1406	**0.0000**

Table 5.8 Authentication performance of dissimilarity-based unknown tea brand classifier with SNR = 40 (gold working electrode)

	BR	DD	GD	LP	LD	MV	MR	RD	TJ	TG	TT	TR
BR	**0.0000**	0.3105	0.2978	0.0082	0.0436	0.0706	0.3105	0.1840	0.1288	0.2185	0.1937	0.0208
DD	0.3032	**0.0055**	0.0779	0.3179	0.3209	0.3324	0.3289	0.3686	0.3605	0.2136	0.4120	0.3004
GD	0.2915	0.1142	**0.0005**	0.3242	0.3243	0.2479	0.1619	0.3098	0.3650	0.1587	0.2947	0.2598
LP	0.0088	0.3238	0.3293	**0.0000**	0.0515	0.1125	0.3596	0.1899	0.1346	0.2293	0.2436	0.0468
LD	0.0431	0.3311	0.3323	0.0505	**0.0000**	0.1252	0.3263	0.2299	0.0311	0.1400	0.2241	0.0581
MV	0.0677	0.3403	0.2535	0.1097	0.1238	**0.0000**	0.1654	0.1442	0.1987	0.2749	0.0604	0.0279
MR	0.3028	0.3540	0.1717	0.3534	0.3188	0.1565	**0.0002**	0.2830	0.3287	0.2157	0.0851	0.2444
RD	0.1827	0.3698	0.3108	0.1882	0.2293	0.1438	0.2889	**0.0000**	0.3156	0.3833	0.2141	0.1259
TJ	0.1282	0.3729	0.3744	0.1337	0.0304	0.1991	0.3355	0.3158	**0.0000**	0.1222	0.2610	0.1409
TG	0.2152	0.2424	0.1731	0.2263	0.1350	0.2718	0.2162	0.3832	0.1169	**0.0001**	0.3202	0.2220
TT	0.1888	0.4217	0.2989	0.2394	0.2211	0.0558	0.0902	0.2115	0.2593	0.3229	**0.0001**	0.1369
TR	0.0200	0.3084	0.2662	0.0455	0.0584	0.0296	0.2531	0.1262	0.1418	0.2262	0.1416	**0.0000**

Table 5.9 Authentication performance of dissimilarity based unknown tea brand classifier with SNR = 100 (platinum working electrode)

	BR	DD	GD	LP	LD	MV	MR	RD	TJ	TG	TT	TR
BR	**0.0000**	0.4192	0.4007	0.0962	0.0964	0.2585	0.5831	0.1015	0.1598	0.0941	0.2422	0.1420
DD	0.4192	**0.0000**	0.1342	0.3065	0.2984	0.3384	0.4248	0.3217	0.2853	0.3227	0.2997	0.3156
GD	0.4007	0.1342	**0.0000**	0.2649	0.2387	0.2062	0.3211	0.3055	0.1913	0.3076	0.1696	0.2714
LP	0.0962	0.3065	0.2649	**0.0000**	0.0513	0.1120	0.5492	0.0696	0.1650	0.0584	0.0871	0.0434
LD	0.0964	0.2984	0.2387	0.0513	**0.0000**	0.1252	0.5166	0.0980	0.0795	0.0940	0.0944	0.1020
MV	0.2585	0.3384	0.2062	0.1120	0.1252	**0.0000**	0.4772	0.1920	0.2282	0.1913	0.0108	0.1112
MR	0.5831	0.4248	0.3211	0.5492	0.5166	0.4772	**0.0000**	0.6064	0.4090	0.6048	0.4661	0.5850
RD	0.1015	0.3217	0.3055	0.0696	0.0980	0.1920	0.6064	**0.0000**	0.2390	0.0011	0.1770	0.0286
TJ	0.1598	0.2853	0.1913	0.1650	0.0795	0.2282	0.4090	0.2390	**0.0000**	0.2319	0.1780	0.2606
TG	0.0941	0.3227	0.3076	0.0584	0.0940	0.1913	0.6048	0.0011	0.2319	**0.0000**	0.1749	0.0251
TT	0.2422	0.2997	0.1696	0.0871	0.0944	0.0108	0.4661	0.1770	0.1780	0.1749	**0.0000**	0.1065
TR	0.1420	0.3156	0.2714	0.0434	0.1020	0.1112	0.5850	0.0286	0.2606	0.0251	0.1065	**0.0000**

Table 5.10 Authentication performance of dissimilarity-based unknown tea brand classifier with SNR = 70 (platinum working electrode)

	BR	DD	GD	LP	LD	MV	MR	RD	TJ	TG	TT	TR
BR	**0.0000**	0.4192	0.4007	0.0962	0.0964	0.2585	0.5831	0.1015	0.1598	0.0941	0.2422	0.1420
DD	0.4192	**0.0000**	0.1341	0.3065	0.2984	0.3383	0.4247	0.3217	0.2852	0.3227	0.2997	0.3156
GD	0.4007	0.1342	**0.0000**	0.2649	0.2387	0.2062	0.3211	0.3055	0.1913	0.3076	0.1695	0.2714
LP	0.0962	0.3065	0.2649	**0.0000**	0.0513	0.1120	0.5492	0.0696	0.1650	0.0584	0.0871	0.0434
LD	0.0964	0.2984	0.2387	0.0513	**0.0000**	0.1252	0.5166	0.0980	0.0795	0.0940	0.0944	0.1020
MV	0.2585	0.3384	0.2062	0.1120	0.1252	**0.0000**	0.4772	0.1920	0.2282	0.1913	0.0108	0.1112
MR	0.5831	0.4248	0.3211	0.5492	0.5166	0.4772	**0.0000**	0.6064	0.4090	0.6048	0.4661	0.5850
RD	0.1014	0.3217	0.3055	0.0696	0.0980	0.1920	0.6064	**0.0000**	0.2390	0.0011	0.1770	0.0286
TJ	0.1598	0.2853	0.1913	0.1650	0.0795	0.2282	0.4090	0.2390	**0.0000**	0.2319	0.1780	0.2606
TG	0.0941	0.3227	0.3076	0.0584	0.0940	0.1913	0.6048	0.0011	0.2319	**0.0000**	0.1749	0.0251
TT	0.2422	0.2997	0.1696	0.0871	0.0944	0.0108	0.4661	0.1770	0.1780	0.1749	**0.0000**	0.1065
TR	0.1420	0.3156	0.2714	0.0434	0.1020	0.1111	0.5850	0.0286	0.2606	0.0251	0.1065	**0.0000**

Table 5.11 Authentication performance of dissimilarity-based unknown tea brand classifier with SNR = 50 (platinum working electrode)

	BR	DD	GD	LP	LD	MV	MR	RD	TJ	TG	TT	TR
BR	**0.0000**	0.4193	0.4008	0.0962	0.0965	0.2586	0.5831	0.1015	0.1598	0.0942	0.2423	0.1420
DD	0.4186	**0.0000**	0.1321	0.3060	0.2976	0.3377	0.4227	0.3214	0.2837	0.3224	0.2989	0.3153
GD	0.4005	0.1345	**0.0000**	0.2646	0.2384	0.2058	0.3210	0.3053	0.1910	0.3074	0.1691	0.2712
LP	0.0962	0.3065	0.2649	**0.0000**	0.0514	0.1120	0.5491	0.0697	0.1649	0.0585	0.0871	0.0435
LD	0.0964	0.2984	0.2387	0.0513	**0.0000**	0.1252	0.5166	0.0980	0.0795	0.0939	0.0943	0.1019
MV	0.2583	0.3382	0.2060	0.1118	0.1251	**0.0000**	0.4772	0.1918	0.2280	0.1911	0.0107	0.1110
MR	0.5824	0.4250	0.3203	0.5482	0.5154	0.4752	**0.0004**	0.6055	0.4079	0.6040	0.4644	0.5839
RD	0.1014	0.3217	0.3055	0.0696	0.0980	0.1920	0.6064	**0.0000**	0.2390	0.0011	0.1771	0.0286
TJ	0.1595	0.2854	0.1916	0.1646	0.0792	0.2279	0.4093	0.2387	**0.0000**	0.2316	0.1777	0.2603
TG	0.0942	0.3228	0.3077	0.0584	0.0941	0.1914	0.6048	0.0011	0.2319	**0.0000**	0.1750	0.0251
TT	0.2420	0.2996	0.1696	0.0869	0.0941	0.0108	0.4662	0.1768	0.1778	0.1746	**0.0000**	0.1063
TR	0.1419	0.3156	0.2714	0.0433	0.1019	0.1112	0.5851	0.0286	0.2605	0.0250	0.1065	**0.0000**

Table 5.12 Authentication performance of dissimilarity-based unknown tea brand classifier with SNR = 40 (platinum working electrode)

	BR	DD	GD	LP	LD	MV	MR	RD	TJ	TG	TT	TR
BR	**0.0000**	0.4195	0.4013	0.0963	0.0971	0.2592	0.5835	0.1012	0.1605	0.0939	0.2430	0.1418
DD	0.4146	**0.0020**	0.1149	0.3015	0.2908	0.3308	0.4057	0.3190	0.2706	0.3199	0.2910	0.3125
GD	0.3980	0.1375	**0.0001**	0.2617	0.2350	0.2025	0.3207	0.3036	0.1871	0.3057	0.1655	0.2693
LP	0.0957	0.3067	0.2656	**0.0000**	0.0516	0.1127	0.5498	0.0690	0.1655	0.0577	0.0879	0.0430
LD	0.0959	0.2986	0.2391	0.0507	**0.0000**	0.1251	0.5168	0.0977	0.0795	0.0936	0.0942	0.1016
MV	0.2576	0.3378	0.2059	0.1109	0.1246	**0.0000**	0.4776	0.1910	0.2279	0.1902	0.0105	0.1103
MR	0.5785	0.4225	0.3123	0.5419	0.5076	0.4600	**#0.0244**	0.5989	0.4015	0.5977	0.4521	0.5754
RD	0.1007	0.3219	0.3059	0.0695	0.0978	0.1927	0.6064	**0.0000**	0.2384	0.0011	0.1776	0.0290
TJ	0.1568	0.2873	0.1938	0.1617	0.0768	0.2254	0.4120	0.2365	**0.0000**	0.2293	0.1754	0.2577
TG	0.0938	0.3229	0.3080	0.0584	0.0941	0.1919	0.6049	0.0011	0.2317	**0.0000**	0.1754	0.0254
TT	0.2409	0.2994	0.1701	0.0858	0.0933	0.0109	0.4671	0.1757	0.1776	0.1734	**0.0000**	0.1053
TR	0.1416	0.3155	0.2714	0.0432	0.1016	0.1113	0.5851	0.0284	0.2601	0.0249	0.1066	**0.0000**

Brookbond (BR), Double Diamond (DD), Godrej (GD), Lipton (LP), Lipton-Darjeeling (LD), Marvel (MV), Maryada (MR), Redlabel (RD), Tajmahal (TJ), Tatagold (TG), Tetly (TT) and Trishul (TR)

 # Misclassification = (1/144) × 100 = 0.69%

5.4.2 *Performance evaluation of the designed dissimilarity-based tea classifier*

Variability was allowed in the range of SNR of 100, 70, 50, and 40 (noise content of 1–2.5%) in the training/template tea samples for simulating any unknown tea brand. For all of the working electrodes, this revealed that there was no misclassification for the simulated tea sample until an SNR of 50 (noise content of 2%). For an SNR of 40 (equivalent noise content of 2.5%) and with all the working electrodes, the resultant dissimilarity values of the order of 10^{-2} were treated as misclassifications (five such cases). A misclassification of 1.38% for using silver and 0.69% for using platinum resulted in the developed classifiers. For the gold electrode system, there was no misclassification. A higher level of noise infiltration enhances the misclassification possibility. The proposed moving window-based pattern-matching efficiency matrices including pool accuracy, pattern-matching efficiency, and overall effectiveness were also calculated, and they were all 100% for all the 12 tea brands at SNRs of 100, 70, and 50 and a maximum of 98.6% at an SNR of 40. This is a sheer revelation of the robustness of the developed tea assessment/classification system, which could function with 100% accuracy up to a variability of 2.0% in unknown tea brands (with respect to their parental brands). The tea samples apart from those 12 brands, which have been considered for developing the classifier, were treated as outcasts. The requisite MATLAB® code for calculating dissimilarity (KL expansion based) was already provided in Chapter 2.

5.5 Design of RPCA-based tea classifier

E-tongue signatures due to various branded tea samples (Brookebond (BR), Double Diamond (DD), Godrej (GD), Lipton (LP), Lipton-Darjeeling (LD), and Marvel (MV)) have been considered as batches of time series data (*batch number* × *variables* × *time*; $I \times J \times K$) with nonstationarity. This can be unfolded as variable-wise ($IK \times J$) or batch-wise ($I \times JK$). Variable-wise (unknown/known tea brands) unfolding has been considered in this book. To adapt to the nonstationarity of the tea data, an efficient recursive PCA (RPCA) is chosen as the machine-learning algorithm in the classifier development. It is not the PCA score but the two statistical matrices that describe the fitness of the PCA model, namely Q threshold, Q_{upper}, and T^2 threshold, T^2_{upper}, which are calculated for every tea brand and updated along a moving window to provide a basis for authentication/classification. The updating of the thresholds has been done using RPCA (model updates on an ever-increasing data set that includes new samples without discarding old ones) with the efficient Lanczos procedure. The recursive PCA model, hence, the recursive update of Q statistics and Q_{upper}, and the monitoring/identifying index using the Lanczos procedure, were already discussed in Chapter 2.

5.5.1 Authentication/classification algorithm of unknown tea brands

The increment in control limit of T^2 statistics (while an unknown sample does not resemble any of the reference/template brands) has not been considered an attribute to the authentication of an unknown sample. Instead of their revelation of appreciable T^2 statistics, the tea brands could not be identified with that attribute because of inadequate distinction among them with respect to T^2 statistics. The low to moderate increment in Q statistics (Q_{upper}) occurs when an unknown sample does not resemble any of the reference/template brands used for classifier development. Q statistics-based (for all three electrodes—gold, silver, and platinum) authentication performances of unknown tea brands when compared with the Q statistics of the six training/reference tea brands are expressed as the difference in their Q_{upper} values and are tabulated. For a specific test tea brand, the minimum entry values (preferably within a range of 0–10^{-3}) among the generated ($Q_{upper,\ diff}$) indicate the resemblance of it to any one of the six template tea brands. Every tea brand was simulated with an SNR of 100, 70, and 50 resembling the unknown test brand with certain degrees of variability. In order to authenticate six test tea brands of six unknown categories, six training data sets are created, and they undergo pattern matching with the six categories of test data sets three times (simulated test data sets possess an SNR of 100, 70, and 50). Each time, the training data set composition remains unchanged, but the composition of the test data for each table keeps on changing with respect to the SNR of the unknown brands of tea. The pattern matching results are tabulated and each of them is of dimension six by six (6×6). The proposed algorithm is as follows:

For e = 1 to 3 (1 = silver; 2 = gold; 3 = platinum)

Step 1: Creation of simulated unknown tea brands in triplicate (each containing three samples) resembling the tea brands considered for the classifier development. The SNRs considered are 100, 70, and 50.

Step 2: Generation of Q statistics for six training tea brands (i = 1 to 6)

i) Considering i = 1. Initially 200 samples are taken (X_k). R_k , b_k, $Q(k)$ (Q value at k instants) and Q_{upper} are calculated and stored.
ii) The data matrix is then augmented sample-wise (X_{k+1}) not ignoring the previous data. b_{k+1} and R_{k+1} are calculated recursively.
iii) Application of Lanczos procedure on symmetric correlation matrix R_{k+1} to yield a symmetric tridiagonal matrix (Γ) having fewer principal components.
iv) Eigenvector decomposition of (Γ) and projection of (Γ) on eigenvector to give rise to the loading matrix (P).
v) Calculation of new score and error, hence $Q(k)$ (that is Q value at k instants) and Q_{upper} using the augmented data matrix X_{k+1} and loading matrix (P) as obtained in step iv.
vi) Repetition of steps ii to v for the specific i^{th} training brand of tea until all (4402 × 3) observations are complete. The maximum $Q(k)$ or Q_{upper} for that brand is determined.
vii) Procedures i to vi are repeated six times for six tea brands. This completes generation of Q statistics for training data set.

Step 3: Generation of Q statistics for six tested tea brands (j = 1 to 6).

i) Considering j = 1 as current snapshot/test data. Initially 200 samples are taken (X_k). R_k, b_k, $Q(k)$ (Q value at k instants) and Q_{upper} are calculated and stored.
ii) The data matrix is then augmented sample-wise (X_{k+1}) not ignoring the previous data. b_{k+1} and R_{k+1} are calculated recursively.
iii) Application of the Lanczos procedure on symmetric correlation matrix R_{k+1} to yield a symmetric tridiagonal matrix (Γ) having fewer principal components.
iv) Eigenvector decomposition of (Γ) and projection of (Γ) on eigenvectors to give rise to the loading matrix (P).
v) Calculation of new score and error, hence $Q(k)$ (that is Q value at k instant) and Q_{upper} using the data matrix X_{k+1} and loading matrix (P) as obtained in step iv.
vi) Repetition of steps ii to v for the specific i^{th} tested brand of tea until all (4402 × 3) observations are complete for that brand. The maximum $Q(k)$ or Q_{upper} for that test brand is determined.
vii) Procedures i to vi are repeated six times for six tested tea brands. Therefore, completion of generation of the six testing data sets with their Q statistics is completed.

viii) It may happen (for online monitoring) that the minimum difference Q_{upper} between the template and test tea brand is detected without repeating procedures i to vi six times as stated in step vii. In that case the algorithm gets terminated.

Step 4: A comparison in Q_{upper} of each category of test data against all six training brands and six absolute differences in Q_{upper} values is generated.
Step 5: After step 4 is repeated three times (SNR 100, 70, and 50), three sets of (6 × 6) matrix (W) are generated, along the row of these three sets are the training tea data sets and along the columns, the testing tea data sets.
Step 6: Unknown sample authentication: Along each column of the generated W matrices, there remains a minimum entry value. The minimum difference Q_{upper} values will authenticate a specific brand.

End (e)

The general code developed using 12 tea brand samples (in triplicate) for RPCA classifier follows.

All tea data sets for testing and on the program F_distribution.m are available on the first author's personal webpage: http://sites.google.com/site/madhusreekundu/

Program 5.1 MATLAB code for recursive PCA classifier

%% The MATLAB® code given below is for developing and testing a recursive PCA classifier for the first tea brands.

```
clc
clear all

format long

snr=input('Enter the signal to noise ratio for the test data');

DD=zeros(6,6);

for ii=1:6

Type=ii;

if Type==1
 fname='brookbond' ;
elseif Type==2
 fname='double-diamond' ;
elseif Type==3
  fname='goodrej' ;
elseif Type==4
 fname='lipton' ;
elseif Type==5
 fname='lipton-darjeeling' ;
elseif Type==6
```

```
 fname='marvel' ;
elseif Type==7
    fname='maryada';
elseif Type==8
    fname='redlabel';
elseif Type==9
    fname='tajmahal';
elseif Type==10
    fname='tatagold';
elseif Type==11
    fname='tetley';
elseif Type==12
    fname='trishul';
end

X1=testing_RPCA(Type,snr);
X1=X1(:,2);

D=[];

for i=1:6

Type1=i;

   if i==1
 fname2='brookbond'
   end

if i==2
 fname2='double-diamond'
end

if i==3
  fname2='goodrej'
end

if i==4
 fname2='lipton'
end

if i==5
 fname2='lipton-darjeeling'
end

if i==6
 fname2='marvel'
end

if i==7
 fname2='maryada';
end

if i==8
 fname2='redlabel';
```

```
end

if i==9
 fname2='tajmahal';
end

if i==10
 fname2='tatagold';
end

if i==11
 fname2='tetley';
end

if i==12
 fname2='trishul';
end

X2=train_RPCA(Type1);
X2=X2(:,2);
Q1=max(X1);
Q2=max(X2);

n=size(X1);
fname4=strcat(fname,'_',fname2,'_compare.fig')
fname5=strcat(fname,'_',fname2);

T=0:1:n(1,1)-1 ;
T=T' ;
hgsave(fname4)

d=abs(Q2-Q1);
DD(ii,i)=d;

[Q1 Q2 abs(Q2-Q1)];

D=[D; Q1 Q2 d];

end

disp( 'Test  Trained  Difference')

D

end

dlmwrite('DD.txt', DD, 'delimiter', '\t', ...
         'precision', '%0.5f');

%%%%%%%%%%%%%%%%%%%%%%%%%%%%%%%%%%%%%%%%%%%%%%%%%%%%%%%%%%%%%%%%%%%%%%%%%%%%%%%
%%%%%%%%%%%%  Implementation of recursive PCA on test data %%%%%%%%%%%%

function Q=testing_RPCA(Type,snr)

% The input to this program is tea category type (1-12)
```

```
% The outputs of this program are Squared predicted error 'Qk ,
Hotelling's
% T2 and upper limit of Qk

[X,fname]=test_data(Type,snr);

t=200 ;
X=ascale(X);
Xnew=X(t+1:end,:);                          % OLD DATA IS NOT IGNORED
size(Xnew)

t2=zeros(length(Xnew),1);                                % SAMPLE WISE
UPDATION
qk=zeros(length(Xnew),1);
Qupr=zeros(length(Xnew),1);
T2upr=zeros(length(Xnew),1);

for i=1:1:length(Xnew)

    if i==1              % initial data block
        xk=X(1:t,:);
        bk=(xk'*ones(length(xk),1))/length(xk);
        sdt=std(xk);
        xks=(xk-ones(length(xk),1)*bk')/(diag(sdt));
        Rk=(xks'*xks)/(length(xk)-1);
        Rk1=Rk;
    elseif i>1                 % for k+1 data blocks
        xk=X(1:t+(i-2),:);           % previous data block
        xk1=X(1:t+i,:);               % increased data matrix
        k=i;
        [bk1,Rk1]=Recursive(xk,xnk1,xk1,bk,Rk,k,sdt);
        bk=bk1;
        Rk=Rk1;
    end

     if i==1
        X1=xk;
     else
        X1=xk1;
     end

    [PI,PL,D,T0]=LANCZOS(Rk1,X1); %P=PL;S=D;
    [P,S,T]=CPV(Rk1,X1,PL,D);
    xnew=Xnew(i,:);                % collecting sample for monitoring
    [Qk]=qstatic(xnew,PL);          % Squared Prediction Error
    qk(i)=Qk;
    [Qupper,T1]=NOC(X1,P,S,Rk1,PI);
    Qupr(i)=Qupper;
    [T2,t2upr]=tstatic(xnew,P,S,T,X1);       % Hotelling's T2
    t2(i)=T2;
    T2upr(i)=t2upr;
    xnk1=xnew;                      % data augmentation
```

```
end

Q=[qk,Qupr , t2, T2upr];

end
%%%%%%%%%%%%%%%%%%%%%%%%%%%%%%%%%%%%%%%%%%%%%%%%%%%%%%%%%%%%%%%%%%%%%%%%%%%%%%%%
%%%%%%%%%%%%  MATLAB code for generating test data   %%%%%%%%%%%%%%%%%%%%%

function [Y,fname]=test_data(Type,snr)
% This program generates the test data set

format long ;

if Type==1
 fname='brookbond' ;
elseif Type==2
 fname='double-diamond' ;
elseif Type==3
  fname='goodrej' ;
elseif Type==4
 fname='lipton' ;
elseif Type==5
 fname='lipton-darjeeling' ;
elseif Type==6
 fname='marvel' ;
elseif Type==7
    fname='maryada';
elseif Type==8
    fname='redlabel';
elseif Type==9
    fname='tajmahal';
elseif Type==10
    fname='tatagold';
elseif Type==11
    fname='tetley';
elseif Type==12
    fname='trishul';
end

fname1=strcat(fname,'_3s_N.txt') ;
fname2=strcat(fname,'_test.txt') ;

X=load(fname1,'-ascii');

y1=awgn(X(:,1),snr,'measured');
y2=awgn(X(:,2),snr,'measured');
y3=awgn(X(:,3),snr,'measured');

Y=[y1 y2 y3];

end

%%%%%%%%%%%%%%%%%%%%%%%%%%%%%%%%%%%%%%%%%%%%%%%%%%%%%%%%%%%%%%%%%%%%%%%%%%%%%%%%
```

```
%%%%%%%%%%% MATLAB code for updating of mean vector and correlation
matrix  %%%%%%%%%%%

function [bk1,Rk1]=Recursive(xk,xnk1,xk1,bk,Rk,k,sdt)

bk1=((length(xk))/length(xk1))*bk+(1/length(xk1))*xnk1'*ones(size(xnk1,1)
,1); % mean update
ssq=zeros(size(xk,2),1);                % xk-previous data block (raw
data)
sv=zeros(size(xk,2),1);

for i=1:1:size(xk,2)                    % xnk1-current data block (raw
data)
    ssq(i)=(norm(xk(:,i)-ones(length(xk),1)*bk(i))^2)/(length(xk)-1);
    sv(i)=sqrt(ssq(i));
end

deltabk1=bk1-bk;
ssq1=zeros(size(xk,2),1);
sv1=zeros(size(xk,2),1);

for j=1:1:size(xk,2)                    % variance update
    ssq1(j)=(length(xk)-1)*ssq(j)+(length(xk)*(deltabk1(j))^2)+norm(xnk
1(j)-ones(size(xnk1,1),1)*(bk1(j)))^2;
    sv1(j)=sqrt(ssq1(j));
end

Xk1=[xk1-ones(length(xk1),1)*bk1']*inv(diag(sdt));       % auto scaled
matrix
e1=diag(sv);
e2=inv(diag(sv1));
rk1=((k-1)*(e2)*e1*Rk*(e2)*e1)/k;                         % updated
correlation matrix
rk2=(e2)*(deltabk1)*deltabk1'*(e2);
rk3=(xnk1'*xnk1)/k;
Rk1=rk1+rk2+rk3;

end
%%%%%%%%%%%%%%%%%%%%%%%%%%%%%%%%%%%%%%%%%%%%%%%%%%%%%%%%%%%%%%%%%%%%%%%%%%%%%
%%%%%%%%%%%%%%%%%  MATLAB code for Lanczos algorithm  %%%%%%%%%%%%%

function [PI,PL,Cd,T0]=LANCZOS(RK,X1)             % RK is correlation
matrix of one batch

r0=(ones(size((RK),1),1))/sqrt(length(RK));
beta0=norm(r0);
beta=zeros(length(RK),1);
r=zeros(length(RK),length(RK));
alpha=zeros(length(RK),1);
% PI=zeros(length(RK),length(RK));
q=zeros(length(RK),length(RK));          % for storage of Lanczos vectors
% U=zeros(length(RK),1);
% --------------- Initial data for Lanczos Unsymmetric method -------------
```

```
p1=(ones(size((RK),1),1))/sqrt(length(RK));
P=zeros(length(RK),length(RK));
Q1=(ones(size((RK),1),1))/sqrt(length(RK));
Q=zeros(length(RK),length(RK));
S=zeros(length(RK),length(RK));
p0=0;
Q0=0;
t=0;
gama=zeros(length(RK),1);
%------------------------------------------------------------------------
if (RK==RK')
    disp('----------------------------')
    disp(':correlation matrix is symmetric')
    disp('----------------------------')
    c=1;
else
    disp('----------------------------')
    disp(':correlation matrix is unsymmetric ')
    disp('----------------------------')
    c=0;
end
b1=1;
q0=0;
% PI(1,:)=[];
% PI(:,1)=[];
% while(tracepi/traceRK<=e)
if c==1
    disp('----------------------------')
    disp(':Symmetric Lanczos Algorithm')
    disp('----------------------------')
    % ----------------Symmetric Lanczos Algorithm -------------------
    for k=1:1:length(RK)
        if k==1  % calculation of symmetric matrix begin
            q(:,k)=r0/beta0;
        else
            q(:,k)=r(:,k-1)/beta(k-1);
        end
            U=RK*q(:,k);
            if k==1
                r(:,k)=U-q0*beta0;
            else
                r(:,k)=U-q(:,k-1)*beta(k-1);
            end
            alpha(k)=q(:,k)'*r(:,k);
            r(:,k)=r(:,k)-alpha(k)*q(:,k);
            sume=0;
    for k1=k:1:1  % reorthogonalization
        sum1=sume+q(:,k1)*(q(:,k1)'*r(:,k));
        sume=sum1;
    end
    r(:,k)=r(:,k)-sum1;
    beta(k)=normest(r(:,k));                               % 2-norm (verify)
%     k=k+1;
    end
```

```
% ---------------- End of calculations of symmetric matrix -----------
% ----------------- Unsymmetric Algorithm -----------------------
elseif c==0
    disp('----------------------------')
    disp(':Unsymmetric Lanczos Algorithm')
    disp('----------------------------')
    s0=p1;r0=Q1;
    while (t<=length(RK))
        if t==0
            beta0=normest(r0);
            gama0=(s0'*r0)/beta0;
            Q(:,t+1)=r0/beta0;
            P(:,t+1)=s0/gama0;
        elseif t>0
            beta(t)=normest(r(t));
            gama(t)=(S(:,t)'*r(:,t))/beta(t);
            Q(:,t+1)=r(:,t)/beta(t);
            P(:,t+1)=S(:,t)/gama(t);
        end
        t=t+1;
        alpha(t)=P(:,t)'*RK*Q(:,t);
        if t==1
        r(:,t)=(RK-alpha(t)*eye(length(RK)))*Q(:,t)-gama0*Q0;
        S(:,t)=(RK-alpha(t)*eye(length(RK)))'*P(:,t)-beta0*p0;
        elseif t>1

r(:,t)=(RK-alpha(t)*eye(length(RK)))*Q(:,t)-gama(t-1)*Q(:,t-1);

S(:,t)=(RK-alpha(t)*eye(length(RK)))'*P(:,t)-beta(t-1)*P(:,t-1);
        end
        r0=r(:,t);
        s0=S(:,t);
        if sum(r0)==0 || sum(s0)==0 || sum(r0)==0 && sum(s0)==0
            disp('--------------------------------')
            disp('1:either r0 or so or both is zero')
            disp(':breakdown occured')
            disp('--------------------------------')
            c=2;
            break;
        elseif sum(r0)~=0 && sum(s0)~=0 && (s0'*r0==0)
            disp('------------------------------------')
            disp(':breakdown occured')
            c=2;
            disp(':there is no information about eigenvalues of
tridiagonal matrix')
            disp(':use Lanczos look ahead algorithm to cure breakdown')
            disp('-----------------------------------------------------')
            break;
        end
    end
    if s0'*r0==0 || sum(r0)==0 || sum(s0)==0
%          cl=2;
        disp('----------------------------')
        disp(':A Look ahead Lanczos Algorithm')
```

```
        disp('---- LAL --------------------')
         % Initialization
%         Pl=zeros(length(RK),length(RK));
%         Ql=zeros(length(RK),length(RK));
%         Beta=zeros(length(RK),1);
        Gama=zeros(length(RK),1);
        B=zeros(length(RK),1);
        Alpha=zeros(length(RK),1);
%         u=1;
        z=zeros(length(RK),1);
        z(1)=1;
%         L=1;
        L=1+size((p0),1);
        R=zeros(length(RK),length(RK)+1);
        Sl=zeros(length(RK),length(RK)+1)';
        R(:,L)=(ones(length(RK),1)); %/sqrt(length(RK));
        Sl(L,:)=(ones(length(RK),1))'; %/sqrt(length(RK));
        ql=zeros(length(RK),length(RK));
        pl=zeros(length(RK),length(RK));
        Rn=zeros(length(RK),1);
        Sn=zeros(length(RK),1);
        w=Sl(L,:)*R(:,L);
        tol=0.0001;
% ------------------------ Look Ahead Lanczos Algorithm ----------------
        for I=1:1:length(RK)
                     % s': variable s is not designated as s' but the
values of s are transposed %
if I==1
            % This method is maintained throughout LAL algorithm%
R(:,L+1)=RK*R(:,L)-q0*z(I)*w;
            Sl(L+1,:)=(Sl(L,:))*RK-w*p0';             % without transpose
            else                                          %this
consideration also holds for P%
                R(:,L+1)=RK*R(:,L)-ql(I-1)*z(I)*w;
                Sl(L+1,:)=Sl(L,:)*RK-w*pl(I-1);
            end
            theta=Sl(L,:)*R(:,L+1);                   %needed inner
products%
            wb=Sl(L+1,:)*R(:,L+1);
%             Rn(L)=norm()
            if L==2
            pi1=w/((norm(R(:,L)))*norm(Sl(L,:)));
            else
                pi1=w/(Rn(I-1)*Sn(I-1));
            end
            pi2=0;
            if theta==0
                if abs(pi1)<tol && pi2<tol      % test for failure
                    disp('---------------------------')
                    disp(':both angles are less than tol')
                    disp(':algorithm will terminate without having any
information about Lanczos vectors')
disp('------------------------------------------------------------
-----------')
```

```
                    break;
                end
            else
                tau1=w/theta;
                tau2=wb/theta;
                if L==2
                rnb1=sqrt(((norm(R(:,L+1)))^2)-(2*tau2*(R(:,L)'*R(:,L+1))
)+(tau2^2*norm(R(:,L))^2));
                snb=sqrt((norm(Sl(L,:)')^2)-2*tau1*(Sl(L,:)*Sl(L+1,:)')+(
tau1^2)*norm(Sl(L+1,:)')^2);
                psi1=theta/(norm(R(:,L))*norm(Sl(L+1,:)));

                else
                    rnb1=sqrt(((norm(R(:,L+1)))^2)-(2*tau2*(R(:,L)'*R(:,L
+1)))+(tau2^2*Rn(I-1)^2));
                    snb=sqrt((Sn(I-1)^2)-2*tau1*(Sl(L,:)*Sl(L+1,:)')+(tau
1^2)*norm(Sl(L+1,:)')^2);
                    psi1=theta/(Rn(I-1)*norm(Sl(L+1,:)));
                end
                psi2=(w*tau2-theta)/(rnb1*snb);
                pi2=min(abs(psi1),abs(psi2));
            end
            biasf=0;          % single step is assumed
            if (abs(pi1))>=((biasf)*pi2);

disp('--------------------------------------------------------------')
                disp(':LAL reduces to standard Lanczos algorithm from
this point onwards')
                disp('------- single step
------------------------------------------')
                if L==2
                beta1=norm(R(:,L))*sqrt(pi1);
                else
                    beta1=Rn(I-1)*sqrt(pi1);
                end
                G1=w/beta1;
                ql(:,I)=R(:,L)/beta1;
                pl(I,:)=(Sl(L,:))/G1;
                Gama(I)=z(I)*G1;
                B(I)=beta1;
                R(:,L+1)=R(:,L+1)/beta1;          % form new residuals
                Sl(L+1,:)=Sl(L+1,:)'/G1;
                Alpha(I)=1/tau1;
                R(:,L+1)=R(:,L+1)-ql(:,I)*Alpha(I);     %
biorthogonalization%
                Sl(L+1,:)=Sl(L+1,:)-Alpha(I)*pl(I,:);
                if L==2
                Rn(I)=sqrt((norm(R(:,L+1))^2)-(2*Alpha(I)*R(:,L)'*R(:,L+1
))+(Alpha(I)^2)*norm(R(:,L))^2)/beta1;
                Sn(I)=sqrt(norm(Sl(L+1,:)')^2-2*Alpha(I)*Sl(L,:)*Sl(L+1,:
)'+(Alpha(I)^2)*norm(Sl(L,:)')^2)/G1;
                else
```

```
                    Rn(I)=sqrt((norm(R(:,L+1))^2)-(2*Alpha(I)*R(:,L)'*R(:
,L+1))+(Alpha(I)^2)*Rn(I-1)^2)/beta1;
                    Sn(I)=sqrt(norm(Sl(L+1,:)')^2-2*Alpha(I)*Sl(L,:)*Sl(L
+1,:)'+(Alpha(I)^2)*Sn(I-1)^2)/G1;
                end
                w=(wb/w)-Alpha(I)^2;
                z(I+1)=[1];
            end                    % end of calculations for single step
            L=L+1;
        end
    end
% ----------------- Unsymmetric Lanczos Algorithm -----------------------
end
e=0.995;
tracePI=0;
traceRK=1;
f=1;
% PI=zeros(f,f);
while ((tracePI/traceRK)<=e)          % termination, i.e.if eigenvalues
of PI
    PI=zeros(f,f);
    for w=1:1:f                      % are approximately equal to eig(RK).
        for l=1:1:f
            if w==l
                if c==2
                    PI(w,l)=Alpha(w);
                else
                PI(w,l)=alpha(w);
                end
            elseif w==(l+1) && l==(w-1)
                if c==2
                    PI(w,l)=B(l);
                else
                    PI(w,l)=beta(l);
                end
            elseif l==w+1 && w==l-1
                if c==1
                PI(w,l)=beta(w);
                elseif c==0
                    PI(w,l)=gama(w);
                elseif c==2
                    PI(w,l)=Gama(w);
                end
            end
        end
    end
    [vrk,drk]=eig(RK);
    [vpi,dpi]=eig(PI);
    tracePI=sum(diag(dpi));                      % sum of eigenvalues
    traceRK=sum(diag(drk));
    f=f+1;
    if f==(length(RK)+1)
        disp('----------------------------')
        disp(':PI is not approximation of RK')
```

```
        disp(':no.of iterations are equal to dimensions of RK')
        disp('-------------------------------------------------')
        break;
    end

end
% if f<=length(PI)
% PI(:,f:end)=[];
% PI(f:end,:)=[];
% end
if f<length(RK)
    disp('----------------------------')
    disp(':dimension of PI matrix is smaller than dim.of RK ')
    disp('----------------------------')
    fprintf('%10d\n',f);
elseif f==length(RK)
    disp('----------------------------')
    disp(':dim.of PI equal to dim.of RK')
    disp('----------------------------')
end
% Further reduction of size of PI by eliminating very low eigenvalues.
[V,D]=eig(PI);
v1=1;
sum2=0;     % -------- Cumulative Percent Variance ------------------
% no.of largest PCs determined by the use of 95% variance criteria%.
while ((sum2/trace(RK))<=0.99)

        if D(v1,v1)~=0
        sum2=sum2+D(v1,v1);
        end

        if v1==length(D)
            disp('----------------------------')
           disp(':max.no.of dim.is reached')
           disp('----------------------------')
           break;
        end

       v1=v1+1;

%          end
%      end
end
% v1=v1-1;
v11=length(PI);
i=1;
[Cv,Cd]=eigs(PI);
if c==1
    PL=q(:,1:f-1)*Cv;
elseif c==0
    PL=Q(:,1:f-1)*Cv;
elseif c==2
    PL=ql(:,1:f-1)*Cv;
end
```

```
T0=X1*PL;
end
%%%%%%%%%%%%%%%%%%%%%%%%%%%%%%%%%%%%%%%%%%%%%%%%%%%%%%%%%%%%%%%%%%%%%%%%%%
%%%%%%%%%%%%%% Matlab code for cumulative percent variance %%%%%%%%%%%%%%

function [P,S,T]=CPV(Rk1,X1,v,d)

sum=0;
i=1;

while (sum/trace(d))<=0.95
    sum=sum+d(i,i);
i=i+1;
end

if i<=length(d)
v(:,i:end)=[];
d(:,i:end)=[];
d(i:end,:)=[];
end

P=v;
S=d;
T=X1*P;

end

%%%%%%%%%%%%%%%%%%%%%%%%%%%%%%%%%%%%%%%%%%%%%%%%%%%%%%%%%%%%%%%%%%%%%%%%%%
%%%%%%%%%%%%%%% MATLAB code for squared prediction error %%%%%%%%%%%%%%%%

function [Q]=qstatic(xnew,P)
tnew=xnew*P*inv(P'*P);
enew=xnew-tnew*P';
Q=enew*enew';
Q=xnew*(eye(length(P*P'))-P*P')*xnew';

%%%%%%%%%%%%%%%%%%%%%%%%%%%%%%%%%%%%%%%%%%%%%%%%%%%%%%%%%%%%%%%%%%%%%%%%%%
%%%%%  MATLAB code for the upper limit of squared prediction error
%%%%%

function [Qupper,T]=NOC(X,P,S,R,PI)
T=X*P;
E=X-T*P';
V1=(E'*E)/(length(X)-1);
theta1=(trace(V1));
theta2=(trace(V1^2));
theta3=(trace(V1^3));
ho=1-((2*theta1*theta3)/(3*(theta2^2)));
nalpha=1.645;
Qupper=theta1*(((nalpha*sqrt(2*theta2*ho^2))/
theta1)+1+((theta2*ho*(ho-1))/theta1^2))^(1/ho);
end
```

```
%%%%%%%%%%%%%%%%%%%%%%%%%%%%%%%%%%%%%%%%%%%%%%%%%%%%%%%%%%%%%%%%%%%%%%%%%%
%%%%%%%%%%%% MATLAB code for Hotelling's T2 statistic and its upper
limit  %%%%%%%%%%%%%%

function [T2,T2upr]=tstatic(xnew,P,D,T,X1)
tnew=xnew*P*inv(P'*P);
% enew=xnew-tnew*P';
size(tnew)
S=(T'*T)/(1);
size(S)
T2=tnew*inv(S)*tnew';
% T2=xnew*P*inv(D)*P'*xnew';
n=length(X1);
m1=length(D);
[F]=F_distribution(0.05,m1,n-m1);
T2upr=[((n*n-1)*m1)/(n*(n-m1))]*F;

%%%%%%%%%%%%%%%%%%%%%%%%%%%%%%%%%%%%%%%%%%%%%%%%%%%%%%%%%%%%%%%%%%%%%%%%%%
%%%%%%%%%% Implementation of recursive PCA on training data
%%%%%%%%%%%

function Q=train_RPCA(Type)
% The input to this program is tea category type (1-12)
% The outputs of this program are squared predicted error 'Qk,
Hotelling's
% T2 and upper limit of Qk

[X,fname]=training_data(Type);

N=size(X);
T=0:1:N(1,1)-1;
plot(T',X)
grid on

t=200 ;
X=ascale(X);
Xnew=X(t+1:end,:);                        % OLD DATA IS NOT IGNORED
size(Xnew)
t2=zeros(length(Xnew),1);                                        % SAMPLE
WISE UPDATE
qk=zeros(length(Xnew),1);
Qupr=zeros(length(Xnew),1);
T2upr=zeros(length(Xnew),1);

for i=1:1:length(Xnew)

    if i==1            % initial data block
        xk=X(1:t,:);
        bk=(xk'*ones(length(xk),1))/length(xk);
        sdt=std(xk);
        xks=(xk-ones(length(xk),1)*bk')/(diag(sdt));
        Rk=(xks'*xks)/(length(xk)-1);
        Rk1=Rk;
```

```
    elseif i>1                    % for k+1 data blocks
        xk=X(1:t+(i-2),:);            % previous data block
        xk1=X(1:t+i,:);               % increased data matrix
        k=i;
        [bk1,Rk1]=Recursive(xk,xnk1,xk1,bk,Rk,k,sdt);
        bk=bk1;
        Rk=Rk1;
    end

    if i==1
        X1=xk;
    else
        X1=xk1;
    end

    [PI,PL,D,T0]=LANCZOS(Rk1,X1); %P=PL;S=D;
    [P,S,T]=CPV(Rk1,X1,PL,D);
    xnew=Xnew(i,:);              % collecting sample for monitoring
    [Qk]=qstatic(xnew,PL);         % Squared Prediction Error
    qk(i)=abs(Qk);
    [Qupper,T1]=NOC(X1,P,S,Rk1,PI);
    Qupr(i)=Qupper;
    [T2,t2upr]=tstatic(xnew,P,S,T,X1);     % Hotelling's T2
    t2(i)=T2;
    T2upr(i)=t2upr;
    xnk1=xnew;

end
Q=[qk , Qupr , t2, T2upr];
end

%%%%%%%%%%%%%%%%%%%%%%%%%%%%%%%%%%%%%%%%%%%%%%%%%%%%%%%%%%%%%%%%
%%%%%%%%%%%%  MATLAB code for generating training data
%%%%%%%%%%%%%%%

function [X,fname]=training_data(Type)

% This program generates data for training

format long ;

if Type==1
 fname='brookbond' ;
elseif Type==2
 fname='double-diamond' ;
elseif Type==3
  fname='goodrej' ;
elseif Type==4
 fname='lipton' ;
elseif Type==5
 fname='lipton-darjeeling' ;
elseif Type==6
 fname='marvel' ;
elseif Type==7
```

```
    fname='maryada';
elseif Type==8
    fname='redlabel';
elseif Type==9
    fname='tajmahal';
elseif Type==10
    fname='tatagold';
elseif Type==11
    fname='tetley';
elseif Type==12
    fname='trishul';
end

fname1=strcat(fname,'_3s_N.txt') ;
X=load(fname1,'-ascii');

End
==============================================================
```

5.5.2 *Performance evaluation of the designed RPCA-based classifier*

Tables 5.13 through 5.21 reveal Q statistics-based (for all three electrodes—gold, silver, and platinum) authentication performances of unknown tea brands while compared with the Q statistics of the six training/reference tea brands expressed as differences in their (Q_{upper}) values. For a specific test tea brand, the minimum entry values (preferably within a range of $0–10^{-3}$) among the columns in Tables 5.13 through 5.21 designate the unknown tea brands resembling any one of the six tea brands considered here for classifier development.

On keen observation, one can find that along the principal diagonal of any of the 12 tables there remains a minimum entry value (compared to the other entries along the row) or a value within a range of $0–10^{-2}$. At that minimum entry value, one can find a match. For the silver working-electrode system and test data of SNR 50, a misclassification rate of 2.7% has been found (Table 5.15). A misclassification rate of 5.5% has been found for the platinum working-electrode system and test data of SNR 50 (Table 5.21). The RPCA classifiers, gold, silver, or platinum electrodes, are able to detect and classify the unknown tea samples pertaining to specific brands with 96% accuracy (misclassification% = (13/324) × 100 = 4.0%), when the variability in those unknown samples is up to 2.0% (SNR = 50) with respect to their parent brands. The developed RPCA-based classifier has the potential to supplant the role of tea tasters in assessing tea quality/grade.

Table 5.13 Authentication performance of RPCA-based unknown tea brand classifier with SNR = 100 (silver working electrode)

	BR	DD	GD	LP	LD	MV
BR	**0.00117**	5.49936	5.00917	1.53056	10.81914	1.18549
DD	5.49061	**0.00758**	0.48261	7.02234	5.32736	6.67727
GD	5.00741	0.49078	**0.00059**	6.53914	5.81056	6.19407
LP	1.53181	7.03000	6.53981	**0.00008**	12.34978	0.34515
LD	10.81944	5.32125	5.81144	12.35117	**0.00148**	12.00611
MV	1.18687	6.68506	6.19487	0.34486	12.00484	**0.00021**

Table 5.14 Authentication performance of RPCA-based unknown tea brand classifier with SNR = 70 (silver working electrode)

	BR	DD	GD	LP	LD	MV
BR	**0.01850**	5.47968	4.98949	1.55024	10.79946	1.20517
DD	5.69234	**#0.19415**	0.68434	7.22407	5.12563	6.87900
GD	5.00100	0.49719	**0.00699**	6.53274	5.81696	6.18767
LP	1.53985	7.03804	6.54785	**0.00812**	12.35782	0.35319
LD	10.81420	5.31601	5.80620	12.34593	**0.00377**	12.00086
MV	1.18697	6.68516	6.19497	0.34477	12.00493	**0.00030**

Table 5.15 Authentication performance of RPCA-based unknown tea brand classifier with SNR = 50 (silver working electrode)

	BR	DD	GD	LP	LD	MV
BR	**#0.15662**	5.34157	4.85138	1.68836	10.66134	1.34329
DD	5.14857	**#0.34962**	0.14057	6.68030	5.66940	6.33523
GD	5.00473	0.49346	**0.00327**	6.53646	5.81324	6.19139
LP	1.53806	7.03625	6.54606	**0.00633**	12.35602	0.35139
LD	8.69871	3.20052	3.69071	10.23045	**#0.7827**	9.88538
MV	0.88849	6.38668	5.89649	0.64324	11.70645	**#0.29818**

Table 5.16 Authentication performance of RPCA-based unknown tea brand classifier with SNR = 100 (gold working electrode)

	BR	DD	GD	LP	LD	MV
BR	**0.000225**	2.743935	4.030219	6.527170	2.950282	1.468625
DD	2.744769	**0.000609**	6.774763	9.271714	5.694826	1.275919
GD	4.030397	6.774557	**0.000403**	2.496548	1.080340	5.499247
LP	6.527004	9.271164	2.497010	**0.000059**	3.576947	7.995854
LD	2.922467	5.666627	1.107527	3.604478	**0.027590**	4.391317
MV	1.468751	1.275409	5.498745	7.995696	4.418808	**0.000099**

Table 5.17 Authentication performance of RPCA-based unknown tea brand classifier with SNR = 70 (gold working electrode)

	BR	DD	GD	LP	LD	MV
BR	**0.007256**	2.751416	4.022739	6.519689	2.942801	1.476106
DD	2.847465	**#0.103305**	6.877460	9.374410	5.797522	1.378615
GD	4.033274	6.777434	**0.003279**	2.493671	1.083217	5.502124
LP	6.530122	9.274282	2.500127	**0.003177**	3.580065	7.998972
LD	2.475345	5.219505	1.554650	4.051600	**#0.474712**	3.944195
MV	1.470325	1.273835	5.500320	7.997270	4.420382	**0.001475**

Table 5.18 Authentication performance of RPCA-based unknown tea brand classifier with SNR = 50 (gold working electrode)

	BR	DD	GD	LP	LD	MV
BR	**0.092110**	2.836270	3.937884	6.434835	2.857947	1.560960
DD	3.457800	**#0.713640**	7.487794	9.984745	6.407857	1.988950
GD	4.051121	6.795281	**0.021127**	2.475824	1.101064	5.519971
LP	6.509309	9.253469	2.479315	**0.017636**	3.559252	7.978159
LD	2.561538	5.305698	1.468456	3.965407	**#0.388519**	4.030388
MV	1.418251	1.325909	5.448245	7.945196	4.368308	**0.050599**

Table 5.19 Authentication performance of RPCA-based unknown tea brand classifier with SNR = 100 (platinum working electrode)

	BR	DD	GD	LP	LD	MV
BR	**0.000143**	5.458275	11.649867	2.849186	0.727702	5.146609
DD	5.459757	**0.001625**	6.189967	8.309086	4.732198	0.313291
GD	11.653679	6.195547	**0.003955**	14.503008	10.926120	6.507213
LP	2.849387	8.307519	14.499111	**0.000058**	3.576946	7.995853
LD	0.848843	4.609289	10.800881	3.698172	**#0.121284**	4.297623
MV	5.146627	0.311505	6.503097	7.995956	4.419068	**0.000161**

Table 5.20 Authentication performance of RPCA-based unknown tea brand classifier with SNR = 70 (platinum working electrode)

	BR	DD	GD	LP	LD	MV
BR	**0.000613**	5.457519	11.649111	2.849942	0.726946	5.145853
DD	5.545786	**0.087654**	6.103938	8.395115	4.818227	0.399320
GD	11.661063	6.202931	**0.011339**	14.510392	10.933504	6.514597
LP	2.851848	8.309980	14.501572	**0.002519**	3.579407	7.998314
LD	0.613616	4.844516	11.036108	3.462945	**#0.113943**	4.532850
MV	5.148779	0.309353	6.500945	7.998108	4.421220	**0.002313**

Table 5.21 Authentication performance of RPCA-based unknown tea brand classifier with SNR = 50 (platinum working electrode)

	BR	DD	GD	LP	LD	MV
BR	**0.028043**	5.430089	11.621681	2.877372	0.699516	5.118423
DD	5.444879	**0.013253**	6.204845	8.294208	4.717320	0.298413
GD	8.331489	2.873357	**#3.318235**	11.180818	7.603930	3.185023
LP	2.821764	8.279896	14.471488	**0.027565**	3.549323	7.968230
LD	5.423275	0.034857	6.226449	8.272604	**#4.695716**	0.276809
MV	5.227013	0.231119	6.422711	8.076342	4.499454	**0.080547**

Abbreviations: Brookbond (BR), Double Diamond (DD), Godrej (GD), Lipton (LP), Lipton-Darjeeling (LD), Marvel (MV), # Entry = overall misclassification = (13/324) × 100 = 4.0%

5.6 Design of FDA-based classifier for commercial tea brands

A tea data set was prepared (collected over a specific working electrode) considering six chosen tea brands (Brookebond, Double Diamond, Godrej, Lipton, Lipton-Darjeeling, and Marvel, each containing three sets of data or three sample runs) available in the Indian marketplace. The classifier designed here is unlike the unsupervised PCA and similarity/dissimilarity-based PCAs. The DDAG method has been adapted here for FDA-based multiclass classification of the previously mentioned tea brands using the decisions generated by the binary classifiers. The DDAG method in essence is similar to the working principle of pairwise classification. For classification among six commercial tea brands, 15 binary classifiers $\left\{\left(\frac{M(M-1)}{2}\right), \text{ where } M \text{ or tea class} = 6\right\}$ have been designed for each of the electrode systems. Given a test instance, the multiclass classification was executed by evaluating decisions of all 15 binary classifiers and assigning the instance to the class which got the highest number of resulting outputs (decision class).

5.6.1 FDA-based authentication/classification algorithm using DDAG

Each of the triplicate runs for a specific brand (among the six brands chosen) contains 4402 features and have been considered for the design of FDA-based binary classifiers. The data set was generated/simulated by considering any sample of a brand (among the six brands considered here and generated using silver, gold, glassy-carbon, and platinum working electrode systems) added up/adulterated with 5–8% of random noise; the members of the set treated as unknown brand samples and used for authentication purposes. In FDA, the number of variables used as the input to the model has to be smaller than the total number of samples (observations/runs here). The scores of the most significant principal components are used as inputs to a binary FDA-based classification model. For authentication of an unknown tea brand using binary classifiers, PCA decomposition of ($3 + 3 + 1 = 7 \times 4402$) data set was made. Finding a discriminant direction (in FDA) along with the DDAG procedure is discussed in Chapter 2. Linear and quadratic kernels are used for finding discriminants. The proposed tea classification/authentication algorithm is as follows:

For e = 1 to 4 (e = silver; gold; platinum; glassy carbon)

Step 1: Creation of simulated unknown tea brand resembling any of Brookebond, Double Diamond, Godrej, Lipton, Lipton-Darjeeling, and Marvel.

Step 2: Numbering of tea brands: Brookebond = 1, Double Diamond = 2, Godrej = 3, Lipton = 4, Lipton-Darjeeling = 5 and Marvel = 6

Step 3: Designation of binary classifiers (15 numbers here, for six tea brands) (M_1–M_2).

$M_1 = 6$ to 2 (a decrement by −1) and $M_2 = 1$ to M_1−1.

Step 4: PCA decomposition of ($3 \times 2 + 1 = 7 \times 4402$) tea data set (one run is simulated as unknown tea brand). Generation of scores corresponding to most significant principal components (PC1 and PC2).

Step 5: Designing FDA-based binary classifiers/authenticator, where inputs have two-dimensional features of seven tea samples (2 brands of tea in triplicate and 1 unknown tea brand sample).

Step 6: Results of binary classifiers are used in DDAG-based membership generation of the previously mentioned six brands (algorithm already proposed in Chapter 2). The results of authentications of test samples generated through binary classifiers are expressed either as not M_1 or not M_2. In this way, every tea brand (six to one) gets a membership value out of their detected classes through 15 binary classifiers. Assign the unknown tea sample to the tea class receiving the highest membership.

End (e)

The MATLAB® code developed for the FDA-based (classify_fisher_ag.m) classification is provided in Program 5.2. After the generation of binary classifier results, one can generate the triangular DDAG graph, and hence generate membership numbers for each category of tea brand present in the database using the algorithm provided in Chapter 2.

Program 5.2 MATLAB code for FDA classifier

```
% This program takes on Ag working electrode-based e-tongue data on six
grades of tea samples ('alltea_ag.txt' of dimension 18×4402). A simulated
unknown sample is created by infiltrating noise 5-8% to any of the tea
data sample. The scores of the first two principal components of the tea
data samples become the input to the binary FDA classifier. For any
unknown tea sample authentication 15 binary classifiers are designed
(according to DDAG) and the unknown sample is tested whether it belongs
to either of the category with respect to whom it is getting examined.
Each of the binary classifications is done using a linear or quadratic
FDA based 'classify' function. The decision boundary is created and the
distance of the sample from the boundary is also determined.

clc
clear all

X=load('alltea_ag.txt','-ascii') ;

% Creation of simulated test sample

y=input('Enter test category:')
k=1+(y-1)*3

R=[X(:,k) X(:,k+1) X(:,k+2)]
R=mean(R,2)
n=size(R)
b=input('Enter bias :')
    r=b*randn(n,1);
R=R+r
%=================================================
for M1=6:-1:2
```

```
for M2=1:M1-1
% M1 and M2 are the two categories of tea samples to
% be used to design the binary classifier.
Z=[];
k1=1+(M1-1)*3
k2=1+(M2-1)*3

 % creation of  Z (7×4402) matrix

Z=[Z X(:,k1) X(:,k1+1) X(:,k1+2) X(:,k2) X(:,k2+1) X(:,k2+2) R]
===========================================================================

% nomenclature of 15 binary classifiers begins

if M1==1
N1=['BR1';'BR2';'BR3']
text2=' BROOKBOND'
end

if M1==2
N1=['DD1';'DD2';'DD3']
text2=' DOUBLE-DIAMOND'

end

if M1==3
N1=['GD1';'GD2';'GD3']
text2=' GODREJ'

end

if M1==4
N1=['LP1';'LP2';'LP3']
text2=' LIPTON'

end

if M1==5
N1=['LD1';'LD2';'LD3']
text2=' LIPTON-DARJEELING'

end

if M1==6
N1=['MV1';'MV2';'MV3']
text2=' MARVEL'
end

if M2==1
N2=['BR1';'BR2';'BR3']
text2=strcat(text2,' and BROOKBOND')

end
```

```
if M2==2
N2=['DD1';'DD2';'DD3']
text2=strcat(text2,' and DOUBLE-DIAMOND')

end

if M2==3
N2=['GD1';'GD2';'GD3']
text2=strcat(text2,' and GODREJ')

end

if M2==4
N2=['LP1';'LP2';'LP3']
text2=strcat(text2,' and LIPTON')

end

if M2==5
N2=['LD1';'LD2';'LD3']
text2=strcat(text2,' and LIPTON-DARJEELING')

end

if M2==6
N2=['MV1';'MV2';'MV3']
text2=strcat(text2,' and MARVEL')

end


% end of nomenclature
==========================================================================

N=[N1 ;N2]

N3=' X'% unknown sample

 % PCA decomposition of z matrix

Z=Z' ;

[pc,score,latent,tsquare] = princomp(Z);
size(pc)

size(score)
data=score(:,1:2)

data1=data(1:6,1:2) ; % generation of 6×2 matrix using scores of Z
matrix, %which will be the input matrix to the FDA classifier

pattern=data(7,:) ; % 7th row of score resembling unknown sample.
```

```
% pattern is a 1×2 matrix, it will also be the input matrix to the LDA

% classifier

% formation of axis

F=[ data1 ; pattern];
L1=min(F(:,1)) ;
L2=min(F(:,2));
L3=max(F(:,1));
L4=max(F(:,2));
L5=L1-0.5 ;
L6=L3+0.5 ;
L7=L2-0.5 ;
L8=L4+0.5 ;

[ L5 L6 L7 L8]

 =============================================% end of axes formation
% start of plotting class 1, class 2 and unknown class data
% on the same figure
% l1=input('Enter category as class-1:')
% l2=input('Enter  category as class-2:')
l1=1 ; l2=2 ;

group=[l1 ; l1; l1; l2 ;l2 ;l2] ;

plot(data1(:,1) , data1(:,2) ,'b+','Linewidth',2)
grid on
axis([ L5 L6 L7 L8])
gname(N)
hold on
plot(pattern(:,1) , pattern(:,2) ,'rv','Linewidth',2)
gname(N3)

pause

test=pattern ;

% decision on kernel to be used
k=input('Enter kernel type:')
if k==1
   type='linear'
end

if k==2
type='quadratic'
end
 ======================================================================

 % LDA classification begins, where inputs are pattern and data1 matrix
% classify.m is a MATLAB inbuilt function
```

```
[C,err,P,logp,coeff] = classify([pattern(1) , pattern(2)],[data1(:,1)
data1(:,2) ],group,type)

% start visualizing the classification
K = coeff(1,2).const;
L = coeff(1,2).linear ;

if k==1
f = sprintf('0 = %g+%g*(x)+%g*(y)',K,L(1,1),L(2,1));

end

if k==2
Q = coeff(1,2).quadratic;
f = sprintf('0 =%g+%g*(x)+%g*(y)+%g*x^2+%g*x.*y+%g*y.^2',K,L,Q(1,1),Q(1,2
)+Q(2,1),Q(2,2))

end
text=num2str(f);
h2 = ezplot(f,[L5, L6 , L7,  L8]);
set(h2,'Color','b','LineWidth',2)
axis([L5 L6 L7 L8])
xlabel('PC-1')
ylabel('PC-2')
title('{\bf Classification with Fisher Discriminant Analysis with AG
electrode}')

if k==1
s1=sign(L(1,1));
if s1==-1
    s11='-';
else
    s11='+' ;
end
s2=sign(L(2,1)) ;

if s2==-1
   s12='-' ;
 else
   s12='+' ;


end

text1=strcat('0=',num2str(K),s11,num2str(abs(L(1,1))),'x',s12,num2str(abs
(L(2,1))),'y');
end

if k==2
s1=sign(L(1,1));
if s1==-1
    s11='-';
else
    s11='+' ;
```

```
end

s2=sign(L(2,1)) ;
if s2==-1
   s12='-' ;
else
   s12='+' ;

end

text1=f
end

text=type ;

text=strcat('Decission boundary:',text)
gtext(text) % mouse placement of text
gtext(text1)
ctext1='class-1' ;
ctext2='class-2' ;
gtext(ctext1)
gtext(ctext2)

fname=strcat(fname,'_',type,'.fig')

hgsave(fname)

if k==1
% Find projection distance
m=-L(1,1)/L(2,1)
k=-K/L(2,1)
y0=pattern(2) ; x0=pattern(1)

A=(y0-k+x0/m) ; B=m+1/m
y=m*(A/B)+k;
x=A/B;
D=sqrt((x-x0)^2+(y-y0)^2)
U=[x0 y0 ; x y]
text4=num2str(D)
text4=strcat('Distance=',text4)

plot(U(:,1) , U(:,2),'r:','Linewidth',4)
gtext(text4)
[C D]
end

if k==2

   C

end
```

5.6.1.1 *Performance evaluation of the designed FDA-based classifier*

Figure 5.8 represents the Ag electrode-based DDAG classifier used for authentication of the unknown Marvel tea brand manifesting the evaluation path. Based on the evaluation path in Figure 5.8, the maximum membership for each brand is determined, and Marvel got the highest membership as a test sample (Table 5.22). Figure 5.9 reveals a representative combination of 15 plots of binary classification using FDA and supporting Figure 5.8. Au electrode-based DDAG has been developed and is presented in Figure 5.10. Table 5.23 reveals the membership value calculated for every brand considering the evaluation path in Figure 5.10. Lipton gets the highest membership; hence, the test sample is unequivocally the Lipton. Figures 5.11 and 5.12 represent the DDAGs for glassy-carbon and platinum electrode-based classifiers, respectively. Based on the evaluation paths in Figures 5.11 and 5.12, the membership values of each class of tea have been determined. Tables 5.24 and 5.25 document the outputs in terms of membership values for each tea category in accordance with the evaluation paths of the designed DDAGs. The authenticated unknown tea brands are Lipton-Darjeeling and Double Diamond, respectively, for the glassy-carbon and platinum electrode-based classification systems.

Choosing the FDA-based binary classification method and utilizing it for multiclass classification has been adapted here for successful authentication as well as classification of commercial tea brands. The present study has handled the multiclass classification problem using a decision tree as a consequence of execution of several binary classifiers, with linear decision boundaries. Multiclass decision functions with possible nonlinear decision boundaries might face overlapping problems and necessitate the incorporation of the kernel trick. When the number of tea brands is limited (6–8), the proposed method is immensely useful without much computational complexity.

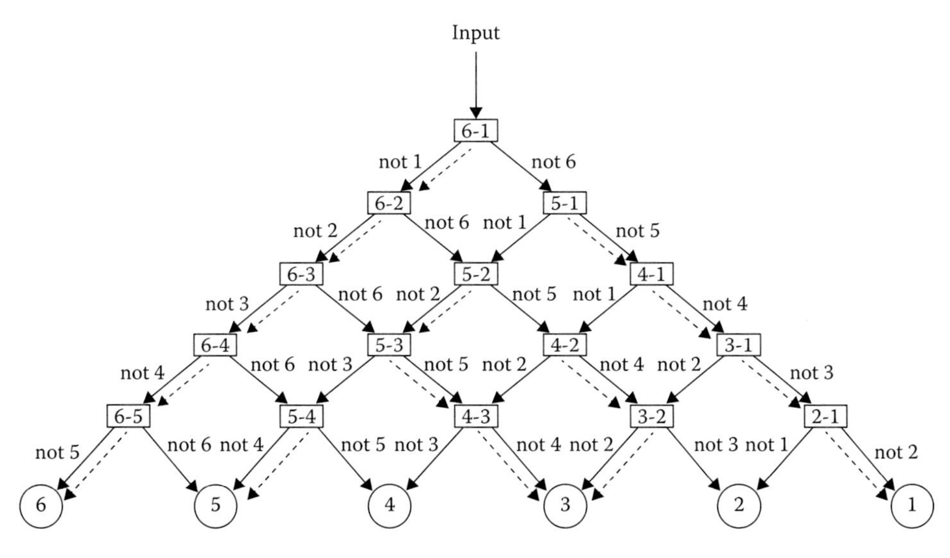

Figure 5.8 DDAG-based authentication of Marvel tea brand (Ag working electrode) using pairwise FDA (- - - -: evaluation path).

Table 5.22 Membership for different tea brands of the DDAG in Figure 5.8

Membership (tea brand) number	Number of detected classes
6 (Marvel)	**5: (6-1); (6-2); (6-3); (6-4); (6-5)**
5 (Lipton-Darjeeling)	2: (5-2) & (5-4)
4 (Lipton)	0:
3 (Godrej)	3: (3-2); (4-3); (5-3)
2 (Double Diamond)	1: (4-2)
1 (Brookbond)	4: (5-1); (4-1); (3-1); (2-1)

Unknown = Marvel

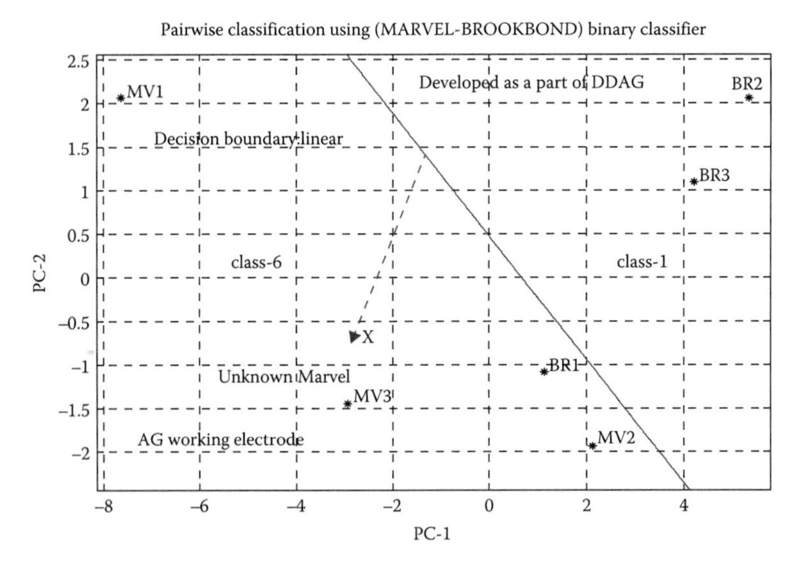

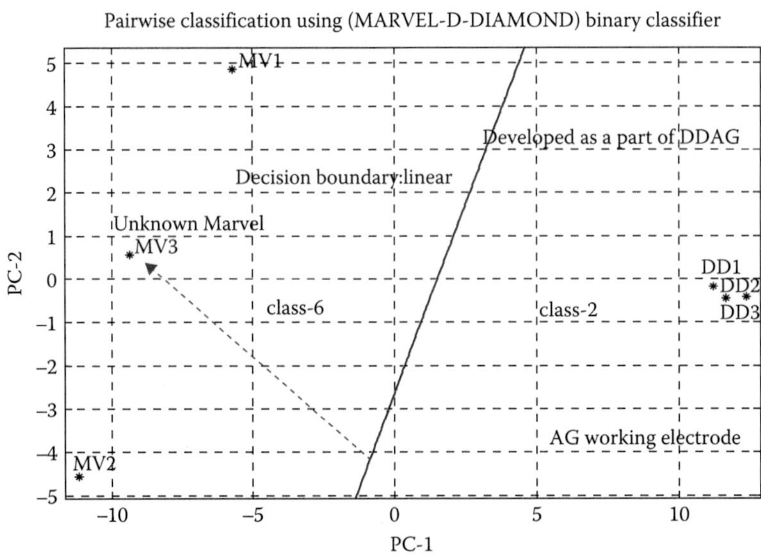

Figure 5.9 Outputs of 15 FDA-based binary classifiers when Marvel is the test tea sample.

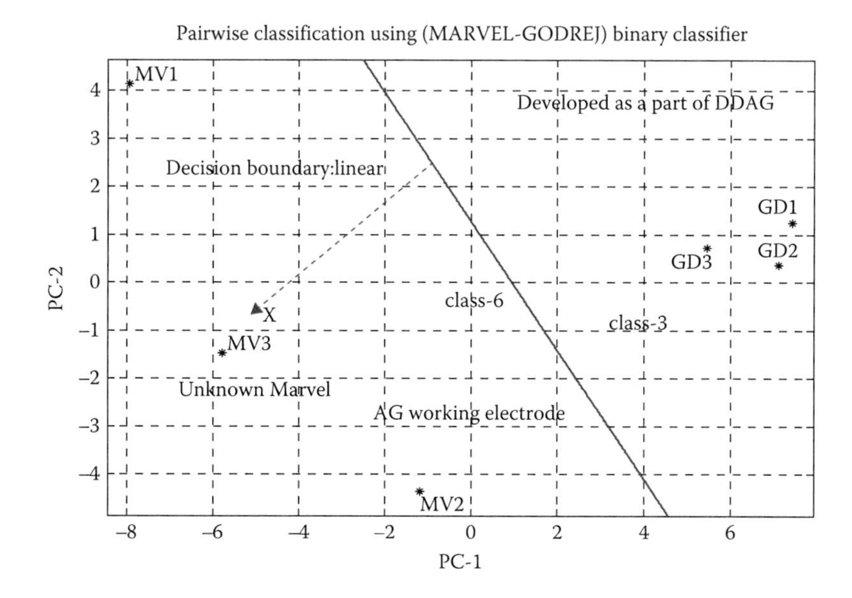

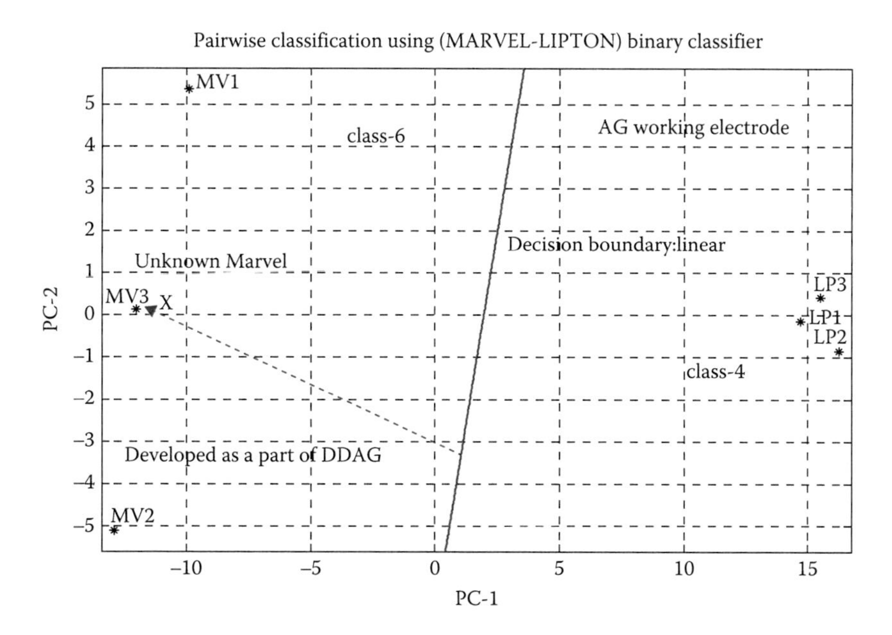

Figure 5.9 (Continued) Outputs of 15 FDA-based binary classifiers when Marvel is the test tea sample.

(Continued)

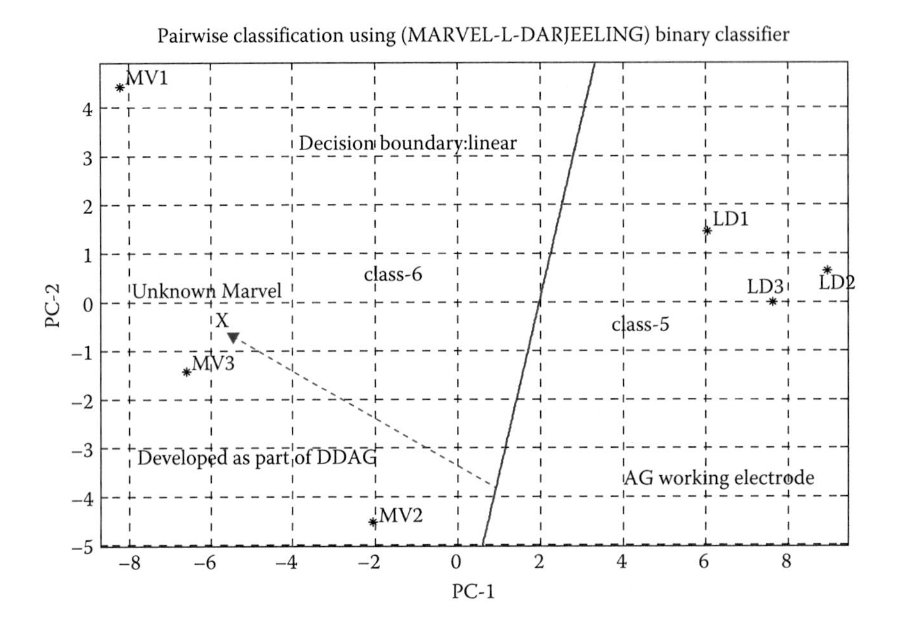

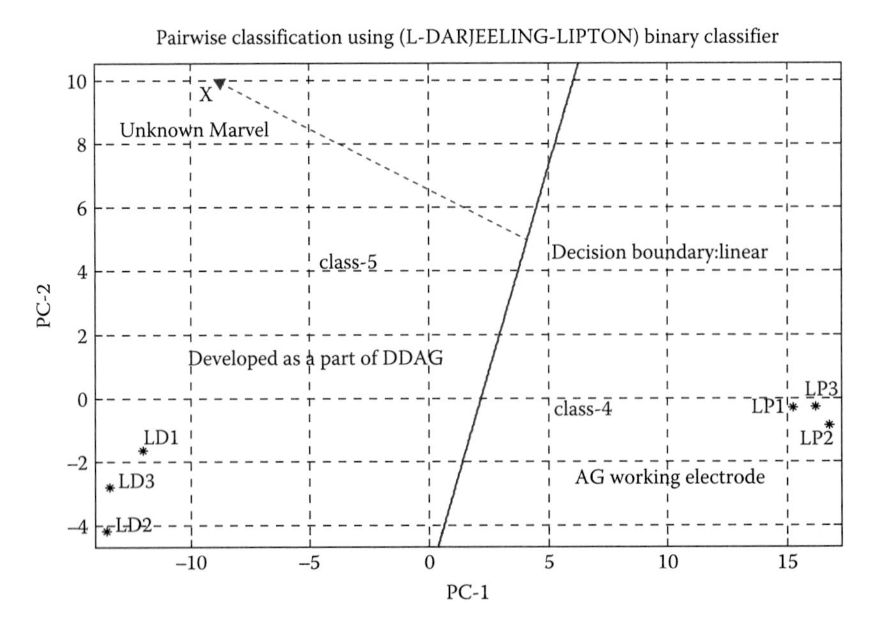

Figure 5.9 **(*Continued*)** Outputs of 15 FDA-based binary classifiers when Marvel is the test tea sample.

(Continued)

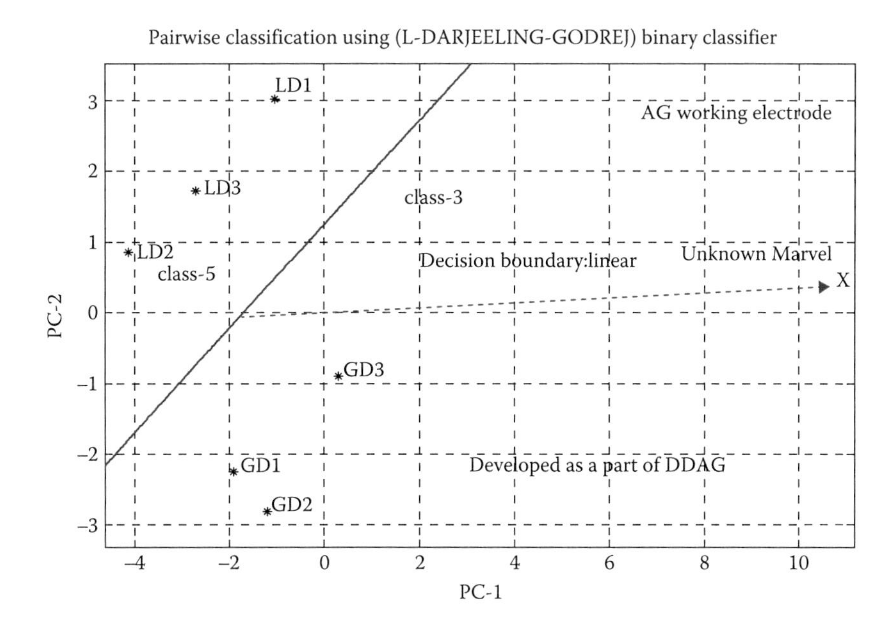

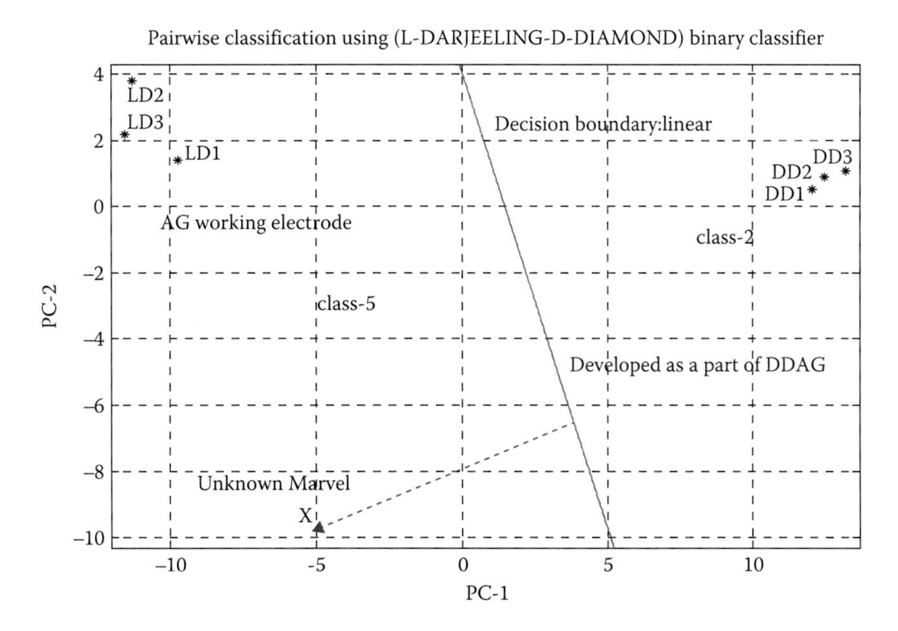

Figure 5.9 **(*Continued*)** Outputs of 15 FDA-based binary classifiers when Marvel is the test tea sample.

(*Continued*)

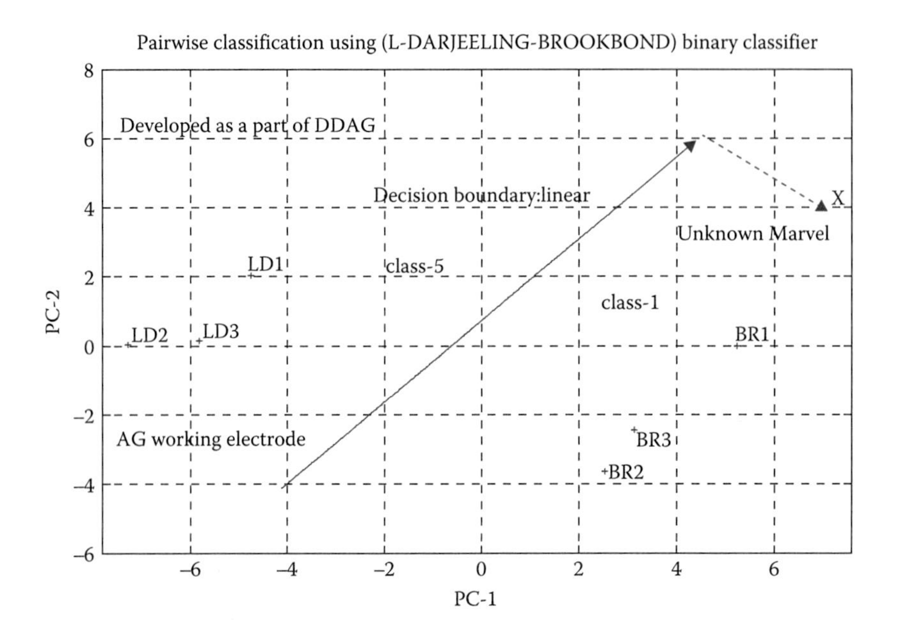

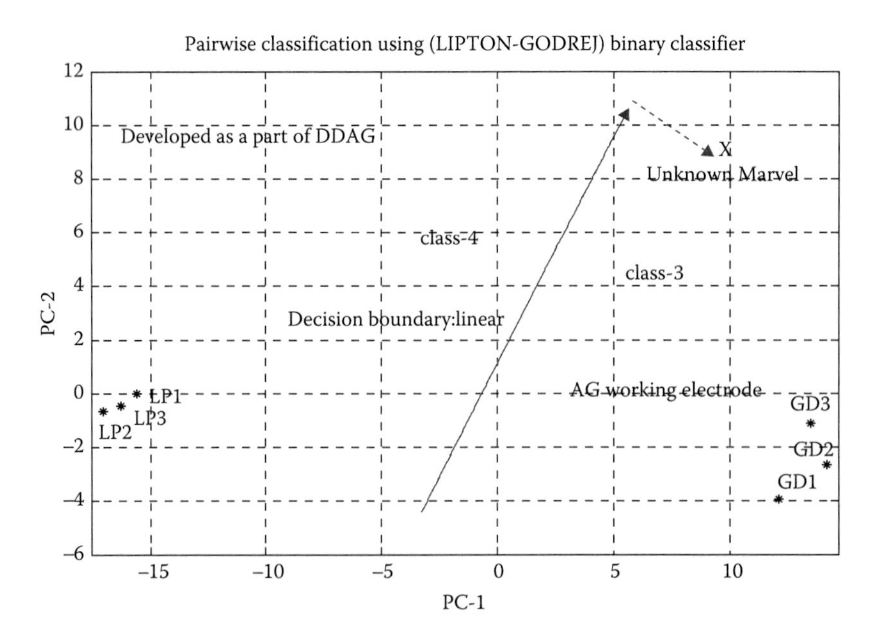

Figure 5.9 (Continued) Outputs of 15 FDA-based binary classifiers when Marvel is the test tea sample.

(Continued)

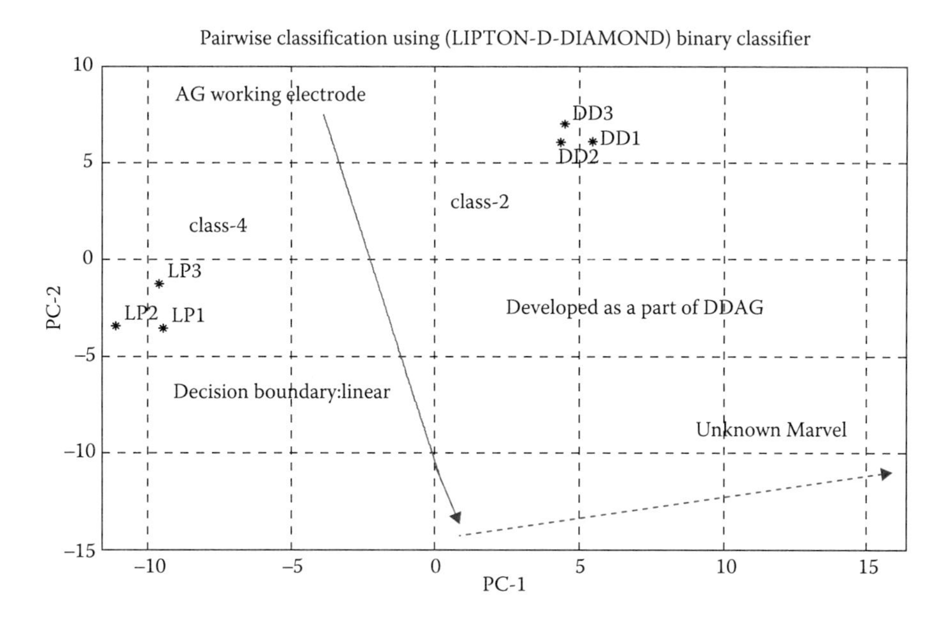

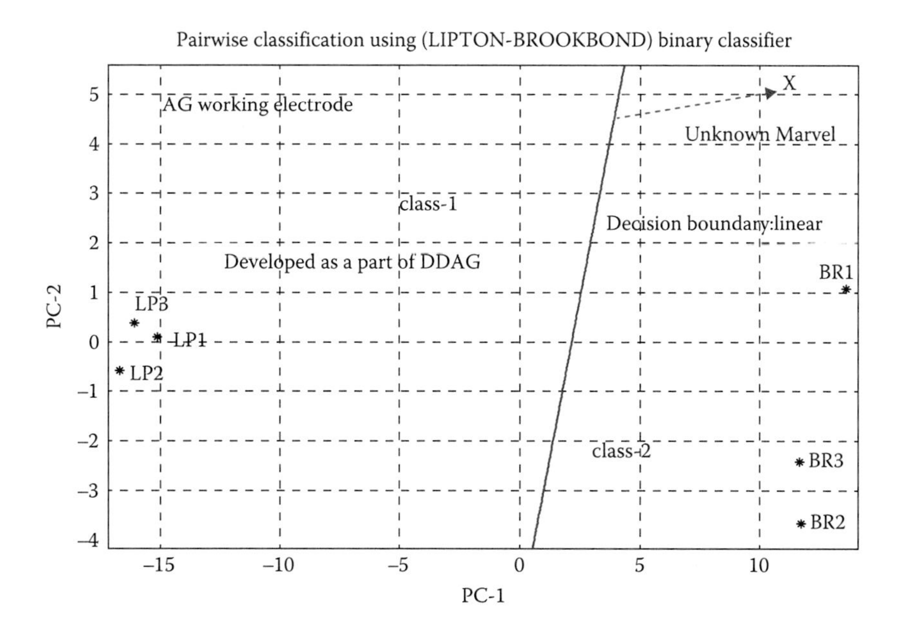

Figure 5.9 (Continued) Outputs of 15 FDA-based binary classifiers when Marvel is the test tea sample.

(Continued)

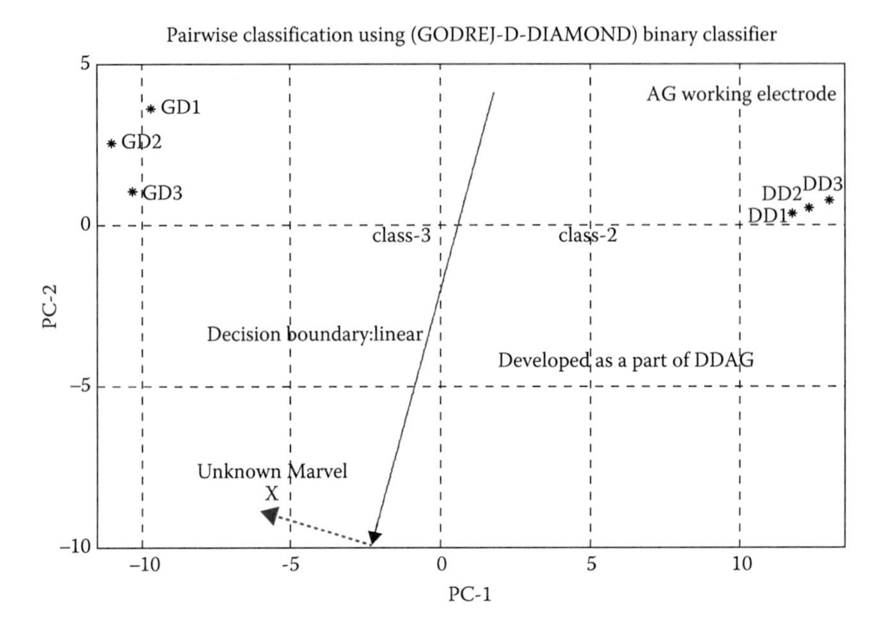

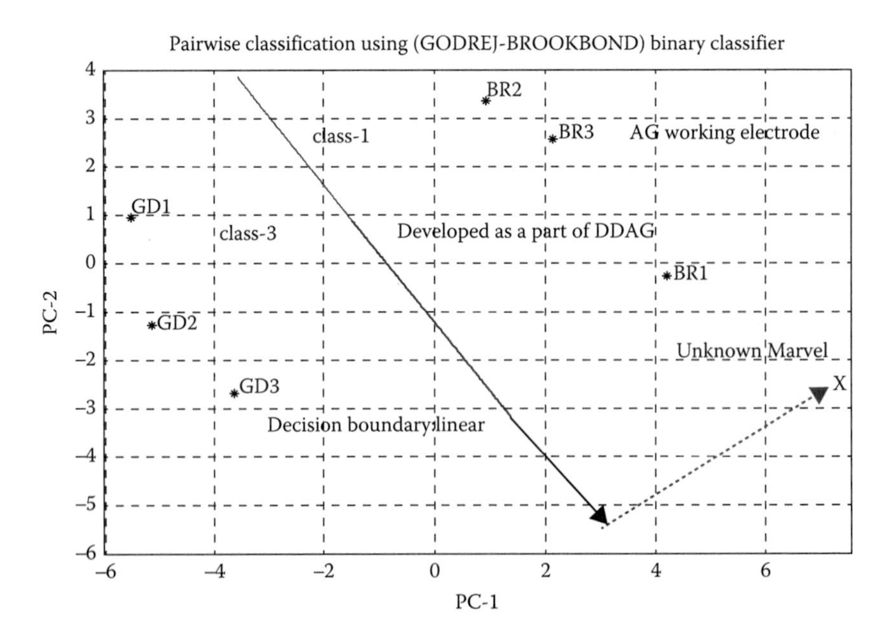

***Figure 5.9* (*Continued*)** Outputs of 15 FDA-based binary classifiers when Marvel is the test tea sample.

(*Continued*)

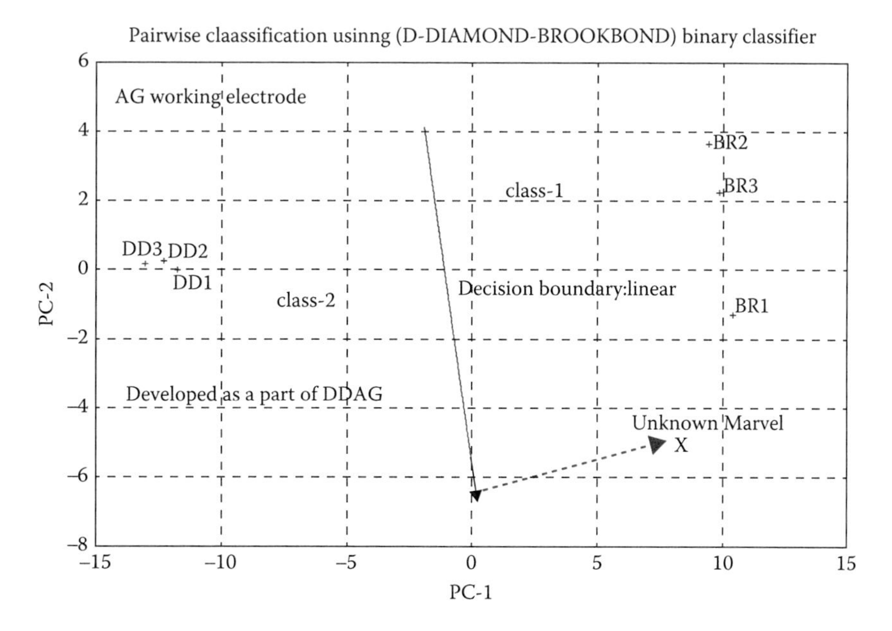

Figure 5.9 (Continued) Outputs of 15 FDA-based binary classifiers when Marvel is the test tea sample (X = unknown test sample; and it is marvel tea brand here).

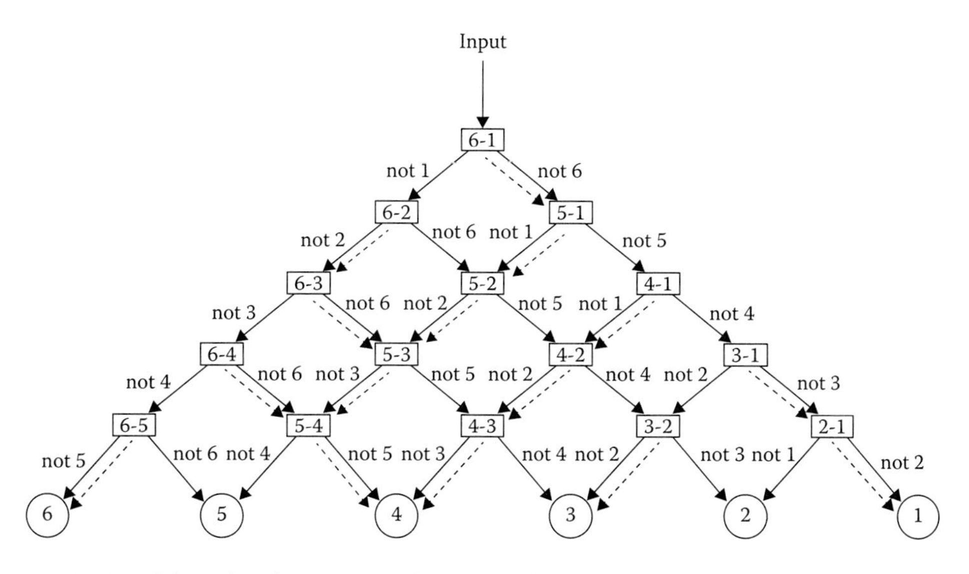

Figure 5.10 DDAG-based authentication of Lipton tea brand (Au working electrode) using pairwise FDA (- - -: evaluation path).

Table 5.23 Membership for different tea brands of the DDAG in Figure 5.10

Membership (tea brand) number	Number of detected classes
6 (Marvel)	2: (6-5); (6-2)
5 (Lipton-Darjeeling)	3: (5-1); (5-2); (5-3)
4 (Lipton)	**5: (4-1); (4-2); (4-3); (5-4); (6-4)**
3 (Godrej)	2: (3-2); (6-3)
2 (Double Diamond)	0:
1 (Brookbond)	3: (6-1); (3-1); (2-1)

Unknown = Lipton

5.7 Comparative classification performance

The designed dissimilarity-based tea brand classifier revealed an utmost error of less than 1.4% (for silver working electrode at an SNR of 40, i.e., a variability of 2.5% with respect to their parent brands). The proposed tea classifier is possibly a suitable one for online tea grading because of its computational simplicity and robust nature.

The designed RPCA classifiers, with all the working electrodes were able to detect and classify the unknown tea samples pertaining to specific brands with 96% accuracy (misclassification% = (13/324) × 100 = 4.0%) up to a variability of 2.0% (SNR = 50) with respect to their parent brands. The developed RPCA-based classifier has the potential to supplant the role of tea tasters in assessing tea quality/grade and for online tea grading.

The designed FDA classifier has evolved to be 100% efficient for unknown test samples containing noise from 5% to 8% with respect to their parent brand. The FDA classifier is a supervised classifier, and the resulted classification efficiency is already expected. However, the multiclass classification of tea brands is implemented as a

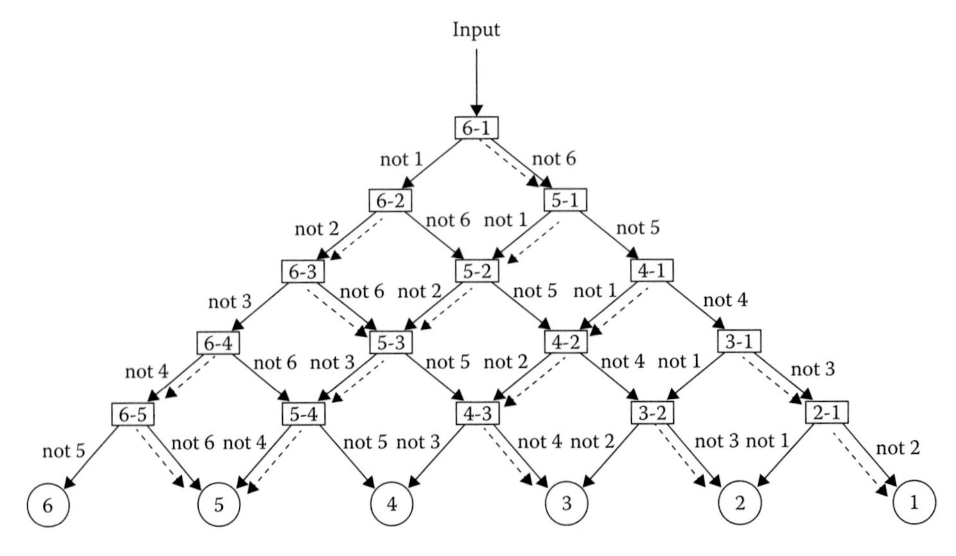

Figure 5.11 DDAG-based authentication of Lipton-Darjeeling tea brand (glassy-carbon working electrode) using pairwise FDA (---: evaluation path).

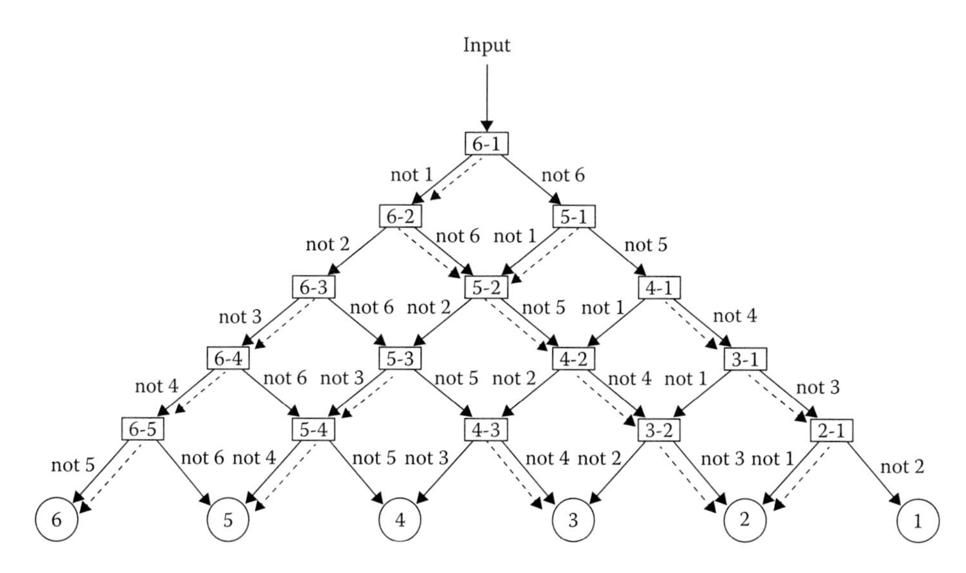

Figure 5.12 DDAG-based authentication of Double Diamond tea brand (Pt working electrode) using pairwise FDA (---: valuation path).

Table 5.24 Membership for different tea brands of the DDAG in Figure 5.11

Membership (tea brand) number	Number of detected classes
6 (Marvel)	2: (6-4); (6-2)
5 (Lipton-Darjeeling)	**5: (5-1); (5-2); (5-3); (5-4); (6-5)**
4 (Lipton)	2: (4-1); (4-2)
3 (Godrej)	2: (4-3); (6-3)
2 (Double Diamond)	1: (3-2)
1 (Brookbond)	3: (6-1); (3-1); (2-1)

Unknown = Lipton-Darjeeling

Table 5.25 Membership for different tea brands of the DDAG in Figure 5.12

Membership (tea brand) number	Number of detected classes
6 (Marvel)	4: (6-1); (6-3); (6-4); (6-5)
5 (Lipton-Darjeeling)	3: (5-1); (5-3); (5-4)
4 (Lipton)	0:
3 (Godrej)	1: (4-3)
2 (Double Diamond)	**5: (6-2); (5-2); (4-2); (3-2); (2-1)**
1 (Brookbond)	2: (4-1); (3-1)

Unknown = Double Diamond

combination of several binary classifications, which makes this category of classification inappropriate for online classification/monitoring because it demands a great deal of CPU time.

5.8 Prototype development principles of commercial tea grader

In principle, an electronic version of beverage classifiers, the commercial tea grader should consist of the following specific steps:

1. Generation of voltammogram using electronic tongue instrumentation.
2. Determination/extraction of suitable features using the voltammogram.
3. Development of a robust authenticator based on these extracted features.

So far the classification of different commercial tea brands using different machine-learning algorithms has considered the previously mentioned steps. The present work uses a potentiostat (Gamry Instruments, USA; reference 600 potentiostat, with a 16-bit A to D converter, maximum current ± 600 ma, minimum current resolution = 20 aA, maximum potential = ± 11 V, minimum voltage resolution 1 μV), which connects the electron transfer event occurring in an electrochemical cell (Bob cell) via the DAS (PHE200™ and PV220™ software packages provided by Gamry) to a PC. The potentiostat system is connected to a PC through a USB interface. DAS is used to create an excitation voltage to the cell and record the return waveform (called a "voltammogram") from the cell to store it in the PC for subsequent feature extraction and analysis using different algorithms that have been developed. The potentiostat contains the hardware related to all sorts of communication between the cell and the PC.

So far, because Step 1 is concerned with capturing electrical signals in the form of a voltammogram, there remains the scope of low cost and indigenous development of the electronic tongue instrumentation with respect to the following avenues:

- Replacing the potentiostat and DAS with indigenous hardware/software modules
- Fabricated/tailored working electrode being used in the electrochemical cell generating the voltammogram

The first proposition can reduce the development cost to a notifiable extent. It is relevant at this juncture to briefly discuss the hardware and working principle of a potentiostat and DAS.

5.8.1 Potentiostat

The controlled variable in a potentiostat is the cell potential, and the measured variable is the cell current. The potentiostat is widely used in electroanalytical techniques to identify, quantify, and characterize a redox-active species and in evaluating thermodynamic and kinetic parameters of electron transfer events. Figure 5.13 reveals the schematic of a potentiostat.

5.8.1.1 Electrometer

The electrometer circuit measures the difference in voltage between the reference and working electrodes. Its output has two major functions: providing the feedback signal in the potentiostat circuit and the signal being measured, whenever the cell voltage is

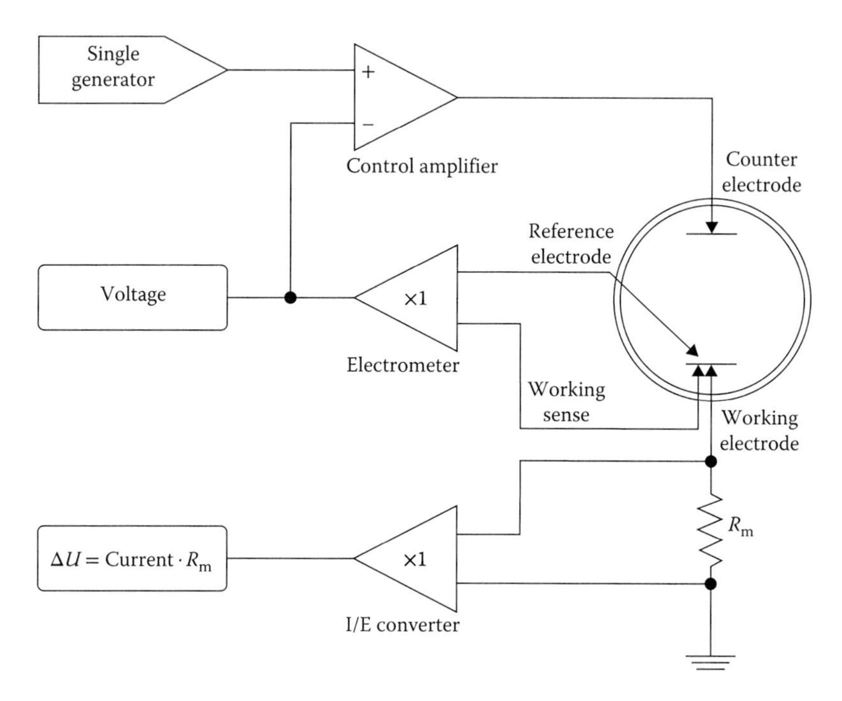

***Figure* 5.13** Schematic of potentiostat. (Source: literature on the potentiostat provided by www.gamry.com/Potentiostats.)

needed. An ideal electrometer has zero input current and infinite input impedance. Current flowing through the reference electrode changes its potential. All modern electrometers have near zero input currents; hence, this effect can usually be ignored. Two important attributes of the electrometer are its bandwidth and its input capacitance. The electrometer bandwidth portrays the AC frequencies the electrometer can measure, driven from a low-impedance source. The electrometer bandwidth must be the highest among the electronic components in the potentiostat. The electrometer input capacitance and the reference electrode resistance together form an RC filter. If this filter's time constant is too large, it can limit the effective bandwidth of the electrometer causing system instabilities. Smaller input capacitance translates into more stable operation and greater tolerance for high-impedance reference electrodes.

5.8.1.2 I/E converter

The current-to-voltage (I/E) converter in Figure 5.13 measures the cell current. It forces the cell current to flow through a current measurement resistor R_m. The voltage drop across R_m is a measure of the cell current. In some experiments, cell current doesn't vary appreciably. In other experiments, such as a corrosion experiment, the current can often vary by seven orders of magnitude. It is not possible to measure current over such a wide range using a single resistor. A number of different R_m resistors can be connected into the I/E circuit under computer control. This ensures measurement of widely varying ranges of currents, with each range being measured using an appropriate resistor. An "I/E autoranging" algorithm is required to select the appropriate resistor values. The "I/E converters" bandwidth depends strongly on its sensitivity. Measurement of small currents requires large R_m values. Undesirable capacitance in the I/E converter forms a resistor-capacitor (RC) filter with R_m, limiting the I/E bandwidth.

5.8.1.3 Control amplifier

A servo control amplifier was used. It maintains the cell voltage and allows current through the cell by generating feedback of the difference between the measured and the desired cell voltage. The measured voltage acts as a negative input to the control amplifier. A positive perturbation in the measured voltage creates a negative control amplifier output. This negative output counteracts the initial perturbation. This is a negative feedback control scheme. Under normal conditions, the cell voltage is controlled to be identical to the signal source voltage. The control amplifier has a limited output capability.

5.8.1.4 Signal circuit

The signal circuit is a computer-controlled voltage generator. Computer-generated number sequences are converted into voltage through the digital-to-analog (D/A) converter. Proper choice of number sequences allows the signal circuit to generate constant voltages, voltage ramps, and even sine waves as its output. A D/A converter makes an analog to digital conversion of a waveform. It contains small voltage steps; the size of the steps is controlled by the resolution of the D/A converter.

5.8.2 DAS

DAS (data acquisition software) is used by the potentiostat to perform an in-depth study of the electrode-electrolyte interface and the mechanisms of electrochemical reactions. It is a useful tool for analyzing electrochemical reaction, sensor development, and small-scale energy storage devices. Measurements and analysis using different electrochemical techniques like voltammetry, chronoamperometry, and chronocoulometry become possible by maneuvering the DAS thoughtfully. DAS provides the opportunity to experiment with a large range of parameters like voltage scan rate, voltage step size, sampling period, etc. PHE200 physical electrochemistry and PV220 pulse voltammetry software packages have been used in the experimentat.

5.8.3 Proposed tea/beverage grading device

The framework of the proposed indigenous voltammetric beverage/tea grading device is as follows:

Figure 5.14 presents the tea-grader prototype schematic (the detailed hardware design with circuitry is beyond the scope of the present book). Following is the architecture summary architecture.

- An array of electrodes is in the electronic tongue test cell, which establishes communication through DAS for generating a voltammogram in PCs.
- The functionality of the potentiostat is implemented by a data acquisition card (DAQ card) and DAS.
- The manufacturer of the DAQ card provides a set of APIs (application program interfaces) that are utilized by the DAS (a software program developed using any high-level language such as LabView, MATLAB®, C#, or Visual Basic) to communicate with the DAQ card. The major functionality of the DAS includes: generation of excitation (voltage) signals, controlling various parameters such as sampling rate, voltage level, etc., and receiving the output generated by the DAQ card and storing it in a file in the PC for later analysis.

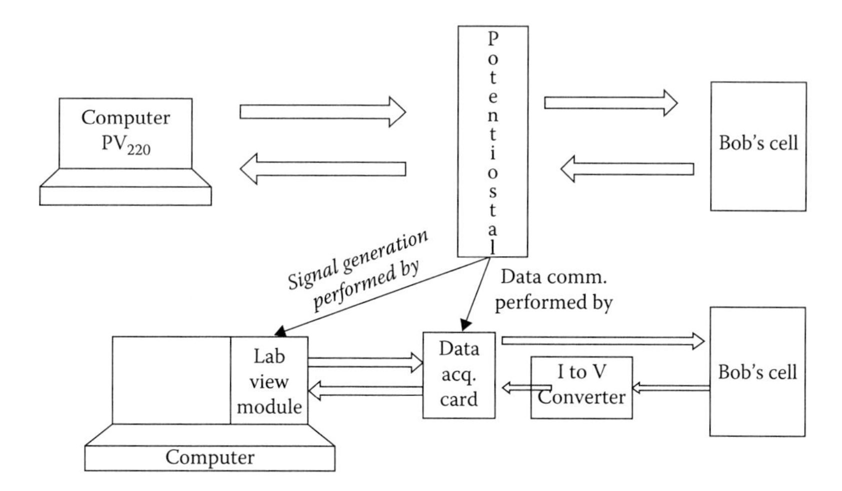

Figure 5.14 Proposed e-tongue prototype schematic.

- The data acquisition card takes care of the communication between the cell and the DAS, such as the generated excitation signal is applied across the electrodes and the current pattern produced is sent back to the DAS.
- Data stored in the PC generate the voltammogram (current/voltage plot).
- The generated voltammogram is then used in feature extraction and in developing PC-based machine-learning algorithms.

The merits and demerits of the proposed tea/beverage grader are as follows:

- Significant economic solution for bulk monitoring.
- Full control/flexibility in changing the excitation signal pattern at any given point of time.
- The process is highly dependent on the accuracy of the DAQ card.

5.9 Summary of deliverables

In conclusion, a few things need to be stated about this chapter's deliverables, which are as follows:

1. An electronic tea grader is proposed using voltammetric electronic-tongue instrumentation.
2. This chapter has exercised some of the novel approaches for chemometric classifier design and development using e-tongue data, which can potentially find application in tea grading. Of them, some are suitable for online and some are for offline quality monitoring/grading, as a replacement for human tea tasters.
3. Grading tea in terms of the various organic and inorganic materials present in it is a complicated task, for which keen observation on the voltammogram developed using e-tongue instrumentation is not enough. Due to the non-stationary behavior of the time series data produced for tea samples, the significance of machine-learning algorithms is paramount.

4. The proposed tea classifiers can only function for the commercial tea brands utilized in their design; all others will be treated as outcasts.
5. All the code developed using MATLAB® is provided.
6. The guidelines regarding the prototype development of a commercial tea grader are presented in this chapter. These principles can be extended for the design of possible low-cost and everyday appliances like beverage quality monitors, adulterated milk detectors, out-of-date food product detectors, and arsenic meters to quantify arsenic in contaminated water.
7. This chapter will benefit engineers engaged in product research and development, as well as entrepreneurs. Hardware design, development, and the implementation needed for these appliances are beyond the scope of this book and could not be included.

References

1. H. Xiao, J. Wang (2008). Discrimination of Xihulongjing tea grade using an electronic tongue. *African Journal of Biotechnology*, vol. 8, 6985–6992.
2. S. Sarkar, N. Bhattacharyya, V. Palakurthi (2011). Taste recognizer by multi sensor electronic tongue: A case study with tea quality classification. In *Proceedings of 2nd International Conference on Emerging Applications of Information Technology*, IEEE Computer Society, pp. 138–141.
3. M. Palit, B. Tudu, P. Dutta, A. Dutta, A. Jana, J. Roy, N. Bhattacharyya, R. Bandyopadhyay, A. Chatterjee (2010). Classification of black tea taste and correlation with tea taster's mark using voltammetric electronic tongue. *IEEE Transactions on Instrumentation and Measurement*, vol. 59, 2230–2239.
4. K. Toko (2000). Taste sensor. *Sensors and Actuators B*, vol. 64, 205–215.
5. A.V. Legin, Y.G. Vlasov, A.M. Rudnitskaya (1996). Cross-sensitivity of chalcogenide glass sensors in solutions of heavy metal ions. *Sensors and Actuators B*, vol. 34, 456–461.
6. A. Legin, A. Rudnitskaya, Yu. Vlasov, C. Di Natale, E. Mazzone, A. D'Amico (2000). Application of electronic tongue for qualitative and quantitative analysis of complex liquid media. *Sensors and Actuators B*, vol. 65, 232–234.
7. A.V. Legin, Y.G. Vlasov, A.M. Rudnitskaya (2002). Electronic tongues: Sensors, systems, applications. *Sensor Update*, vol. 10, 143–188.
8. F K.Toko (1996). Taste sensor with global selectivity. *Materials Science and Engineering: C Materials Biology Applied*, vol. 4(2), 69–82.
9. K.Toko (2004). Measurement of taste and smell using biomimetic sensor. In *17th IEEE International Conference on Micro Electro Mechanical Systems 2004 (MEMS)*. IEEE, Maastricht, the Netherlands, pp. 201–207.
10. M. Habara, H. Ikezaki, K. Toko (2004). Study of sweet taste evaluation using taste sensor with lipid/polymer membranes. *Biosensors and Bioelectronics*, vol. 19(12), 1559–1563.
11. K. Toko (1998). A taste sensor. *Measurement Science and Technology*, vol. 9, 1919–1936.
12. A. Riul, Jr., A.M.G. Soto, S.V. Mello, S. Bone, D.M. Taylor, L.H.C. Mattoso (2003). An electronic tongue using polypyrrole and polyaniline. *Synthetic Metals*, vol. 132, 109–116.
13. A. Riul, Jr., R.R. Malmegrim, F.J. Fonseca, L.H.C. Mattoso (2003). Nano-assembled films for taste sensor application. *Artificial Organs*, vol. 27, 469–472.
14. A. Riul, Jr., R.R. Malmegrim, F.J. Fonseca, L.H.C. Mattoso (2003). An artificial taste sensor based on conducting polymers. *Biosensors and Bioelectronics*, vol. 18, 1365–1369.
15. E. Borato, A. Riul, Jr., M. Ferreira, O.N. Oliveira, Jr., L.H.C. Mattoso (2004). Exploiting the versatility of taste sensors based on impedance spectroscopy. *Instrumentation Science and Technology*, vol. 32(1), 21–30.
16. M. Ferreira, C.J.L. Constantino, A. Riul, Jr., K. Wohnrath, R.F. Aroca, J.A. Giacometti, O.N. Oliveira, Jr., L.H.C. Mattoso (2003). Preparation, characterization and taste sensing properties of Langmuir–Blodgett Films from mixtures of polyaniline and a ruthenium complex. *Polymer*, vol. 44, 4205–4211.

17. F. Winquist, S. Holmin, C. Krantz-Rulcker, P. Wide, I. Lundstrom (2000). A hybrid electronic tongue. *Analytica Chimica Acta*, vol. 406, 147–157.
18. A. D'Amico, C. Di Natale, R. Paolesse (2000). Portraits of gasses and liquids by arrays of non-specific chemical sensors: Trends and perspectives. *Sensor and Actuators B: Chemical*, vol. 68, 324–330.
19. C. Di Natale, R. Paolesse, A. MacAgnano, A. Mantini, A. D'Amico, M. Ubigli, A. Legin, L. Lvova, A. Rudnitskaya, Y. Vlasov (2000). Application of a combined artificial olfaction and taste system to the quantification of relevant compounds in red wine. *Sensor and Actuators B: Chemical*, vol. 69, 342–347.
20. C. Di Natale, R. Paolesse, M. Burgio, E. Martinelli, G. Pennazza, A. D'Amico (2004). Application of metalloporphyrins-based gas and liquid sensor arrays to the analysis of red wine. *Analytica Chimica Acta*, vol. 513, 49–56.
21. C. Di Natale, R. Paolesse, A. Macagnano, A. Mantini, A. D'Amico, A. Legin, L. Lvova, A. Rudnitskaya, Y. Vlasov (2000). Electronic nose and electronic tongue integration for improved classification of clinical and food samples. *Sensor and Actuators B: Chemical*, vol. 64, 15–21.
22. R.N. Bleibaum, H. Stone, T. Tan, S. Labreche, E. Saint-Martin, S. Isz (2002). Comparison of sensory and consumer results with electronic nose and tongue sensors for apple juices. *Food Quality and Preference*, vol. 13, 409–422.
23. M. Kataoka, K. Yoshida, Y. Miyanaga, E. Tsuji, E. Tokuyama, T. Uchida (2005). Evaluation of the taste and smell of bottled nutritive drinks. *International Journal of Pharmaceutics*, vol. 305, 13–21.
24. S. Takagi, K. Toko, K. Wada, T. Ohki (2001). Quantification of suppression of bitterness using an electronic tongue. *Journal of Pharmaceutical Science*, vol. 90, 2042–2048.
25. J.D. Kim, H.G. Byun, D.J. Kim, Y.K. Ham, W.S. Jung, C.O. Yoon (2006). A simple taste analyzing system design for visual and quantitative analysis of different tastes using multi-array chemical sensors and pattern recognition techniques. *Talanta*, vol. 70, 546–555.
26. B.A. McKinley (2008). ISFET and fiber optic sensor technologies: In vivo experience for critical care monitoring. *Chemical Reviews*, vol. 108, 826–844.
27. L.E. Gilabert, M. Peris (2010). Review: Highlights in recent applications of electronic tongues in food analysis. *Analytica Chimica Acta*, vol. 665, 15–25.
28. G. Pioggia, F. Di Francesco, A. Marchetti, M. Ferro, R. Leardi, A. Ahluwalia (2007). A composite sensor array impedentiometric electronic tongue: Part II. Discrimination of basic tastes. *Biosensors and Bioelectronics*, vol. 22(11), 2624–2628.
29. S. Buratti, S. Benedetti, M. Scampicchio, E.C. Pangerod (2004). Characterization and classification of Italian Barbera wines by using an electronic nose and an amperometric electronic tongue. *Analytica Chimica Acta*, vol. 525, 133–139.
30. V. Parra, A.A. Arrieta, J.B. Fernandez-Escudero, M.L. Rodriguez-Mendez, J.A. De Saja (2006). Electronic tongue based on chemically modified electrodes and voltammetry for the detection of adulterations in wines. *Sensors and Actuators B*, vol. 118, 448–453.
31. S. Buratti, D. Ballabio, S. Benedetti, M.S. Cosio (2007). Prediction of Italian red wine sensorial descriptors from electronic nose, electronic tongue and spectrophotometric measurements by means of Genetic Algorithm regression models. *Food Chemistry*, vol. 100, 211–218.
32. A. Legin, A. Rudnitskaya, L. Lvova, Y. Vlasov, C. Di Natale, A. D'Amico (2003). Evaluation of Italian wine by the electronic tongue: Recognition, quantitative analysis and correlation with human sensory perception. *Analytica Chimica Acta*, vol. 484, 33–44.
33. R.B. Bjorklund, C. Magnusson, P. Martensson, F. Winquist, C. Krantz-Rülcker (2009). Continuous monitoring of yoghurt fermentation using a noble metal electrode array. *International Journal of Food Science and Technology*, vol. 44, 635–640.
34. L.A. Dias, A.M. Peres, M. Vilas-Boas, M.A. Rocha, L. Estevinho, A.A.S.C. Machado (2008). An electronic tongue for honey classification. *Microchimica Acta*, vol. 163, 97–102.
35. S. Iiyama, M. Yahiro, K. Toko (2000). Measurements of soy sauce using taste sensors. *Sensors and Actuators B*, vol. 66, 205–206.
36. L. Moreno, A. Merlos, N. Abramova, C. Jimenez, A. Bratov (2006) A multi-sensor array used as an electronic tongue for mineral water analysis. *Sensors and Actuators B*, vol. 116, 130–134.

37. J. Gallardo, S. Alegret, R. Munoz, M. de Roman, L. Leija, M. del Valle (2003). An electronic tongue using potentiometric all-solid-state PVC-membrane sensors for the simultaneous quantification of ammonium and potassium ionsin water. *Analytical Bioanalytical Chemistry*, vol. 377, 248–256.
38. L.A. Dias, A.M. Peres, A.C.A. Veloso, F.S. Reis, M. Vilas-Boasa, A.A.S.C. Machado (2009). An electronic tongue taste evaluation: identification of goat milk adulteration with bovine milk. *Sensors and Actuators B*, vol. 136, 209–217.
39. T.R.L.C. Paixao, M. Bertotti (2009) Fabrication of disposable voltammetric electronic tongues by using Prussian Blue films electrodeposited onto CD-R gold surfaces and recognition of milk adulteration. *Sensors and Actuators B*, vol. 137, 266–273.
40. P. Ciosek, R. Maminska, A. Dybko, W. Wroblewski (2007). Potentiometric electronic tongue based on integrated array of microelectrodes. *Sensors and Actuators B*, vol. 127, 8–14.
41. P. Ciosek, W. Wroblewski (2008). Miniaturized electronic tongue with an integrated reference microelectrode for the recognition of milk samples. *Talanta*, vol. 76, 548–556.
42. P. Ciosek, Z. Brzozka, W. Wroblewski (2006). Electronic tongue for flowthrough analysis of beverages. *Sensors and Actuators B*, vol. 188, 454–460.
43. F. Winquist, R. Bjorklund, C. Krantz-Rulcker, I. Lundstrom, K. Ostergren, T. Skoglund (2005). An electronic tongue in the dairy industry. *Sensors and Actuators B*, vol. 111–112, 299–304.
44. P. Ciosek, E. Augustyniak, W. Wroblewski (2004). Polymeric membrane ion selective and cross-sensitive electrode-based electronic tongue for qualitative analysis of beverages. *Analyst*, vol. 129, pp.639–644.
45. L. Lvova, S.S. Kim, A. Legin, Y. Vlasov, J.S. Yang, G.S. Cha, H. Nam (2002). All-solid-state electronic tongue and its application for beverage analysis. *Analytica Chimica Acta*, vol. 468, 303–314.
46. F. Winquist, C. Krantz-Rulcker, P. Wide, I. Lundstrom (1998). Monitoring of freshness of milk by an electronic tongue on the basis of voltammetry. *Measurement Science and Technology*, vol. 9, 1937–1946.
47. A. Legin, A. Rudnitskaya, Y. Vlasov, C. Di Natale, F. Davide, A. D'Amico (1997). Tasting of beverages using an electronic tongue. *Sensors and Actuators B*, vol. 44, 291–296.
48. C. Soderstrom, F. Winquist, C. Krantz-Rulcker (2003) Recognition of six microbial species with an electronic tongue. *Sensors and Actuators B*, vol. 89, 248–255.
49. M.J. Gismera, S. Arias, M.T. Sevilla, J.R. Procopio (2009). Simultaneous quantification of heavy metals using a solid state potentiometric sensor array. *Electroanalysis*, vol. 21(8), 979–987.
50. D. Calvo, A. Duran, M. del-Valle (2007). Use of sequential injection analysis to construct an electronic-tongue: Application to multidetermination employing the transient response of a potentiometric sensor array. *Analytica Chimica Acta*, vol. 600, 97–104.
51. A. Rudnitskaya, A. Legin, B. Seleznev, D. Kirsanov, Y. Vlasov (2008). Detection of ultra-low activities of heavy metal ions by an array of potentiometric chemical sensors. *Microchimica Acta*, vol. 163, 71–80.
52. M.J. Gismera, S. Arias, M.T. Sevilla (2009). Simultaneous quantification of heavy metals using a solid state potentiometric Sensor Array. *Electroanalysis*, vol. 21(8), 979–987.
53. A. Legin, D. Kirsanov, A. Rudnitskaya, V. Babain (2008). Cross-sensitive rare earth metal ion sensors based on extraction systems. *Sensors and Actuators B*, vol. 131, 29–36.
54. C. Soderstrom, H. Boren, F. Winquist, C. Krantz-Rulcker (2003). Use of an electronic tongue to analyze mold growth in liquid media. *International Journal of Food Microbiology*, vol. 83, 253–261.
55. C. Soderstrom, F. Winquist, C. Krantz-Rulcker (2003). Recognition of six microbial species with an electronic tongue. *Sensors and Actuators B: Chemical*, vol. 89, 248–255.
56. H. Johnson, O. Karlsson, F. Winquist, C. Krantz-Rulcker, L.G. Ekedahl (2003). Predicting microbial growth in pulp using an electronic tongue. *Nordic Pulp & Paper Research Journal*, vol. 18, 134–140.
57. A. Riul Jr., C.A.R. Dantas, C.M. Miyazakiand, O.N. Oliveira (2010). Recent advances in electronic tongues. *Analyst*, vol. 35(10), 2453–2744.
58. P. Ciosek, W. Wroblewski (2007). Sensor arrays for liquid sensing – electronic tongue systems. *Analyst*, vol. 132, 963–978.

chapter six

Water quality monitoring: Design of an automated classification and authentication tool

Water quality can be monitored with various motives and purposes; however, a broad-based framework of monitoring, improvised under variable circumstances to cater to those varying needs, should be the target to be achieved. Whatever be the purpose, the first and foremost requirement in water quality monitoring is the characterization of water, which can be accomplished either by laboratory-based analytical instruments or using some *simple, reliable, sensitive, and inexpensive* equipment for quick field measurement like an electronic tongue (ET). An ET system using electrochemical analysis provides the opportunity to quantify and qualify water samples required for various applications/purposes and this chapter focuses on an electrochemical/ET system-based characterization of water. A brief account on the research and development efforts related to water quality monitoring using a multisensor array (specifically ET) will provide a proper context and perspective to the aim and scope of the present chapter.

Use of potentiometric electronic tongues (PETs) in classifying natural waters and drinking water has been reported by Martinez-Manez et al. [1] and Garcia-Breijo et al. [2] respectively. Martinez-Manez et al. [1] performed a qualitative analysis of the different natural waters, including seven mineral waters, tap water, and post-osmosis water, using fuzzy ARTMAP neural networks (a synthesis of fuzzy logic and *adaptive resonance-theory* neural networks) with a classification rate of higher than 93%. Garcia-Breijo et al. [2] classified drinking water based on neural networks algorithms using PET data. They also implemented different pattern-matching algorithms including fuzzy ARTMAP, a multilayer feed-forward (MLFF) network, and linear discriminant analysis (LDA) to make a comparative assessment on the recognition rate of an unknown test sample and drinking water discriminating performance. Moreno et al. [3] used an array of ISFET (ion-sensitive field-effect transistor)-based sensors, containing different membranes sensitive to distinct ions, combined with conductivity and redox-potential measurement in mineral water assessment. They classified different kinds of mineral waters and quantified some ions, such as Ca^{+2}, Na^{+}, K^{+}, and Cl^{-}, using hierarchical clustering analysis (HCA) and principal component analysis (PCA). The most important groups of potentiometric sensors are the ion-selective electrodes (ISEs), which are able to measure a potential dependent on the concentration of the target ions present in the solution. The problem with this type of sensor is the fouling of membranes or sensitive surfaces by chemicals or biological agents.

Winquist et al. [4] developed a voltammetric e-tongue (VET) device consisting of an array of four noble metals (platinum (Pt), gold (Au), rhodium (Rh) and iridium (Ir)) as working electrodes, embedded in epoxy resin inside a stainless steel cylinder; the steel cylinder itself was used as an auxiliary electrode. By applying a pulse sequence in this VET, different intensity versus time curves for each electrode was obtained. Krantz-Rulcker et al. [5] and Olsson et al. [6] successfully applied this VET to monitor the efficiency of the tap water filtration process and to detect and quantify surfactants during cleaning stages in automatic washing machines. Kundu et al. [7] used BIS-certified (Bureau of Indian Standards) drinking water samples with the VET to design a PLS (partial least squares)-based classifier, which was also used to authenticate unknown water samples. Kundu et al. [8] also proposed an authentication method involving a hybrid of slantlet-transform (ST) and an artificial neural network (ANN) in commercial mineral water brand classification and authentication, and it revealed encouraging efficiency. Although there are many applications employing VETs, their ability to identify, classify, and quantify gets restricted because of the usage of only noble metal electrodes. Noble metal electrodes are able to detect redox-active species showing reversible or irreversible redox processes in a solution (Faradic systems). It has been found that these electrodes indirectly respond to the concentration of ionic or neutral species (not redox-active) because their presence brings about changes in output current signatures. These changes are related to viscosity, the electrical resistance of the solution, and the capacitance of the electrical double layer [9]. The VET's ability to classify and quantify is significantly enhanced by arraying noble metals with non-noble metal electrodes like nickel or cobalt [10]. In fact, this combination of metals has been proved to be successful in detecting and quantifying anions, which may be a consequence of the chemical reactions occurring between the anion present in the solution and the metal cations generated by electrode surface oxidation. The formation of complex species can also play an important role in the detecting ammonia, amines, or nitrites, among others [9].

Ninenty-seven percent of the earth's water content is saline, leaving only 3% of water as fresh water. Most of the fresh water is locked in icecaps and glaciers (69%) and as groundwater (30%), while the lakes and rivers (jointly) account for only 0.3% of the earth's total fresh water. A simple way to compensate for the fresh water scarcity is to reuse the water after treating wastewater. The primary concerns in recycling treated water are public health and soil quality preservation (toward prospective irrigation). The level of pollutants in the treated wastewater should be monitored and controlled. The analytical methods are laborious, time consuming, and not suitable for bulk monitoring. Continuous monitoring and control of these pollutants demands significant investments because of the usage of online analyzers in wastewater treatment plants (WWTP) (for monitoring levels of ammonium, nitrate, phosphate, biochemical oxygen demand [BOD], chemical oxygen demand [COD], etc.). Generally, control in a WWTP relies on the information obtained from low-cost sensors (pH, redox, conductivity) in order to roughly/indirectly estimate the concentration of pollutants. Based on these indirect measures and on operators' experiences, decisions are made to amend the water treatment processes in response to diurnal/seasonal/occasional fluctuations of the pollutants in the wastewater. Application of an e-tongue device has provided a new dimension to the wastewater treatment (WWT) research. Di Natale et al. [11] presented a multicomponent analysis of polluted water using a potentiometric sensor array, containing several chalcogenide glass sensors. Samples of water from the Neva River (a river flowing through the city of St. Petersburg) had been artificially polluted with ionic metals in order to simulate generic industrial waste pollution. Nonlinear least squares and neural networks were used as machine-learning components of the proposed e-tongue. A flow injection VET was used to characterize the paper mill wastewaters by Gutes et al. [12]. They analyzed wastewater and predicted chemical oxygen demand, conductivity, and pH using a PCA and ANN. Campos et al. [13] reported a VET that had been used to characterize influent and effluent waters of a WWTP treating domestic wastewater. The proposed system consists of an array of eight metallic electrodes: Au, Pt, Ir, Rh, Ag, Cu, Ni, and Co and their combined use as working electrodes in voltammetric experiments. With this system, a prediction of typical pollution parameters measured in WWTPs such as soluble chemical oxygen demand (SCOD), soluble biological oxygen demand (SBOD), ammonia, orthophosphate, sulfate, acetic acid, and alkalinity was achieved using PLS analysis. WWTPs with e-tongue technology to monitor and recognize abnormal operating conditions is a possibility one can harness [14]. In an anticipation of possible contingences, the ET-based strategy will allow WWTPs to take corrective measures.

Contamination of soil and ground water due to heavy metal is one of the major concerns in today's world. Because of the high degree of toxicity, arsenic, cadmium, chromium, lead, and mercury rank among the priority metals that are sensitive to public health. They are released by natural and anthropogenic activities. Their detection and remediation in ground water are of the utmost importance because billions of people, as well as organisms, are victims of the pollution they cause. There has been an urgent need to develop *simple, reliable, sensitive, and inexpensive* equipment for quick field measurement like ET-based arsenic detection and quantification [15]. Extensive research and development (R&D) initiatives have been taken over the last few years in order to detect and quantify arsenic [16–19], chromium [20–22], copper, and cadmium [23] using electrochemical methods like anodic stripping voltammetry, cyclic voltammetry, and adsorption stripping voltammetry. There has always been a gap between detection and contaminant abatement initiation. Bulk monitoring and *in situ* detection is the sole way to bridge the gap, and an ET system is evolving as a rational choice.

Apart from the already mentioned three categories of water monitoring, the water monitoring can be done on pathogen-contaminated water using ET instrumentation instead of biochemical analysis of contaminated water. In this case, tailoring the working electrode is an important concern. Studies have revealed that the voltammetric ET can assess the growth of mold and bacteria as well as different strains of molds. The deterioration of milk due to microbial growth has been correlated with colony-forming units using e-tongue methods [24].

Summarily, the following has been observed in the literature:

1. Natural and drinking water classification with respect to their mineral content, alkalinity, pH, conductivity, surfactant content, etc., using multisensory data generated by an electronic tongue has been the active research area over the last decade.
2. Recycled water quality monitoring with respect to the pollutants and detection of abnormal operating conditions in WWTPs are the areas that have witnessed increasing application of electronic tongue technology.
3. Detection and monitoring of heavy metals (arsenic, chromium, copper, antimony, etc.) in drinking and ground water using an ET is another significant area of research related to public health and irrigation.
4. Both PET and VET are deployed in search of an alternative to lab-based analytical techniques.
5. The data analysis/pattern recognition techniques so far used as machine-learning components of the ET device are predominantly PCA, PLS, LDA and ANN/wavelet-ANN/fuzzy ANN.

In view of these observations, this chapter aims to present a voltametric ET system design for water sample classification and authentication, which eventually can be extended to a general water-monitoring framework.

This chapter delivers a design of a commercial mineral/drinking water classifier/authenticator. The BIS-certified mineral water samples (Aquafina, Bisleri, Dolphin, Kingfisher, McDowell, and Oasis) are considered unique representatives of specific types, manufactured by the different manufacturers irrespective of the location of the manufacturing unit in India. These mineral water brands are supplied and distributed by different vendors all over the country. So the sample collected from any pack of water bottles from different stores should have same prescribed/unique quality or a minimal degree of variability with respect to their sources. Usually, labels on mineral water bottles mention major ionic compounds, and that can also easily help in characterization of water samples. However, there are several water samples available that do not have labels or have inaccurate labels on them. The mineral water brands under consideration are from famous mineral water manufacturing brands; hence, it is expected that there will not be much difference in their ionic compositions. This makes the authentication problem even harder. In this situation, a simple voltammetric measurement is not enough to clearly identify different samples; hence, the necessity of an automated authentication tool can be justified. Time series signals for six commercial mineral water brands are collected using electronic tongue instrumentation. For each category of water, captured output signals using two types of working electrodes (Ag and Pt) are used to design PLS, Sammon's nonlinear mapping (NLM), and SVM (support vector machine)-based classification and authentication systems. In short, a general water-monitoring framework has been delineated.

6.1 Experimentation for e-tongue signature generation due to various mineral water brands

The chemical composition of commercial mineral water samples generally available in the Indian market with a representative brand, Bisleri, is provided in Table 6.1. It was not possible to ascertain the exact ingredients of all other BIS-certified mineral water brands because of their proprietary nature. A volume of 100 mL was collected from each of the chosen water brands, and four stock solutions were prepared for each brand. By using these four stock solutions, samples were prepared for experimentation. A mixture of 0.1 M citric acid and 0.2 M Na_2HPO_4 was then added to maintain a pH of 6, which prevents changes in the pH of a solution during measurement. During the experiment, the temperature of all samples was maintained at 20–23 °C. Applying the input pulse waveform, the current transients were measured for all water samples with two different electrodes: silver and platinum (Ag and Pt). Four samples for each of the eight water brands were used for experimentation, resulting in 4402 × 4 features for every water brand for a specific working electrode. The evolved data matrix was as large as (4402 × 2 × 24). Figures 6.1 and 6.2 show the common input voltage waveform applied and the normalized output current for some representative water samples for the Ag working electrode configuration.

Some representative correlograms of partial autocorrelation functions (PACF) and autocorrelation functions (ACF) of them are presented in Figures 6.3 and 6.4, respectively. Platinum as the working electrode revealed moderate to large nonstationary behavior for

Table 6.1 Characteristics of Indian bottled water samples

Parameters	pH	TDS (mg/L)	Hardness (mg/L)	HCO3– (mg/L)	Cl– (mg/L)	NO3– (mg/L)	SO4– (mg/L)	Na+ (mg/L)	K+ (mg/L)	Mg++ (mg/L)	Ca++ (mg/L)
For Indian commercial mineral water brands	7.1–7.3	150–170	60–70	55–65	2.06–28.3	0.3–10.7	0.1–11.5	1–40	0.02–2.04	0.03–8	0.2–25
Bisleri	7.2	160	66.1	58	22	2	19.3	-	-	7.8	13.6

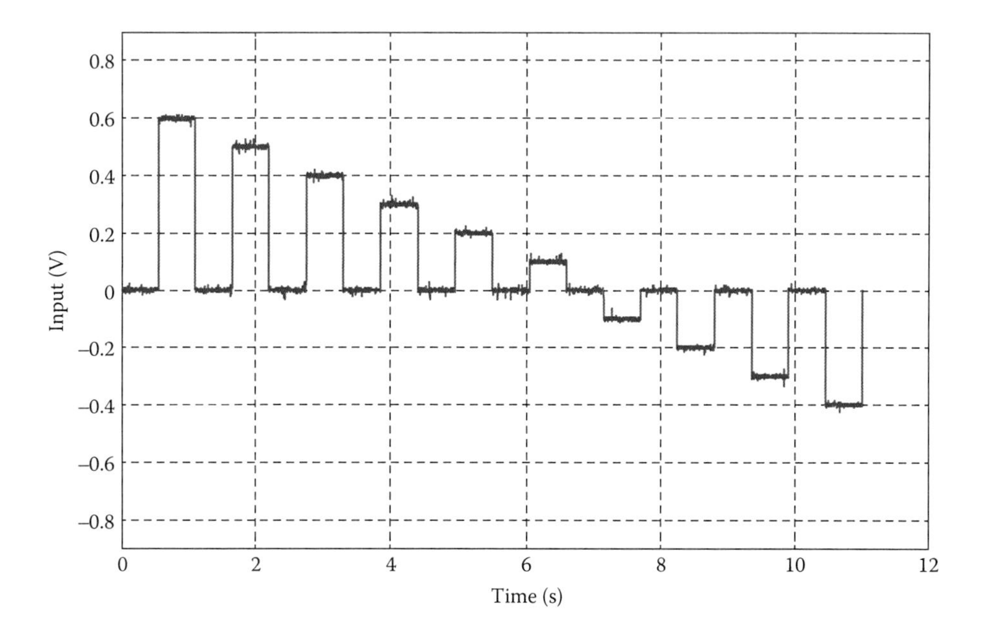

Figure 6.1 Input waveform applied to e-tongue.

e-tongue data using any of the mineral water brands as evidenced by the ACF and PACF functions. The Ljung-Box Q-statistic and Engle's ARCH (autoregressive conditional heteroskedastic effects) tests also supported the nonstationary behavior. Data collected using a silver electrode did not show any nonstationary behavior.

6.2 Design of PLS-based mineral water authenticator/classifier

This section is devoted to designing a water authenticator/classifier using PLS as a machine-learning component. A water data set of six brands (*viz.* Aquafina, Bisleri, Dolphin, Kingfisher, McDowell, and Oasis [each containing three samples]) containing 4402 features for each of them is considered for the design of a PLS-based classifier. The data set is generated/simulated by considering any sample of a brand (among the six chosen brands and generated using both the electrodes) added up/adulterated with 5%–10% of random noise; the generated samples are treated as unknown brand samples and used for authentication purposes. The generated fourth sample of every brand is also treated as an unknown test data set. The incorporation of up to 10% noise level for a sample is tested specifically to justify the applicability of the developed classifier in a rigorous manner. The PLS-based classifier is maneuvered in such a way that the classifier can accommodate the uncertainties in the measurements of unknown test samples. The time series features of six different classes of water samples (each containing three subsamples) are represented in the form of a predictor X matrix and a corresponding response Y matrix indicating the water class is generated in the form of a unit matrix. After the PLS regression, each of the regressed Y vectors of the response matrix are given a class membership. The designed PLS classifier is then used for predicting Ys representing unknown sample classes corresponding to unknown X samples.

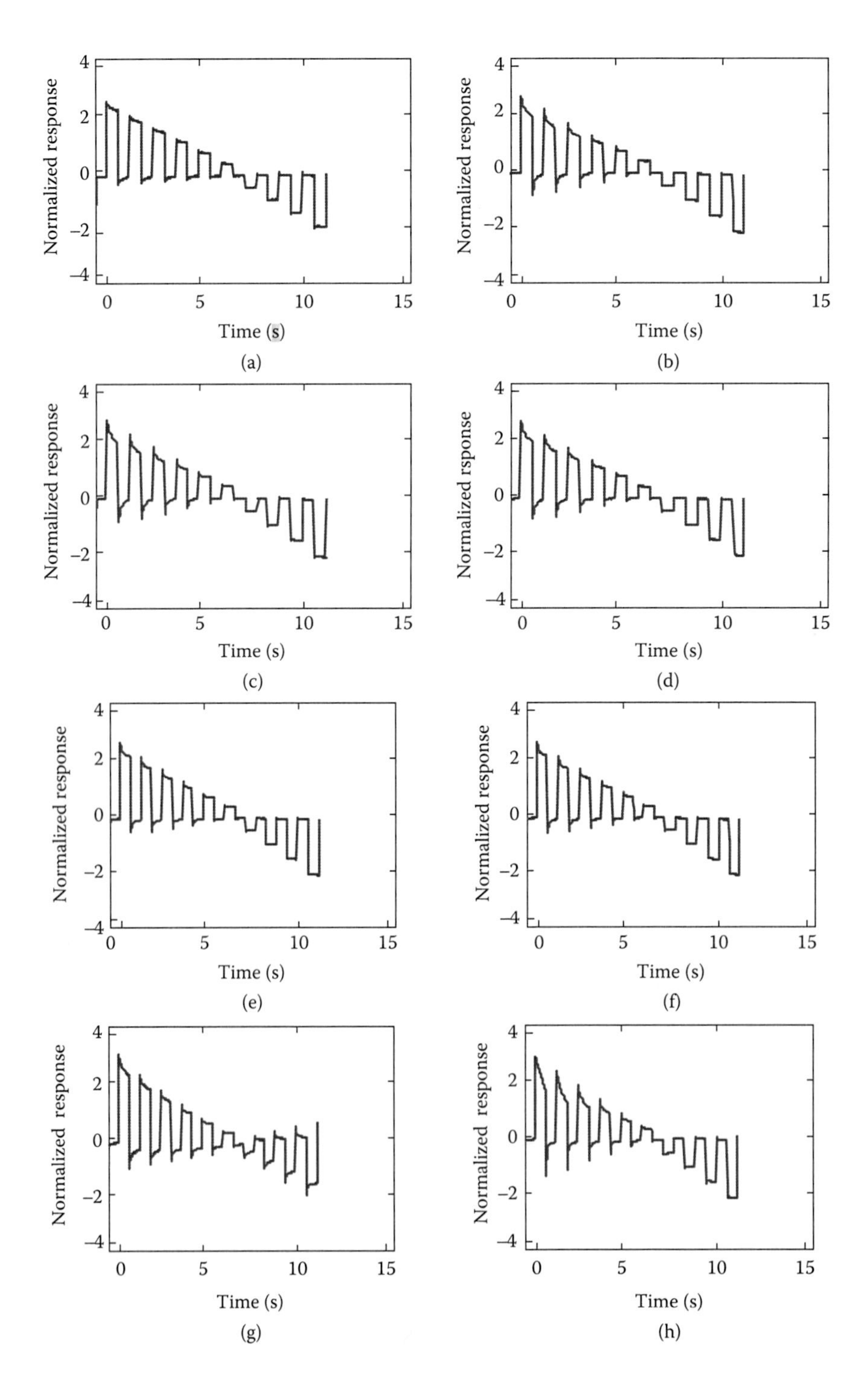

Figure 6.2 Normalized output current signatures for various water brands using a silver (Ag) working electrode: (a) Aquafina, (b) Bisleri, (c) Dolphin, (d) Kingfisher, (e) McDowell, (f) Oasis, (g) distilled water, and (h) tap water.

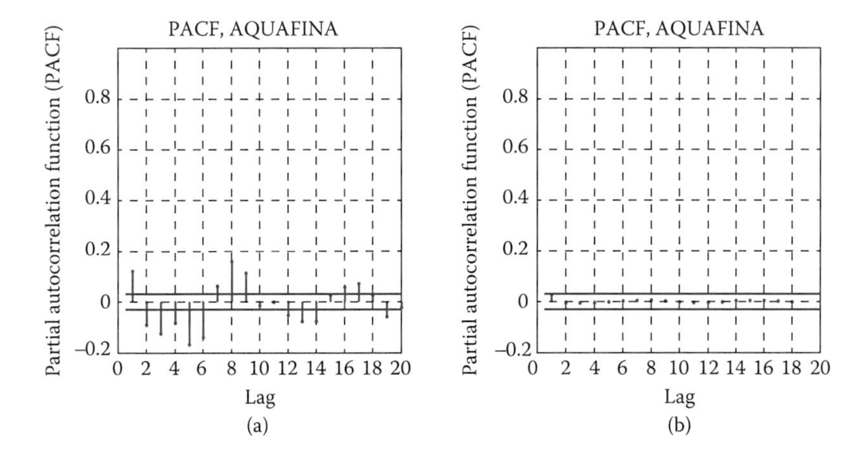

Figure 6.3 Partial autocorrelation functions for representative water brands: (a) Aquafina (platinum electrode) and (b) Bisleri (silver electrode).

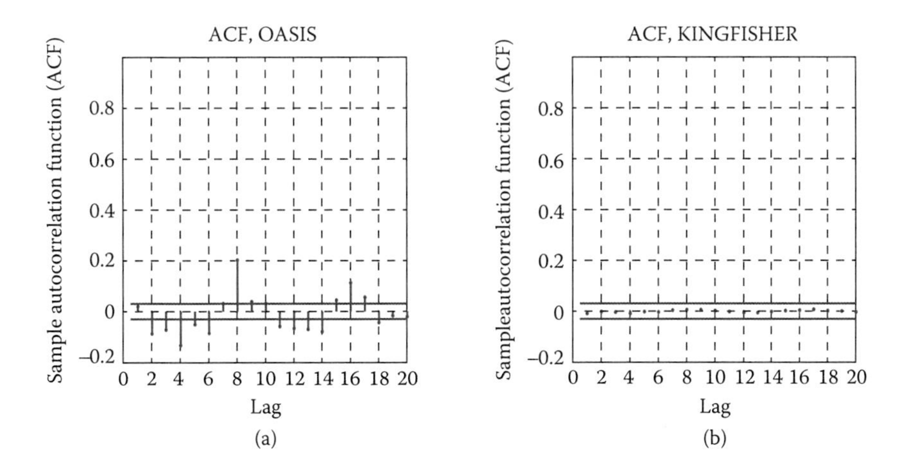

Figure 6.4 Autocorrelation functions for representative water brands: (a) Oasis (platinum electrode) and (b) Kingfisher (silver electrode).

6.2.1 Authentication/classification algorithm

The algorithm of the proposed PLS-based classifier operated in two modes, which are as follows:

Mode 1: Assigning Sample Classes

1. Formation of training/predictor matrix $X_{18\times4402}$: Out of the row vectors, row vector 1, 2, 3 = sample class 1; row vectors 4, 5, 6 = sample class 2; row vectors 7, 8, 9 = sample class 3; row vectors 10, 11, 12 = sample class 4; row vectors 13, 14, 15 = sample class 5; and row vectors 16, 17, 18 = sample class 6
2. $X = X^T$

3. Assigning response matrix $Y_{18\times18}$ as unit matrix.
4. Relating X and Y by a PLS regression; hence, determining the matrix of the regression coefficient and loading matrices corresponding to X and Y data.
5. Formation of test X matrix of dimension $(k \times 4402)$, $\in$ sample classes ≤ 6. $X = X^T$.
6. Prediction of k numbers of (1×18) Y vectors corresponding to k numbers of test vectors of the test X dataset using the developed model in step 4.
7. Determination of k numbers of (1×18) dimensional abs $(Y-1)$ vectors.
8. Detection of an outlier: A sample not among Aquafina, Bisleri, Dolphin, Kingfisher, McDowell, and Oasis. Detection of the minimum entry value among all 18 columns for each of k numbers of (1×18) dimensional abs $(Y-1)$ vectors. If any of the minimum entry from the 18 columns (abs $(Y-1)$) for those k numbers of test vectors is > (±20% of 1.0), most likely the sample class corresponding to that k is an outcast.
9. Generation of class membership: Detection of the minimum entry value among all the columns for each of the k numbers of (1×18) dimensional abs $(Y-1)$ vectors is synonymous to finding out the entry close to 1 among all the columns for each of the k numbers of (1×18) dimensional PLS regressed abs (Y) vectors. The column number corresponding to the minimum entry is the class identification number of that k^{th} test vector. In this way, all k numbers of vectors of the test X data set occupied a unique class membership ranging from 1 to 18.

If the

$$\begin{gathered}\text{identification number} = 1,\ 2,\ 3,\ \text{detected class} = 1\\ \text{identification number} = 4,\ 5,\ 6,\ \text{detected class} = 2\\ \text{identification number} = 7,\ 8,\ 9,\ \text{detected class} = 3\\ \text{identification number} = 10,\ 11,\ 12,\ \text{detected class} = 4\\ \text{identification number} = 13,\ 14,\ 15,\ \text{detected class} = 5\\ \text{identification number} = 16,\ 17,\ 18,\ \text{detected class} = 6\end{gathered}$$

No two k vectors had the same membership/identification number or class because the authenticator in *mode 1* operated with 100% efficiency.

Mode 2: Authentication of Unknown Sample

1. Formation of test X matrix of dimension $(k \times 4402)$, $k \in$ sample classes ≤ 7, where the seventh class is an unknown class. $X = X^T$.
2. Prediction of k (1×19) Y vectors corresponding to k test vectors of the X test data set using the developed model in step 4 of *mode 1*.
3. Determination of k (1×19) dimensional abs $(Y-1)$ vectors.
4. Detection of outlier: Detection of the minimum entry value among all 19 columns (abs $(Y-1)$) for $k = 7$. If any of the minimum entry among them is > (± 20% of 1), most likely the sample class corresponding to $k = 7$ is an outcast. (This was proved with tap water.)
5. Authentication of unknown sample: This was decided from the minimum entry values among all 19 columns (abs $(Y-1)$) for $k = 7$. This very minimum entry is the repetition (within ± 0.001) of any one of the other 18 minimum entries of the classifier, hence it possesses a repeating class identification number ranging from 1 to 18. These

identification numbers correspond to any one of the classes among 1–6, as already discussed in *mode 1*.

A PLS-based water authentication/classification code is provided in Program 6.1.

6.2.2 *Performance evaluation of PLS-based authentication and classification*

Table 6.2 presents the membership values (minimum entry values) corresponding to the identified water classes for both the electrode systems. Figures 6.5 through 6.8 reveal some of the encouraging authentication performances based on experimental results of Ag and Pt electrodes, respectively. Any unknown sample taken for authentication, apart from the samples which have been used for development of the classifier, is not going to be detected by the developed PLS-based classifier. The designed classifiers performed without misclassification. So far as the performances are concerned, the Pt and Ag electrode-based classifiers made no distinction.

Table 6.2 Membership values of identified water classes using Pt and Ag electrode-based PLS authentication

Electrode system	Aquafina	Bisleri	Dolphin	Kingfisher	McDowell	Oasis	Unknown
Pt electrode	0.1776	0.0009	0.0148	0.0124	0.0250	0.0098	0.0148
Ag electrode	0.1823	0.0954	0.1440	0.0899	0.1287	0.0017	0.1287

For **Pt** electrode system: Dolphin as unknown sample; for **Ag** electrode system: McDowell as unknown sample.

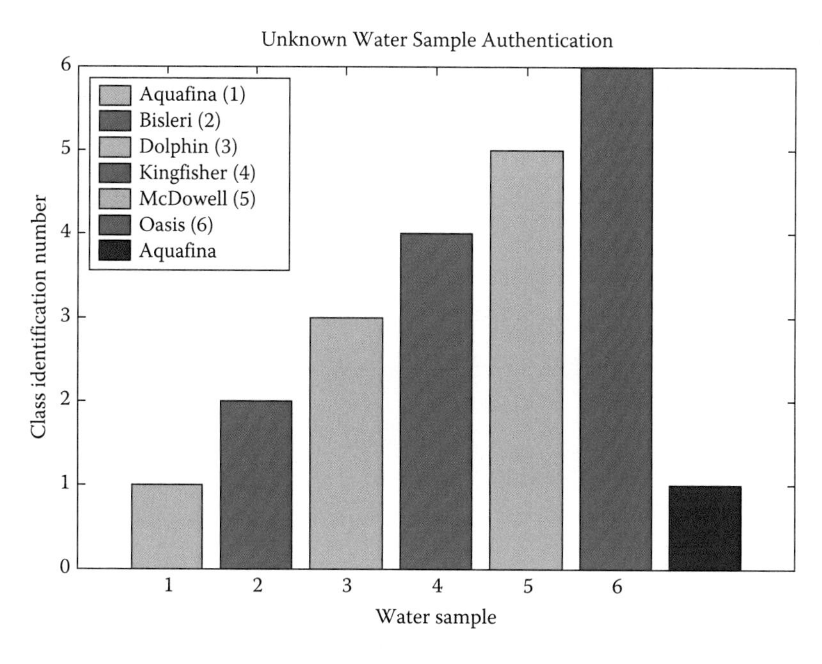

Figure 6.5 Authentication performance of Ag electrode-based PLS classifier (Aquafina as unknown sample).

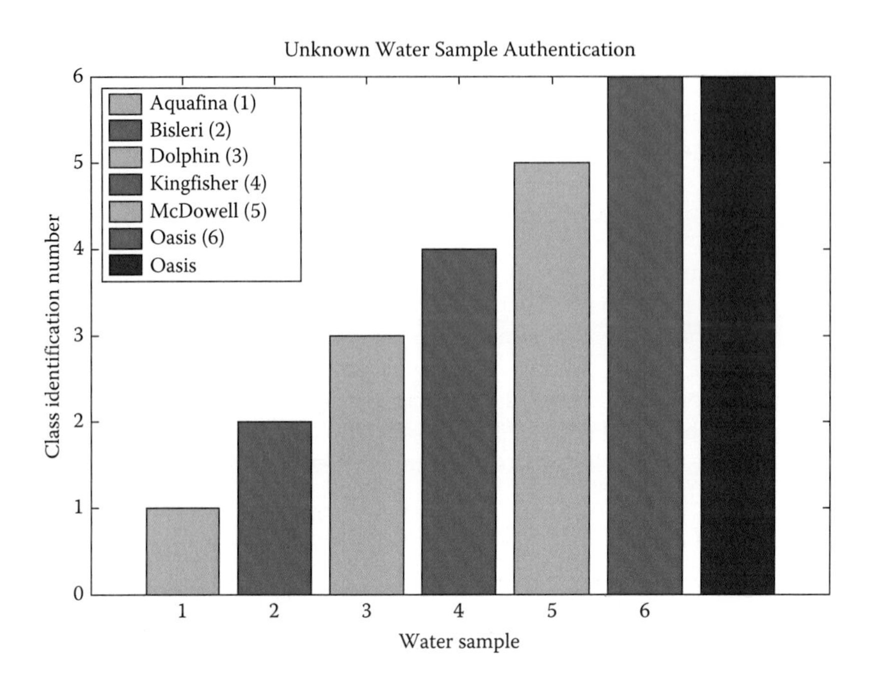

Figure 6.6 Authentication performance of Ag electrode-based PLS classifier (Oasis as unknown sample).

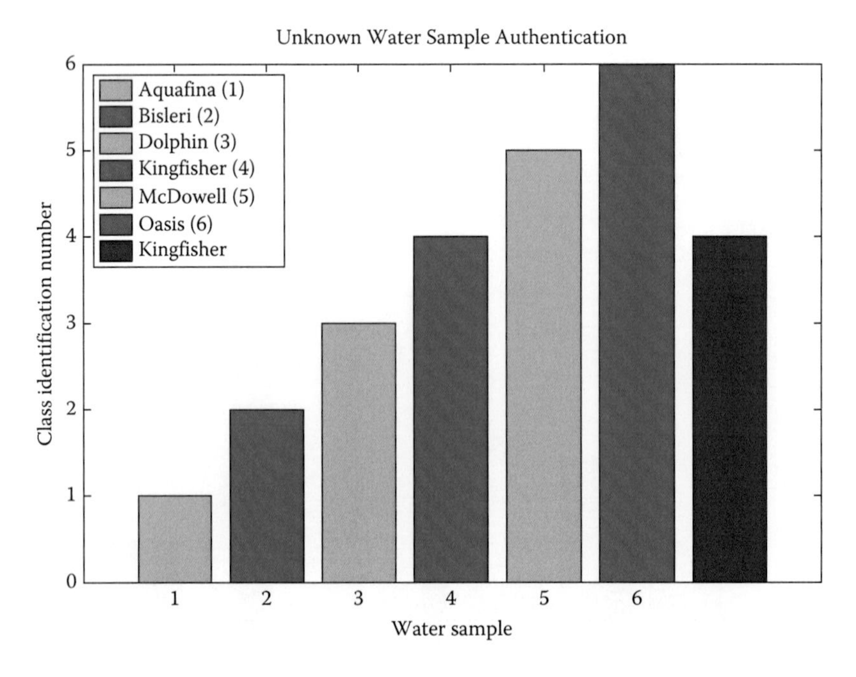

Figure 6.7 Authentication performance of Pt electrode-based PLS classifier (Kingfisher as unknown sample).

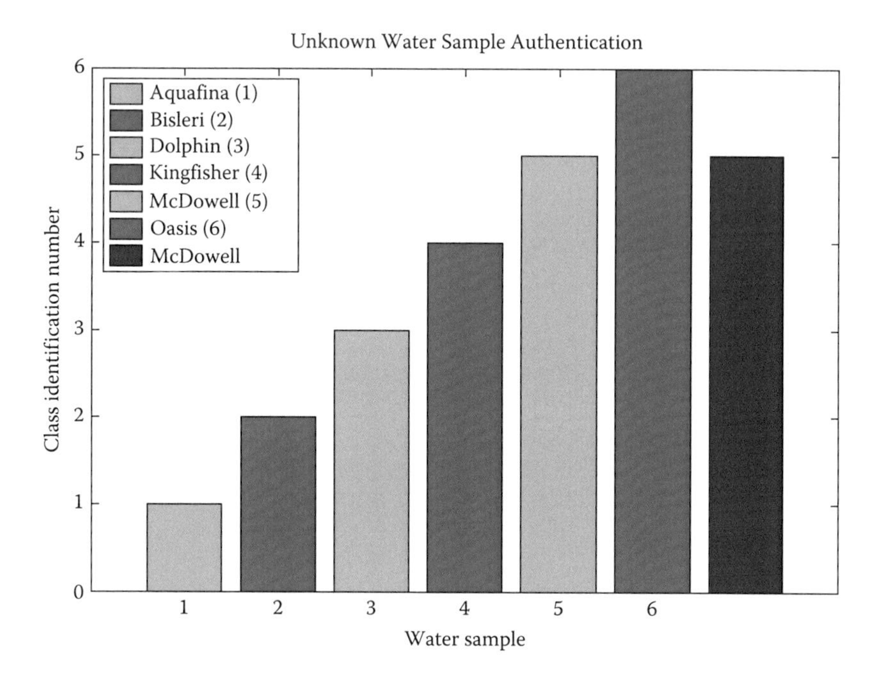

Figure 6.8 Authentication performance of Pt electrode-based PLS classifier (McDowell as unknown sample).

Program 6.1 MATLAB® Code for PLS-based authentication/classification

```
% This code is implemented using the PLS function, PLS.m (provided in
chapter 2).This program is developed for authentication/classification of
tea/water/fruit juice/wine/milk samples.
% Input of normalized liquid samples in triplicates.
clc;
clear all

BR=load('BROOKEBOND_3S_N.txt','_ascii');
DD=load('DOUBLEDAIMOND_3S_N.txt','_ascii');
LP=load('LIPTON_3S_N.txt','_ascii');
LD=load('LIPTON-DARJEELING_3S_N.txt','_ascii');
MR=load('MARVEL_3S_N.txt','_ascii');
MRD=load('MARYADA_3S_N.txt','_ascii');
RD=load('REDLABEL_3S_N.txt','_ascii');
TJ=load('TAJMAHAL_3S_N.txt','_ascii');
TG=load('TATAGOLD_3S_N.txt','_ascii');
TET=load('TETLEY_3S_N.txt','_ascii');

A=[BR(:,1:3) DD(:,1:3) LP(:,1:3) LD(:,1:3) MR(:,1:3) MRD(:,1:3) RD(:,1:3)
TJ(:,1:3) TG(:,1:3) TET(:,1:3) RD(:,1)+R];

A=A';
N1=size(A);
```

```
x1=A(1,:);
x2=A(2,:);
x3=A(3,:);
x4=A(4,:);
x5=A(5,:);
x6=A(6,:);
x7=A(7,:);
x8=A(8,:);
x9=A(9,:);
x10=A(10,:);
x11=A(11,:);
x12=A(12,:);
x13=A(13,:);
x14=A(14,:);
x15=A(15,:);
x16=A(16,:);
x17=A(17,:);
x18=A(18,:);
x19=A(19,:);
x20=A(20,:);
x21=A(21,:);
x22=A(22,:);
x23=A(23,:);
x24=A(24,:);
x25=A(25,:);
x26=A(26,:);
x27=A(27,:);
x28=A(28,:);
x29=A(29,:);
x30=A(30,:);
x31=A(31,:);

% Combined training data with normalization, as (31×44402) predictor
matrix
X=[x1;x2;x3;x4;x5;x6;x7;x8;x9;x10;x11;x12;x13;x14;x15;x16;x17;x18;x19;x20;
x21;x22;x23;x24;x25;x26;x27;x28;x29;x30;x31];
N2=size(X);
  %% Creation of test data as (ii×4402) matrix/predicted matrix
BB=[BR(:,3) DD(:,3) LP(:,3) LD(:,3) MR(:,3) MRD(:,3) RD(:,3) TJ(:,3)
TG(:,3) TET(:,3) RD(:,3)];
BB=BB';
R=0.25*rand(1,4402);
z1=BB(1,:)+R;
z2=BB(2,:)+R;
z3=BB(3,:)+R;
z4=BB(4,:)+R;
z5=BB(5,:)+R;
z6=BB(6,:)+R;
z7=BB(7,:)+R;
z8=BB(8,:)+R;
z9=BB(9,:)+R;
z10=BB(10,:)+R;
z11=BB(11,:)+R;
```

```
data1=[z1;z2;z3;z4;z5;z6;z7;z8;z9;z10;z11];
N3=size(data1)

  %>>>>>>>>>>>>>>>>>>>>>>>>>>>>>>>>>>>>>>>>>>>>>>>>>>>>>>>>>>>>
%
% % % % Define class indicator as Y
  Y = kron(eye(31),ones(1,1));
% %
    ymean = mean(Y);
    ystd = std(Y);
%
  Y = (Y - ymean(ones(31,1),:))./ystd(ones(31,1),:);
% %
% % % % % Tolerance for 90 percent score
tol = (1-0.95) * 1 * 31;
% % % Perform PLS
[T,P,U,Q,B] = pls(X,Y,tol);

B
% % % Results
  fprintf('Number of components retained: %i\n',size(B,1))
%
%
% % % % Testing classes, all unknown classes were tested
X1 = (data1(1,:));
X2 = (data1(2,:));
X3 = (data1(3,:));
X4 = (data1(4,:));
X5 = (data1(5,:));
X6 = (data1(6,:));
X7 = (data1(7,:));
X8 = (data1(8,:));
X9 = (data1(9,:));
X10 = (data1(10,:));
X11 = (data1(11,:));
%
%
% % predicted classes
Y1 = X1 * (P*B*Q');
Y2 = X2 * (P*B*Q');
Y3 = X3 * (P*B*Q');
Y4 = X4 * (P*B*Q');
Y5 = X5 * (P*B*Q');
Y6 = X6 * (P*B*Q');
Y7 = X7 * (P*B*Q');
Y8 = X8 * (P*B*Q');
Y9 = X9 * (P*B*Q');
Y10 = X10 * (P*B*Q');
Y11 = X11 * (P*B*Q');

% %
Y1 = Y1 .* ystd(ones(1,1),:) + ymean(ones(1,1),:);
Y2 = Y2 .* ystd(ones(1,1),:) + ymean(ones(1,1),:);
Y3 = Y3 .* ystd(ones(1,1),:) + ymean(ones(1,1),:);
```

```
Y4 = Y4 .* ystd(ones(1,1),:) + ymean(ones(1,1),:);
Y5 = Y5 .* ystd(ones(1,1),:) + ymean(ones(1,1),:);
Y6 = Y6 .* ystd(ones(1,1),:) + ymean(ones(1,1),:);
Y7 = Y7 .* ystd(ones(1,1),:) + ymean(ones(1,1),:);
Y8 = Y8 .* ystd(ones(1,1),:) + ymean(ones(1,1),:);
Y9 = Y9 .* ystd(ones(1,1),:) + ymean(ones(1,1),:);
Y10 = Y10 .* ystd(ones(1,1),:) + ymean(ones(1,1),:);
Y11 = Y11 .* ystd(ones(1,1),:) + ymean(ones(1,1),:);

yclass=[Y1' Y2' Y3' Y4' Y5' Y6' Y7' Y8' Y9' Y10' Y11' ];
% %
% % % % Class is determined from matrix or column vector of (abs(Y1-1))
along
% %the second dimension or column of the matrix which is most close to 1


[dum1,aclassid]=min(abs(Y1-1),[],2);

[dum2,bclassid]=min(abs(Y2-1),[],2);

[dum3,cclassid]=min(abs(Y3-1),[],2);

[dum4,dclassid]=min(abs(Y4-1),[],2);

[dum5,eclassid]=min(abs(Y5-1),[],2);

[dum6,fclassid]=min(abs(Y6-1),[],2);

[dum7,gclassid]=min(abs(Y7-1),[],2);

[dum8,hclassid]=min(abs(Y8-1),[],2);

[dum9,iclassid]=min(abs(Y9-1),[],2);

[dum10,jclassid]=min(abs(Y10-1),[],2);

[dum11,kclassid]=min(abs(Y11-1),[],2);

%>>>>>>>>>>>>>>>>>>>>>>>>>>>>>>>>>>>>>>>>>>>>>..

if (dum11==dum1)
kclassid=1
end
if (dum11==dum2)
kclassid=2
end
if (dum11==dum3)
kclassid=3
end

if (dum11==dum4)
kclassid=4
end
```

```
if (dum11==dum5)
kclassid=5
end

if (dum11==dum6)
kclassid=6
end
if (dum11==dum7)
kclassid=7
end
if (dum11==dum8)
kclassid=8
end

if (dum11==dum9)
kclassid=9
end
if (dum11==dum10)
kclassid=10
end

kclassid

dum=[dum1' dum2' dum3' dum4' dum5' dum6' dum7' dum8' dum9' dum10' dum11']


if (aclassid==1) || (aclassid==2)||(aclassid==3)||(aclassid==31)
aclassid=1
end
if (bclassid==4) || (bclassid==5)||(bclassid==6)||(bclassid==31)
bclassid=2
end
if (cclassid==7) || (cclassid==8)||(cclassid==9)||(cclassid==31)
cclassid=3
end
if (dclassid==10) || (dclassid==11)||(dclassid==12)||(dclassid==31)
dclassid=4
end
if (eclassid==13) || (eclassid==14)||(eclassid==15)||(eclassid==31)
eclassid=5
end
if (fclassid==16) || (fclassid==17)||(fclassid==18)||(fclassid==31)
fclassid=6
end

if (gclassid==19) || (gclassid==20)||(gclassid==21)||(gclassid==31)
gclassid=7
end
if (hclassid==22) || (hclassid==23)||(hclassid==24)||(hclassid==31)
hclassid=8
end
if (iclassid==25) || (iclassid==26)||(iclassid==27)||(iclassid==31)
iclassid=9
end
```

```
if (jclassid==28) || (jclassid==29)||(jclassid==30)||(jclassid==31)
jclassid=10
end
%>>>>>>>>>>>>>>>>>>>>>>>>>>>>>>>>>>>>>Formation of bar graphs. >>>>>>>>


bar(1,aclassid,'b');
hold on
bar(2,bclassid,'r');
% hold on
bar(3,cclassid,'b');
bar(4,dclassid,'r');
bar(5,eclassid,'b');
bar(6,fclassid,'r');
bar(7,gclassid,'b');
bar(8,hclassid,'r');
bar(9,iclassid,'b');
bar(10,jclassid,'r');
bar(11,kclassid,'b');
% %
hold off
% % % % %
```

6.3 Cross-correlation-based PCA and feature enhancement in classification

The simultaneous authentication of two unknown samples of a similar brand, but from a variable source, has been a motivation behind this section [25]. The unknown samples to be authenticated must be among the brands used to develop classifiers. The possibility of the misclassification or misidentification of the second unknown sample in the same category by using ordinary machine-learning algorithms always remains. The nearest neighborhood method is decided upon by a very low range/scale of within-cluster distances, which often produces a false neighborhood and, hence, misclassification. Therefore, the scaling up of the within-cluster distance might help in correctly detecting the two unknown samples of the same brand with a certain degree of variability with respect to their sources. A "variable source" means water samples of any particular brand collected from different plants or batches. This is simulated by adding noise to the parent sample of any brand, showing the degree of departure between the same brand of water samples from variable batches and plants. The present work proposes a soft scale-up of intercluster as well as intracluster distance using cross-correlation coefficients between the input and output signatures of an e-tongue instead of using e-tongue outputs alone. Figures 6.9 and 6.10 show the representative input voltages, output current waveforms, and cross-correlation coefficients for representative water brands with Ag and Pt electrodes, respectively.

6.3.1 Feature extraction

In the present and subsequent section two types of PCA and Sammon's classifiers are designed. For an ordinary PCA and Sammon's classifier, time series generated and simulated by considering any sample of any specific brand (of the six brands considered herein and generated for both the electrode systems) appended/adulterated with 5–10% of random

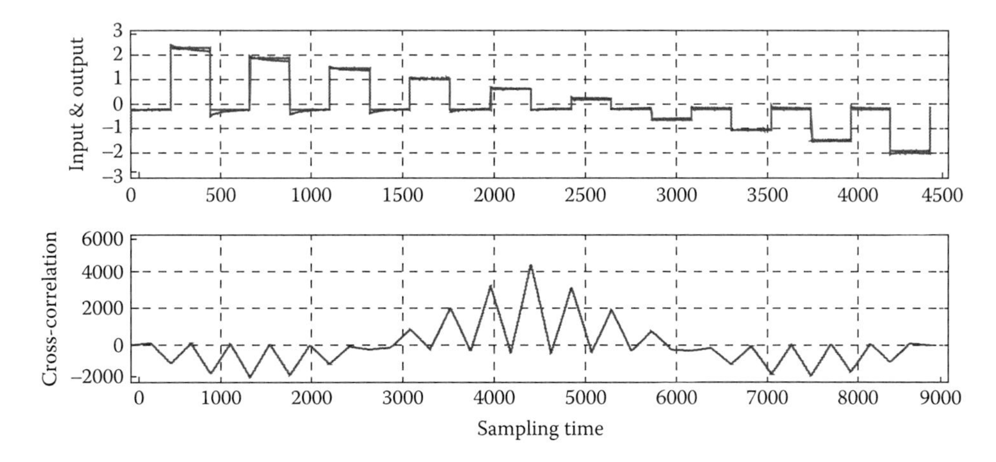

Figure 6.9 Input, output, and cross-correlation waveforms for Aquafina using Ag as working electrode.

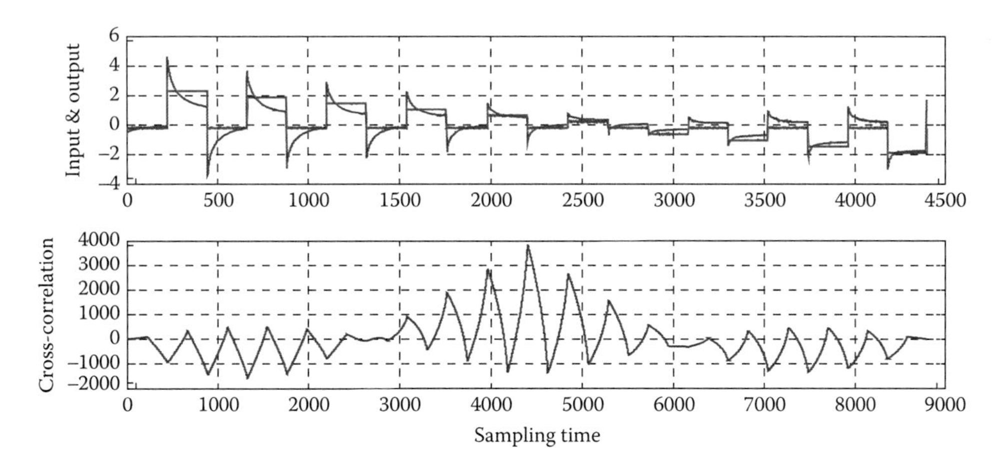

Figure 6.10 Input, output, and cross-correlation waveforms for Oasis using Pt as working electrode.

noise are considered to be the fifth replicate run/sample. The fourth sample of any specific brand appended with random noise, and the fifth sample of that brand of water, are treated as unknown brands and used for authentication purposes.

The water data set containing four replicate runs (e-tongue signatures in the form of time series) for each brand of water has been used for finding cross-correlation coefficients and used in the design of an enhanced PCA and Sammon's classifier. Time series data in the form of cross-correlation coefficients are derived by taking the mean output currents of the first three replicate runs of each brand and the corresponding average input voltage. The second cross-correlation coefficient series is generated between the fourth replicate run/sample output current and the corresponding input voltage. The third series of cross-correlation coefficients for each brand was done by adding 5–10% of random noise to their respective first series of cross-correlation coefficients. The second and third series of cross-correlation coefficients of any particular brand are treated as attributes of two unknown samples of the same brand and used for simultaneous authentication purposes.

A water data set of six brands (each containing five replicate runs/samples) containing 4402 features is considered for the design of a PCA-based classifier. While using the PCA classifier for authentication purposes, it may so happen that two unknown samples are shown to closely resemble any one of the brands among Aquafina, Bisleri, Dolphin, Kingfisher, McDowell, and Oasis. The scores are generated for 20 (6 × 3 known samples + 2 unknown samples of any of the aforementioned brands) water samples that have 4402 features. For authentication using the enhanced PCA classifier, scores are generated for (6 × 1) known + (1 + 1) unknown samples of any of these brands in terms of cross-correlation series (8 × 8803) cross-correlation coefficient matrices.

In order to demonstrate the ordinary Sammon's classification, a water data set of six brands (each containing five replicate runs/samples) containing 4402 features is considered or (30 × 4402) as design data set are considered. The MATLAB code for the generation of a cross-correlated time series is provided in Chapter 2.

6.3.2 *Performance evaluation of a PCA and an enhanced PCA classifier*

In designing an ordinary PCA classifier for the silver working electrode, the first two principal components alone captured 91% of the variance of the data set. The scores for the first and second principal components (PC1 and PC2) discriminates the dataset into seven distinct clusters (any five clusters among Aquafina, Bisleri, Dolphin, Kingfisher, McDowell, and Oasis and two unknown clusters closely resembling any one of the aforementioned brands with a certain degree of variability) at a time. For the platinum electrode, the first two principal components alone captured 93% of the variance of the data set. Figures 6.11 and 6.12 show the representative authentication performances of

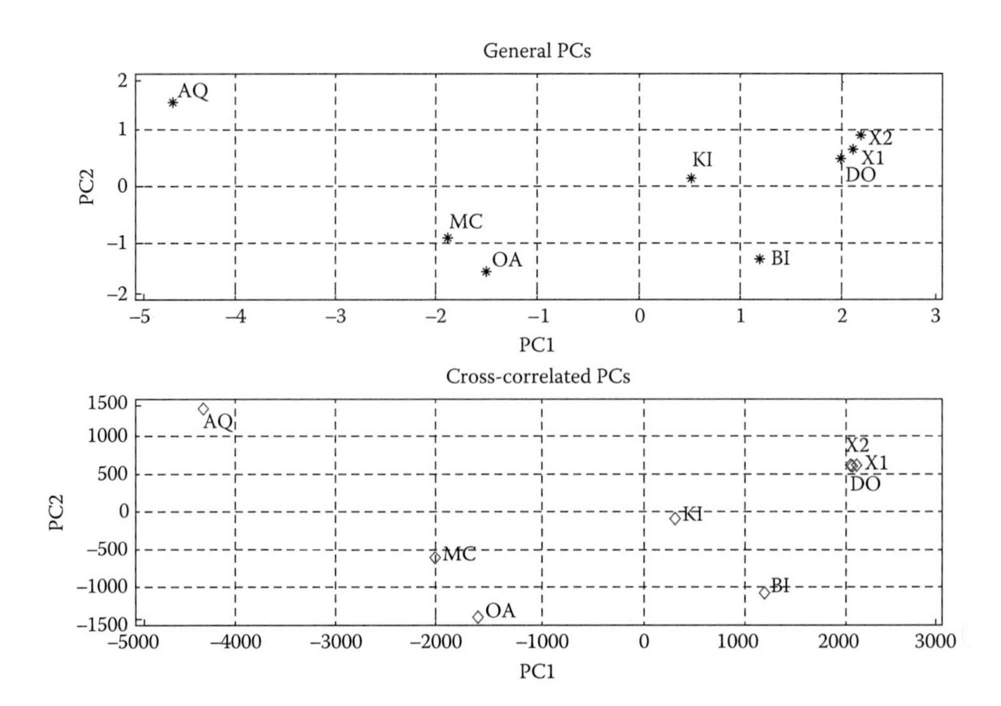

Figure 6.11 Cluster center distances between two unknown Dolphin samples and others in PCA classifiers using silver as working electrode.

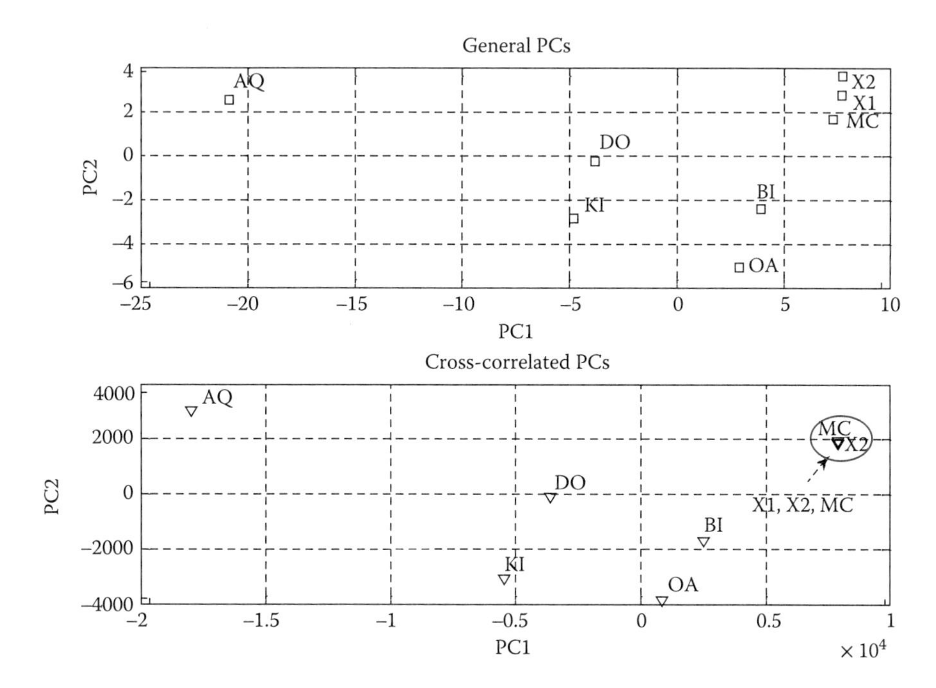

Figure 6.12 Cluster center distances between two unknown McDowell samples and others in PCA classifiers using platinum as working electrode.

the cross-correlation-based PCA classifier using silver and platinum electrodes, respectively, along with the simple PCA-based classification using output current signatures of the mineral water samples. It is evident from the figures that for cross-correlated PCAs, there is an enormous increment of inter- as well as intracluster boundaries, which can be viewed as soft scaling-up of separation margins and making the classification among different water brands simpler. The authentication performance of both PCA classifiers for each brand containing more than one simulated test sample was determined by calculating the Euclidean distances between the coordinates of any unknown sample and the centers of the remaining seven clusters. Thus, one (24 × 7) distance matrix was generated for each electrode system for the two types of PCA classifiers. Tables 6.3 and 6.4 represent the distance matrices for the silver and platinum electrodes, respectively. The minimum value entry in Tables 6.3 and 6.4 reflects the separation distance between the unknown sample and the sample cluster it resembles. The second minimal entry resembles the same brand with a low degree of variability. From the distance matrices, it is clear that for the cross-correlation-based PCA, the enhanced within-cluster boundary and intercluster distances are prevalent, while classification is much better than that by ordinary PCA classifiers. Therefore, the enhanced PCA classifiers hardly carry out any misclassification (Tables 6.3 and 6.4). The misclassification rate in the e-tongue with the platinum electrode is much smaller (2.4%) than that with silver as a working electrode (17.3%). When the intercluster distance of an unknown sample with other samples (excluding the two samples it resembles) is smaller than two minimal entries (responsible for the two unknown samples similar to any of the water brands), this is termed misclassification. The PCA and enhanced classifier code is provided in Program 6.2.

Table 6.3 Distance among known and unknown class centroids in PCA classifiers with silver as working electrode

DIST	AQ1	AQ2	BI	DO	KF	MC	OA
When unknown sample is similar to AQ1							
M-1	0.0283	*6.6448*	**{6.3486}**	6.6885	**{5.3635}**	**{3.3013}**	**{3.8102}**
M-2	21.6	*93.9*	6060.4	6461.0	4883.8	3037.2	3880.4
When unknown sample is similar to AQ2							
M-1	*6.6448*	6.6306	9.2601	9.3870	8.5556	7.5879	7.8535
M-2	*93.9*	76.0	6153.3	6553.8	4977.7	3122.2	3962.1

When unknown sample is similar to BI1

	AQ	BI1	BI2	DO	KF	MC	OA
M-1	**{6.2994}**	0.0364	*6.6961*	**{0.4169}**	**{1.0068}**	**{3.1632}**	**{2.6619}**
M-2	6081.5	22.1	*191.4*	1870.9	1317.7	3300.8	2854.5

When unknown sample is similar to BI2

	AQ	BI1	BI2	DO	KF	MC	OA
M-1	9.3119	*6.6961*	6.6660	6.8846	6.9220	7.4630	7.2104
M-2	6231.2	*191.4*	182.5	1987.8	1503.1	3414.8	2927.5

When unknown sample is similar to DO1

	AQ	BI	DO1	DO2	KF	MC	OA
M-1	6.7325	**{0.7620}**	0.0283	*6.5583*	**{1.4764}**	**{3.8129}**	**{3.3988}**
M-2	6495.8	1915.9	21.1	*65.7*	1916.2	4309.6	4220.3

When unknown sample is similar to DO2

	AQ	BI	DO1	DO2	KF	MC	OA
M-1	9.4385	6.9230	*6.5583*	6.5476	6.8455	7.8875	7.7610
M-2	6556.1	1962.4	*65.7*	{85.2}	1981.5	4375.2	4284.6

When unknown sample is similar to KF1

	AQ	BI	DO	KF1	KF2	MC	OA
M-1	{5.3760}	{0.8005}	{1.3317}	0.0827	*6.5932*	{2.3269}	{1.8738}
M-2	4930.5	1284.2	1877.8	46.6	*173.1*	2443.6	2325.8

When unknown sample is similar to KF2

	AQ	BI	DO	KF1	KF2	MC	OA
M-1	8.4193	6.7885	6.6477	*6.5932*	6.5338	7.1143	7.0418
M-2	4857.4	1430.7	1842.1	*173.1*	127.0	2460.1	2403.2

When unknown sample is similar to MC1

	AQ	BI	DO	KF	MC1	MC2	OA
M-1	{2.5907}	{3.5037}	{4.0526}	{2.6560}	0.5189	*6.8538*	{0.9215}
M-2	3124.8	3204.4	4227.3	2358.8	86.4	*348.5*	826.1

Table 6.3 (Continued) Distance among known and unknown class centroids in PCA classifiers with silver as working electrode

When unknown sample is similar to MC2

	AQ	BI	DO	KF	MC1	MC2	OA
M-1	7.3998	7.3560	7.8325	7.1643	***6.8538***	**<u>6.4952</u>**	**{6.4923}**
M-2	3183.8	3370.6	4501.8	2624.9	***348.5***	**<u>331.3</u>**	711.7

When unknown sample is similar to OA1

	AQ	BI	DO	KF	MC	OA1	OA2
M-1	**{3.4828}**	**{2.7248}**	**{3.3140}**	**{1.9088}**	**{0.4081}**	**<u>0.0697</u>**	***6.5258***
M-2	3854.7	2884.1	4230.5	2372.2	862.4	**<u>46.2</u>**	***77.8***

When unknown sample is similar to OA2

	AQ	BI	DO	KF	MC	OA1	OA2
M-1	7.8889	7.1709	7.7610	7.1682	6.6170	***6.5258***	**6.4821**
M-2	3806.7	2867.3	4186.3	2322.7	798.9	***77.8***	**{92.5}**

M-1 = PCA based on e-tongue output, M-2 = PCA based on cross-correlation

#Rate of misclassification with ordinary PCA classifier = $\{29/(24 \times 7) \times 100 = 17.3\%\}$

#Rate of misclassification with enhanced PCA classifier = $\{2/(24 \times 7) \times 100 = 1.1\%\}$

<u>Bold and underlined data</u> means lowest entry in the distance matrix resembling a distance that an unknown brand maintains with one of the water brands considered to design the classifier.

Bold and italics data means second lowest entry in the distance matrix. The second sample of the same unknown brand (with a certain degree of variability with respect to their sources) maintains that distance with one of the water brands considered to design the classifier.

{} data means the entry corresponds to misclassification.

Table 6.4 Distance among known and unknown class centroids in PCA classifiers with platinum as working electrode

When unknown sample is similar to AQ1

DIST	AQ1	AQ2	BI	DO	KF	MC	OA
M-1	**<u>0.1137</u>**	***1.1740***	25.2249	17.3988	17.0272	28.3442	24.8455
M-2	**<u>40</u>**	***87***	20960	14722	13929	25862	20001

When unknown sample is similar to AQ2

	AQ1	AQ2	BI	DO	KF	MC	OA
M-1	***1.1740***	**<u>1.220</u>**	25.6526	17.7969	17.6117	28.5728	25.4029
M-2	***87***	**<u>52</u>**	20965	14726	13955	25852	20018

When unknown sample is similar to BI1

	AQ	BI1	BI2	DO	KF	MC	OA
M-1	25.5186	**<u>2.7996</u>**	***2.8453***	9.0037	8.8441	7.8532	**{0.8175}**
M-2	20854	**<u>122</u>**	***129***	6178	7952	6557	2680

Table 6.4 (Continued) Distance among known and unknown class centroids in PCA classifiers with platinum as working electrode

When unknown sample is similar to BI2

	AQ	BI1	BI2	DO	KF	MC	OA
M-1	25.1790	***2.8453***	**0.0498**	8.0587	8.7906	5.3379	2.8807
M-2	20977	***129***	**31**	6302	8049	6502	2689

When unknown sample is similar to DO1

	AQ	BI	DO1	DO2	KF	MC	OA
M-1	17.2195	8.1705	**0.0500**	***5.0819***	{2.1291}	10.9820	8.1589
M-2	14653	6330	**25**	***157***	3581	11649	5927

When unknown sample is similar to DO2

	AQ	BI	DO1	DO2	KF	MC	OA
M-1	18.2839	10.4667	***5.0819***	{5.0885}	7.1472	11.5441	11.3655
M-2	14679	6297	***157***	**151**	3444	11682	5831

When unknown sample is similar to KF1

	AQ	BI	DO	KF1	KF2	MC	OA
M-1	17.3435	9.3824	4.3135	**1.2249**	***0.1695***	13.9012	8.0512
M-2	13969	8180	3739	**124**	***249***	14346	6412

When unknown sample is similar to KF2

	AQ	BI	DO	KF1	KF2	MC	OA
M-1	17.1861	9.5432	4.3564	***0.1695***	**1.2532**	14.0485	8.2204
M-2	13778	8428	3913	***249***	**337**	14591	6651

When unknown sample is similar to MC1

	AQ	BI	DO	KF	MC1	MC2	OA
M-1	28.2234	5.2980	11.3558	12.9344	**0.0369**	***2.8554***	7.9827
M-2	25875	6506	11970	14233	**15**	***101***	9110

When unknown sample is similar to MC2

	AQ	BI	DO	KF	MC1	MC2	OA
M-1	28.7976	7.8973	12.6322	14.5889	***2.8554***	**2.8883**	10.6618
M-2	25948	6524	11726	14272	***101***	**100**	9117

When unknown sample is similar to OA1

	AQ	BI	DO	KF	MC1	OA1	OA2
M-1	24.7952	3.0951	8.5243	8.1551	8.2727	**0.0508**	***2.4095***
M-2	20007	2754	5937	6349	9140	**37**	***74***

Table 6.4 (Continued) Distance among known and unknown class centroids in PCA classifiers with platinum as working electrode

When unknown sample is similar to OA2							
	AQ	BI	DO	KF	MC1	OA1	OA2
M-1	25.4952	5.4388	10.1484	9.1995	10.4160	**{2.4095}**	**2.4560**
M-2	20006	2826	5964	6334	9205	***74***	**67**

M-1 = PCA based on e-tongue output, M-2 = PCA based on cross-correlation

#Rate of misclassification with ordinary PCA classifier = $\{4/(24 \times 7) \times 100 = 2.4\%\}$

No misclassification in enhanced PCA.

Bold and underlined data means lowest entry in the distance matrix resembling a distance that an unknown brand maintains with one of the water brands considered to design the classifier.

Bold and italics data means second lowest entry in the distance matrix. The second sample of the same unknown brand (with a certain degree of variability with respect to their sources) maintains that distance with one of the water brands considered to design the classifier.

{} data means the entry corresponds to misclassification.

6.4 *Design of cross-correlation-based enhanced Sammon's classifier for mineral water authentication/classification*

The present section proposes Sammon's NLM-based classification and authentication of mineral water samples. The cross-correlation coefficients among the input and output signature of an e-tongue are considered for determining the principal component features instead of using only the output signature of an e-tongue.

6.4.1 *Performance evaluation of the designed classifier*

The reference water data set for designing the classifier has a dimension of (18 × 4402) (six water brands in triplicate samples, each containing 4402 current signatures). The test data matrix (2 × 4402) is created by taking two simulated unknown samples of each category at a time (i.e., Aquafina, or Bisleri, and so on); this is then combined with the reference data set to give rise to a (20 × 4402) matrix. This combined data set is then reduced and projected to two dimensions after Sammon's transformation. As mentioned earlier, Sammon's NLM theory is based on a point mapping of N L dimensional vectors from the L space to a lower dimensional space d by minimizing the stress function "E" using the gradient descent method. The error is a function of the $d \times N$ variables. In the enhanced Sammon's classifier, scores are generated for (8 × 8803) cross-correlation coefficient matrices. The unknown samples are then classified according to the K-nearest neighborhood classification rule. Figures 6.13 and 6.14 show the representative authentication performances shown by the Sammon's, and cross-correlation-based Sammon's, classifiers using the silver and platinum electrodes, respectively. It is revealed that the enhanced Sammon's classifier performs much better than its ordinary version as far as inter- and intracluster distances are concerned. The class centroid distances of the 12 simulated unknown samples (two unknown samples for each brand at a time) with respect to the remaining seven categories of mineral waters are computed (shown in Tables 6.5 and 6.6) for the silver-based and platinum-based Sammon's classifiers as a measure of their authentication. The within-class as well as interclass boundaries in the enhanced Sammon's classifiers are found to be greater than the ordinary Sammon's classifiers. The ordinary platinum-based Sammon's classifier has 2.97% misclassification in comparison with 7.7% for the silver-based classifier.

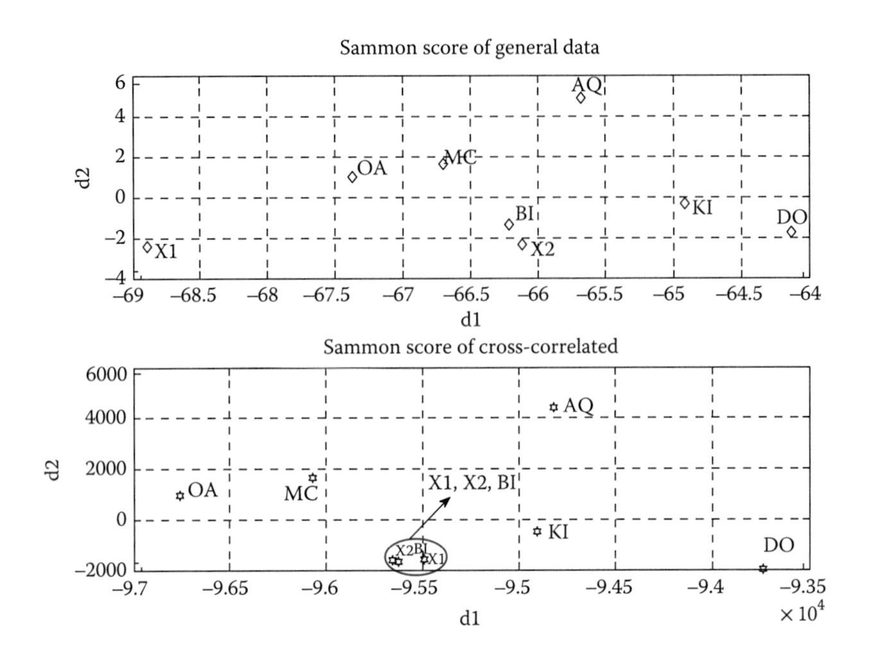

Figure 6.13 Cluster center distances between two unknown Bisleri samples and others in Sammon's classifiers using silver as working electrode.

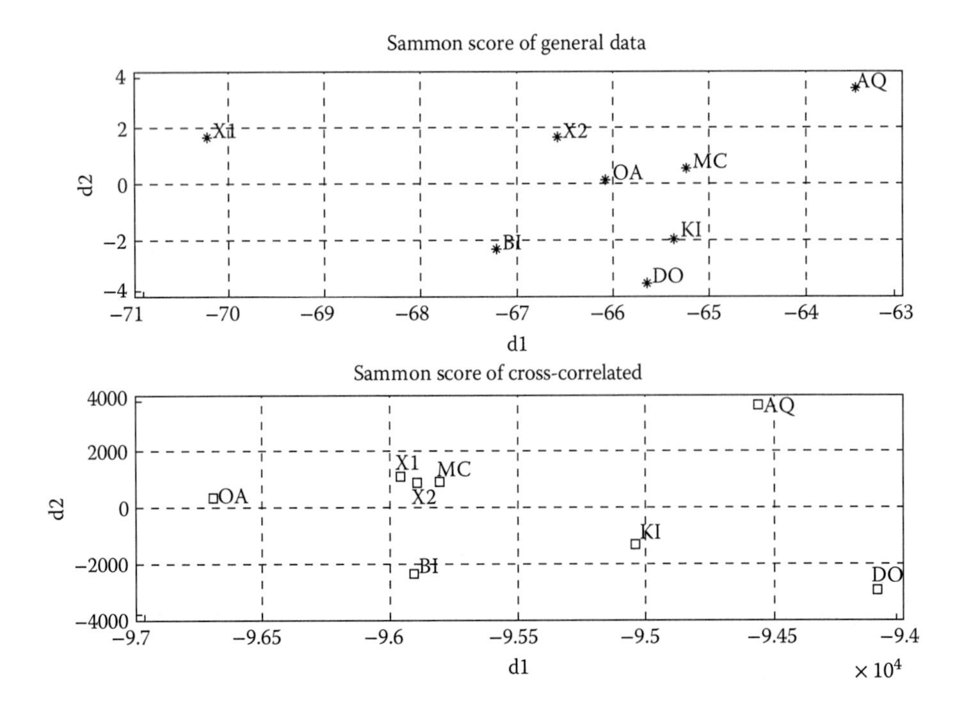

Figure 6.14 Cluster center distances between two unknown McDowell samples and others in Sammon's classifiers using platinum as working electrode.

Table 6.5 Distance among known and unknown class centroids in Sammon's classifiers with silver as working electrode

DIST	AQ1	AQ2	BI	DO	KF	MC	OA
When unknown sample is similar to AQ1							
M-1	**4.0826**	***5.2201***	8.2145	9.3462	7.7852	5.2417	5.5065
M-2	**545**	***3262***	59632	63097	47574	29756	39050
When unknown sample is similar to AQ2							
M-1	***5.2201***	**1.1572**	6.8250	6.6617	5.2606	3.9955	4.9385
M-2	***3262***	**2888**	60977	65263	49243	30133	39160
When unknown sample is similar to BI1							
	AQ	BI1	BI2	DO	KF	MC	OA
M-1	7.9738	**2.8937**	***2.7787***	4.8120	4.5002	4.6229	3.7598
M-2	6029.2	**163**	***166.3***	1798.7	1240.2	3294.9	2846.4
When unknown sample is similar to BI2							
	AQ	BI1	BI2	DO	KF	MC	OA
M-1	7.2024	***2.7787***	**0.9898**	**{2.0686}**	**{2.3334}**	4.0056	3.5598
M-2	6083.8	***166.3***	**64.7**	1950.5	1353.5	3304.9	2809.8
When unknown sample is similar to DO1							
	AQ	BI	DO1	DO2	KF	MC	OA
M-1	9.1196	**{5.6900}**	**6.2688**	***6.8578***	7.2781	8.1174	8.6201
M-2	6801.9	2029.4	**403.9**	***416***	2186	4517.5	4473.1
When unknown sample is similar to DO2							
	AQ	BI	DO1	DO2	KF	MC	OA
M-1	7.5565	**{2.3304}**	***6.8578***	**0.6066**	**{1.6802}**	**{4.4089}**	**{4.0676}**
M-2	6399.1	1963.8	***416***	**33.2**	1879.0	4285.4	4204.4
When unknown sample is similar to KF1							
	AQ	BI	DO	KF1	KF2	MC	OA
M-1	10.2098	6.2977	{4.4904}	**5.3005**	***6.0221***	7.6631	7.8614
M-2	4634.3	1523.0	2215.5	**372.6**	***440.6***	2151.0	2071.2
When unknown sample is similar to KF2							
	AQ	BI	DO	KF1	KF2	MC	OA
M-1	6.0239	6.9711	6.1272	***6.0221***	**5.0836**	***{5.4964}***	6.3341
M-2	4844.2	1527.5	1824.8	***440.6***	**314.5**	2533.8	2510.0
When unknown sample is similar to MC1							
	AQ	BI	DO	KF	MC1	MC2	OA
M-1	6.6841	5.1413	7.0510	6.0761	**4.7807**	***3.9705***	4.0046
M-2	3055.0	3485.0	4576.9	2696.2	**344.3**	***341.7***	927.9

Table 6.5 (Continued) Distance among known and unknown class centroids in Sammon's classifiers with silver as working electrode

When unknown sample is similar to MC2

	AQ	BI	DO	KF	MC1	MC2	OA
M-1	**{3.2678}**	**{3.9036}**	4.9177	**{3.3954}**	***3.9705***	**<u>1.0881</u>**	**{1.2597}**
M-2	2982.0	3444.2	4322.8	2456.9	***341.7***	**<u>65.3</u>**	1049.3

When unknown sample is similar to OA1

	AQ	BI	DO	KF	MC	OA1	OA2
M-1	7.7260	5.4421	7.4220	6.7366	6.0337	**<u>5.3194</u>**	***3.9694***
M-2	4206.1	2824.7	4316.2	2490.0	1347.6	**<u>438.9</u>**	***468.9***

When unknown sample is similar to OA2

	AQ	BI	DO	KF	MC	OA1	OA2
M-1	4.2850	3.3438	4.8682	3.5656	2.0755	***3.9694***	**<u>1.4757</u>**
M-2	3790.5	2888.3	4208.5	2330.6	844.9	***468.9***	**<u>85.9</u>**

M-1 = Sammon based on e-tongue output, M-2 = Sammon based on cross-correlation

#Rate of misclassification with ordinary Sammon = $\{13/(24 \times 7) \times 100 = 7.7\%\}$

No misclassification in enhanced Sammon.

<u>Bold and underlined data</u> means lowest entry in the distance matrix resembling a distance that an unknown brand maintains with one of the water brands considered to design the classifier.

Bold and italics data means second lowest entry in the distance matrix. The second sample of the same unknown brand (with a certain degree of variability with respect to their sources) maintains that distance with one of the water brands considered to design the classifier.

{} data means the entry corresponds to misclassification.

Table 6.6 Distance among known and unknown class centroids in Sammon's classifiers with platinum as working electrode

When unknown sample is similar to AQ1

DIST	AQ1	AQ2	BI	DO	KF	MC	OA
M-1	**<u>2.7406</u>**	***5.4178***	25.9964	17.6402	17.8730	28.4306	26.2006
M-2	**<u>213</u>**	***293***	21125	14765	13826	26283	20184

When unknown sample is similar to AQ2

	AQ1	AQ2	BI	DO	KF	MC	OA
M-1	***5.4178***	**<u>2.8223</u>**	25.9699	17.9804	17.0364	29.3674	25.4489
M-2	***293***	**<u>164</u>**	21023	14682	13672	26227	20048

When unknown sample is similar to BI1

	AQ	BI1	BI2	DO	KF	MC	OA
M-1	27.8604	**<u>3.7707</u>**	***4.7926***	11.1066	12.5524	7.7072	**{4.2789}**
M-2	21036	**<u>347</u>**	***345***	6707	8462	6122	3139

Table 6.6 (Continued) Distance among known and unknown class centroids in Sammon's classifiers with platinum as working electrode

When unknown sample is similar to BI2

	AQ	BI1	BI2	DO	KF	MC	OA
M-1	24.886	***4.7926***	**1.1028**	8.1729	8.4216	5.4210	**{4.2147}**
M-2	21102	***345***	**110**	6546	8178	6450	2800

When unknown sample is similar to DO1

	AQ	BI	DO1	DO2	KF	MC	OA
M-1	17.7458	8.0560	**0.6093**	***4.0195***	**{3.4687}**	11.6901	8.5121
M-2	14647	6580	**25**	***248***	3894	11743	6348

When unknown sample is similar to DO2

	AQ	BI	DO1	DO2	KF	MC	OA
M-1	17.8060	9.8612	***4.0195***	**3.9406**	7.2802	11.6222	11.4378
M-2	14841	6372	***248***	**246**	3809	11615	6102

When unknown sample is similar to KF1

	AQ	BI	DO	KF1	KF2	MC	OA
M-1	17.8939	10.2240	6.7438	**3.2800**	***3.7370***	15.2311	8.7646
M-2	13742	8291	4103	**293**	***302***	14451	6671

When unknown sample is similar to KF2

	AQ	BI	DO	KF1	KF2	MC	OA
M-1	16.1287	9.8479	**{3.4623}**	***3.7370***	**1.1572**	13.8673	9.6713
M-2	13695	8187	3817	***302***	**65**	14297	6660

When unknown sample is similar to MC1

	AQ	BI	DO	KF	MC1	MC2	OA
M-1	27.6690	8.1701	11.5830	14.4239	**5.5757**	***4.4402***	11.5961
M-2	26731	6216	11674	14017	**482**	***495***	8761

When unknown sample is similar to MC2

	AQ	BI	DO	KF	MC1	MC2	OA
M-1	29.1848	5.6677	12.1577	14.0032	***4.4402***	**1.1358**	8.6840
M-2	26898	6475	11723	14172	***495***	**61**	9077

When unknown sample is similar to OA1

	AQ	BI	DO	KF	MC1	OA1	OA2
M-1	26.0940	6.7726	11.3382	9.9218	11.6220	**3.6969**	***3.7230***
M-2	20030	2737	6145	6514	9122	**258**	***289***

Table 6.6 (Continued) Distance among known and unknown class centroids in Sammon's classifiers with platinum as working electrode

When unknown sample is similar to OA2

	AQ	BI	DO	KF	MC1	OA1	OA2
M-1	24.4691	**{3.3736}**	8.3325	7.9983	8.6632	***3.7230***	**<u>1.2034</u>**
M-2	20238	2870	6418	6683	9185	***289***	**<u>77</u>**

M-1 = Sammon based on e-tongue output, M-2 = Sammon based on cross-correlation

#Rate of misclassification with ordinary Sammon = $\{5/(24\times7)\times100 = 2.97\%\}$

No misclassification in enhanced Sammon.

Bold and underlined data means lowest entry in the distance matrix resembling a distance that an unknown brand maintains with one of the water brands considered to design the classifier.

Bold and italics data means second lowest entry in the distance matrix. The second sample of the same unknown brand (with a certain degree of variability with respect to their sources) maintains that distance with one of the water brands considered to design the classifier.

{} data means the entry corresponds to misclassification.

The enhanced Sammon's classifier shows no misclassification irrespective of the working electrode. From the tables as well as from the figures, the major scale-up in the inter- and intracluster boundaries among different water brands is prominent, which is a very crucial step, while simultaneous authentication of more than one unknown sample of a specific brand is an agenda. The PCA and Sammon's NLM-based classification/authentication code is included in Program 6.2.

Program 6.2 MATLAB Code for Sammon and PCA Classifier

```
%This is a program developed for PCA and Sammon's NLM-based
classification of tea/water/fruit juice/wine/milk sample. In an ordinary
PCA and Sammon's classifier, the input feature matrix is (18×4402) for
classification and (19×4402) for authentication. For enhanced classifiers
an (8 x 8803) cross-correlation coefficient matrix becomes the input
(feature) matrix for the same purpose.

%=========================================================================

clc
clear all
fname='allwater.txt' ;
X=load (fname,'-ascii');
X=X(:,1:18);

N
=['AQ1';'AQ2';'AQ3';'BI1';'BI2';'BI3';'DO1';'DO2';'DO3';'KI1';'KI2';'KI3'
;'MC1';'MC2';'MC3';'OA1';'OA2';'OA3';'X ';'AQ';'BI';'DO';'KI';'OA';'MC'];

T=input('Enter type:')

if T==1
  X1=X(:,1:3); S=' Aquafina ';
end
```

```
if T==2
  X1=X(:,4:6); S= 'Bisleri ';
end
if T==3
  X1=X(:,7:9); S=' Dolphin ' ;
end

if T==4
  X1=X(:,10:12); S='Kingfisher' ;
end
if T==5
  X1=X(:,13:15); S=' McDowell' ;
end

if T==6
  X1=X(:,16:18); S=' Oasis' ;
end

Xm=mean(X1,2);

xr=0.035*randn(4402,1);

X=[X (Xm+xr)];
size(X)

X=X'; %input feature matrix

pause

[pc,score,latent,tsquare] = princomp(X);
size(pc)
size(score)
data=score(:,1:2);

c1=[];

for k=1: 3 :18
  c1=[c1; mean(data(k:(k+2),:),1)] ;

end

c11=[c1;data(19,:)]

C1=pdist(c1,'mahal');
C1=squareform(C1)

data1=[data ; c1];

%============end of ordinary PCA-based
classifier=========================
% Sammon's classifier begin===========================
  opts = sammon;
opts.Display = 'iter';
opts.TolFun = 1e-12;
```

```
opts.Initialisation = 'pca';

x=score;

[y, E] = sammon(x, 2, opts);

c2=[];

for k=1: 3 :18
c2=[c2; mean(y(k:(k+2),:),1) ];

end
c22=[c2;y(19,:)];

C2=pdist(c2,'mahal');
C2=squareform(C2)

y1=[y;c2];

subplot(2,1,1)
plot(y1(:,1) , y1(:,2),'ro'); msg='Sammon Mapping Score'
xlabel('Y1') ; ylabel('Y2');Title(msg)
grid on
gname(N)

%==========================================================
subplot(2,1,2)
plot(data1(:,1), data1(:,2),'+');msg='PCA Score'
grid on
xlabel('PC-1');ylabel('PC-2');
gname(N)
%================================================================

Y1=pdist(data,'mahal');
Y3=squareform(Y1);

Y2=pdist(y,'mahal');
Y4=squareform(Y2);

[Y1' Y2' abs(Y1'-Y2')];

%=================================================
```

The PCA and Sammon's NLM-based classifiers, thus developed, may be utilized for the simultaneous authentication of the two unknown samples resembling the same water brand (six such, i.e., resembling any of the six water brands considered) with hardly any misclassification. If the unknown sample taken (e.g., tap water or river water) for authentication is different from the samples used for the development of the classifier, it would not be included in any of the proposed clusters, and thus, it would have been treated as an outcast. As already mentioned, all the water samples (BIS standard) used for the ET experimentation are unique. This effort seems to be a unique one with a view to online water classification and authentication in water quality assessment and monitoring. Platinum-based classifiers perform well in comparison with silver-based classifiers. Enhanced Sammon's classifiers perform better in comparison with ordinary

PCA classifiers, because PCA is an unsupervised technique. The use of cross-correlation coefficients instead of output e-tongue signatures magnifies the intercluster and intracluster distances, thereby facilitating the classification as well as the authentication. The MATLAB code for Sammon's mapping is included in Chapter 2.

6.5 Design of water classifier based on a support vector machine

This section is devoted to the classification of mineral water samples using an SVM-based classifier. The multiclass classification has been designed as a combination of several binary classifiers governed by a DDAG (decision directed acyclic graph) algorithm.

6.5.1 Mineral water sample classification using an SVM

A water data set of six brands (Aquafina, Bisleri, Kingfisher, Oasis, Dolphin, and McDowell; each containing three samples) containing 4402 features is considered for the design of an SVM-based classifier. A data set is generated/simulated by considering any sample of a brand (among the six brands considered and generated using both silver and platinum electrode systems) added up/adulterated with 5–10% of random noise; these samples are treated as unknown brand samples and used for authentication purposes. The principal components of the design data set (21 × 4402) for each electrode system are used as inputs to the SVM-based classification. An RBF (radial basis function), a kernel function with scaling factor (sigma) ranging from 0.6 to 1, has been used for drawing the decision boundaries in the designed binary classifiers. A linear decision boundary was not found as satisfactory as the RBF kernel-based one. The DDAG method has been adapted here for SVM-based multiclass classification of the previously mentioned water brands. For each electrode system, 15 binary classifiers $\left\{\left(\frac{M(M-1)}{2}\right), \text{ where } M = \text{water class} = 6\right\}$ are designed, and the results of those classifiers are the decision classes. Given a test instance, the multiclass classification is then executed by evaluating all 15 binary classifiers and assigning the instance to the class, which gets the highest number of resulting outputs (decision class) of the binary classifiers. In essence, the multiclass classification using SVM is as follows:

1. SVM-based binary classification
2. Application of DDAG algorithm (already provided in Chapter 2) to generate membership against all water brands considered binary classification decisions leading to multiclass classification

Figure 6.15 represents the DDAGSVM (DDAG based on SVM) classifier used for authentication of the representative unknown Dolphin water brand (data collected over an Ag electrode) manifesting the evaluation path. Based on the evaluation path of Figure 6.15, the maximum membership for each brand is determined, and Dolphin gets the highest membership of five (Table 6.7). Similarly, a Pt electrode-based DDAGSVM classifier has been developed and is presented in Figure 6.16. Table 6.8 reveals the membership value calculated for every brand. Kingfisher gets the highest membership; hence, the test sample is Kingfisher. For unknown sample authentication, SVM is implemented on the water database including the simulated unknown water brand. The MATLAB code for an SVM-based water authenticator/classifier is provided in Program 6.3.

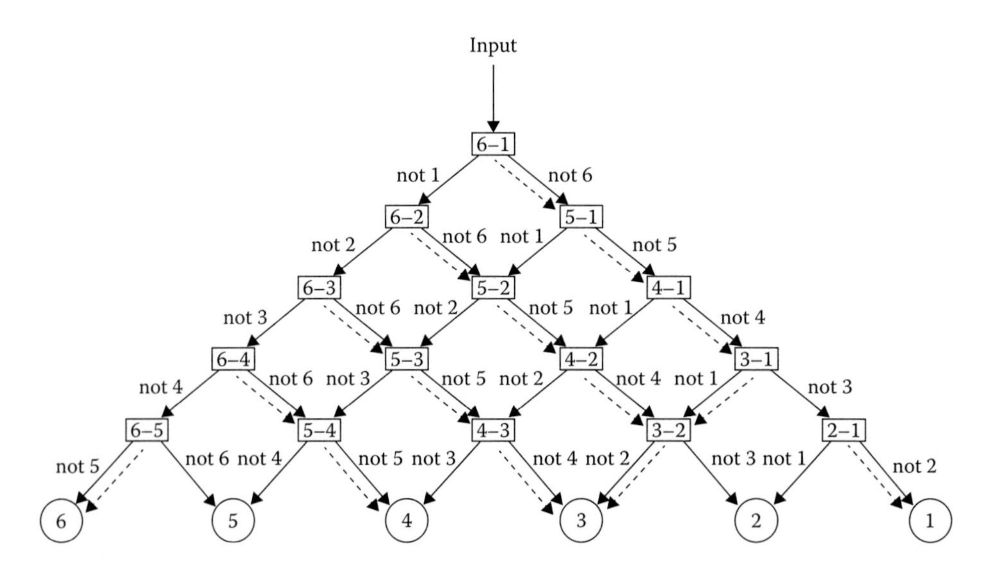

Figure 6.15 DDAG-based authentication of Dolphin water brand (Ag working electrode) using pairwise SVM (---: evaluation path).

Table 6.7 Membership for different water brands of the DDAG in Figure 6.15

Membership (water brand) number	Number of detected classes
6 (Oasis)	1: (6-5)
5 (McDowell)	0:
4 (Kingfisher)	2: (6-4); (5-4)
3 (Dolphin)	**5: (6-3); (5-3); (4-3); (3-2); (3-1)**
2 (Bisleri)	3: (6-2); (5-2); (4-2)
1 (Aquafina)	4: (6-1); (5-1); (4-1); (2-1)

Unknown = Dolphin

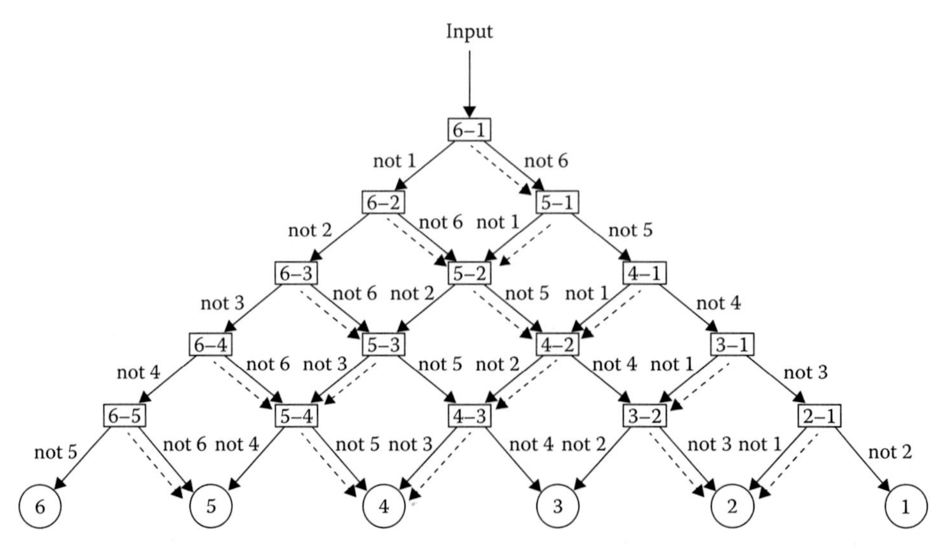

Figure 6.16 DDAG-based authentication of Kingfisher water brand (Pt working electrode) using pairwise SVM (---: evaluation path).

Table 6.8 Membership for different water brands of the DDAG in Figure 6.16

Membership (water brand) number	Number of detected classes
6 (Oasis)	0:
5 (McDowell)	3: (5-1); (5-3); (6-5)
4 (Kingfisher)	**5: (4-1); (4-2); (4-3); (5-4); (6-4)**
3 (Dolphin)	2: (3-1); (6-3)
2 (Bisleri)	4: (6-2); (5-2); (3-2); (2-1)
1 (Aquafina)	1: (6-1)

Unknown = Kingfisher

Program 6.3 MATLAB code for SVM-based water authenticator/classifier

```
% This program works in the same fashion as that of the FDA-based classifier
discussed in Chapter 5 except it works with the MATLAB inbuilt functions
svmclassify and svmstruct. It is applicable for water/tea/fruit juice/
milk samples for their classification and authentication

%===================================================================

clc
clear all

S=['BR1';'BR2';'BR3';'DD1';'DD2';'DD3';'GD1';'GD2';'GD3';'LP1';'LP2';
'LP3';'LD1';'LD2';'LD3';'MV1';'MV2';'MV3']
S1=['X '] % unknown sample

X=load('alltea_au.txt','-ascii') ; % data matrix is based on (gold
% working electrode) e-tongue output signals of dimension (18×4402)

y=input('Enter test category:')
k=1+(y-1)*3

R=[X(:,k) X(:,k+1) X(:,k+2)]
R=mean(R,2)
n=size(R)
b=input('Enter bias :')
   r=b*randn(n,1);
R=R+r % simulation of unknown test sample

%=============================================================
  %creation of input feature matrix of dimension (7×4402) for
authentication
% for binary classifier

s1=input('Enter 1st type:')
s2=input('Enter 2nd type:')

Z=[];
k1=1+(s1-1)*3
k2=1+(s2-1)*3

%
```

```
Z=[Z X(:,k1) X(:,k1+1) X(:,k1+2) X(:,k2) X(:,k2+1) X(:,k2+2) R]

%===starting of nomenclature for binary classifier========

if s1==1
N1=['BR1';'BR2';'BR3']
text2=' BROKBOND'
end

if s1==2
N1=['DD1';'DD2';'DD3']
text2=' DOUBLE-DIAMOND'
end

if s1==3
N1=['GD1';'GD2';'GD3']
text2=' GODREJ'
end

if s1==4
N1=['LP1';'LP2';'LP3']
text2=' LIPTON'
end

if s1==5
N1=['LD1';'LD2';'LD3']
text2=' LIPTON-DARJEELING'
end

if s1==6
N1=['MV1';'MV2';'MV3']
text2=' MARVEL'
end

if s2==1
N2=['BR1';'BR2';'BR3']
text2=strcat(text2,' and BROKBOND')
end

if s2==2
N2=['DD1';'DD2';'DD3']
text2=strcat(text2,' and DOUBLE-DIAMOND')
end

if s2==3
N2=['GD1';'GD2';'GD3']
text2=strcat(text2,' and GODREJ')
end

if s2==4
N2=['LP1';'LP2';'LP3']
text2=strcat(text2,' and LIPTON')
end
if s2==5
```

```
N2=['LD1';'LD2';'LD3'] text2=strcat(text2,' and LIPTON-DARJEELING')
end

if s2==6
N2=['MV1';'MV2';'MV3']
text2=strcat(text2,' and MARVEL')
end
%==========================end of nomenclature===================

N=[N1 ;N2]
N3=' X'
  %=====================PCA decomposition of Z matrix================
Z=Z' ;
[pc,score,latent,tsquare] = princomp(Z);
size(pc)
size(score)
data=score(:,1:2)
data1=data(1:6,1:2) ; % matrix of dimension 6×2 corresponding to two
% different classes in a binary

pattern=data(7,:) ; % matrix of dimension 1×2 corresponding to unknown
test %category

  % starting of generation of axes
F=[ data1 ; pattern];
L1=min(F(:,1)) ;
L2=min(F(:,2));
L3=max(F(:,1));
L4=max(F(:,2));
L5=L1-0.5 ;
L6=L3+0.5 ;
L7=L2-0.5 ;
L8=L4+0.5 ;

[ L5 L6 L7 L8]

l1=input('Enter category as class-1:')
l2=input('Enter category as class-2:')


group=[l1 ; l1; l1; l2 ;l2 ;l2] ;

plot(data1(:,1) , data1(:,2) ,'b+','Linewidth',2)
grid on
axis([ L5 L6 L7 L8])
gname(N)
hold on
plot(pattern(:,1) , pattern(:,2) ,'rv','Linewidth',2)
gname(N3)
xlabel('PC1') ; ylabel('PC2')
title('pairwise SVM analysis')

  pause
  %===============================starting SVM classificationtest=pattern ;
```

```
k=input('Enter kernel type:')
if k==1
type='linear'
end

if k==2
  type='RBF'
end

sigma=input('Enter sigma_value:')
SVMStruct=svmtrain(data1,group,'Kernel_Function','rbf','RBF_Sigma',sigma,
'boxconstraint',Inf,'showplot','true')
c=svmclassify(SVMStruct,pattern)

%================for editing the current figure==================
text=type ;

gtext(text)
ctext1='class-1' ;
ctext2='class-2' ;
gtext(ctext1) % displaying text on the figure with mousegtext(ctext2)
if y==1
fname='brokbond_'

end

if y==2
  fname='double-diamond_'
end
if y==3
  fname='godrej_'

end
if y==4
  fname='lipton_'
end
if y==5
  fname='lipton-darjeeling_'
end
if y==6
  fname='marvel_'

end
fname1=strcat(fname,text2,'_au.fig')

hgsave(fname1) % saves the current figure to a file named filename1

%%%%%%%%%%%%%%%%%%%%%%%%%%%%%%%%%
```

6.6 *Relative performance evaluation of the designed classifiers*

In this chapter, PLS, cross-correlation-based enhanced PCA and Sammon's NLM and SVM-based water classifiers have been designed. PLS-based commercial mineral water classification, as well as authentication, has revealed 100% successful classification rates like

SVM-based classification. However, SVM-based classifiers cannot be ranked as equally efficient to PLS-based ones because of the higher computation load in SVM-based classification. The multiclass classification in SVM-based classification is accomplished as a combination of several binary classification decisions. Multiclass SVM classification as a combination of several binary classifications cannot be recommended for online water quality monitoring. The classifiers designed using PLS and SVM made no distinction in their performance with respect to the working electrodes used in generating the e-tongue signal.

Enhanced Sammon's classifiers perform better in comparison with ordinary PCA classifiers. The ordinary platinum-based Sammon's classifier has 2.97% misclassification in comparison with 7.7% for the silver-based classifier. The enhanced Sammon's classifier shows no misclassification irrespective of the working electrode. The enhanced PCA classifiers hardly carry out any misclassification (1.1%). The misclassification rate in the ordinary PCA classifier with the platinum electrode is much smaller (2.4%) than that with silver as a working electrode (17.3%). The enhanced Sammon's classifier can be a potential candidate for online water quality monitoring like a PLS classifier.

Whatever be the machine-learning component is adapted, all the designed classifiers presented in this chapter are applicable for authentication of an unknown sample belonging to the water brands used in the design of the classifiers. Water samples pertaining to other brands will be treated as outcasts (cannot be recognized).

6.7 *Extended water monitor framework*

The proposed framework is based on electrochemical qualification and quantification of water with respect to the ingredients/harmful ingredients present in it. The proposed framework is a modular device and supposed to cater to the following needs:

1. Natural and drinking water classification with respect to their mineral content, alkalinity, pH, conductivity, surfactant content, etc.
2. Recycled water quality monitoring with respect to the pollutants and detection of abnormal operating conditions in WWTPs.
3. Detection and monitoring of heavy metals (arsenic, chromium, copper, antimony, etc.) in drinking and ground water.

The voltametric e-tongue device is supposed to be the soul of the proposed monitoring framework, tailored for different purposes and adapting different kinds of voltametric techniques like pulse voltammetry, cyclic voltammetry, and anodic stripping voltammetry. Three of the functional blocks of the monitors are as follows:

1. Generation of voltammogram using electronic tongue instrumentation
2. Determination/extraction of suitable features using the voltammogram
3. Development of a robust authenticator/quantifier/classifier based on these extracted features

It is step 1, the ET instrumentation devised different for varying purposes of water monitoring. There always remains the potential for indigenous development with respect to the following:

- Using fabricated/tailored working, counter, and reference electrode assembly in the electrochemical cell for generating the voltammogram
- Replacing potentiostat and DAS with indigenous potentiostats

The indigenous potentiostats have already been elaborated on in Chapter 5.

Step 2, feature extraction using the voltammogram, can be performed using PCA decomposition/wavelet transformation (in order to compress the signal and extract features in the form of wavelet coefficients). The raw signals are normalized, filtered, and processed with various transforms for extracting features. The signals are checked with respect to their nonstationarity by checking their time-bound variation in mean, variance (whether heteroskedasticity prevails), kurtosis, skewness, ACF, PACF, etc. The presence of nonstationarity influences the algorithms developed in step 3. In the water classification/authentication and quantification of pollutants present in it (step 3), the various algorithms developed for tea grading in Chapter 5 are also applicable in this chapter along with those specific algorithims proposed in this chapter.

If the problem is the detection of heavy metal content in ground water/fresh water and its prediction using electrochemical means, step I can be either cyclic voltammetry/stripping voltammetry or adsorption stripping voltammetry-based signal generation. Therefore, step 1 is a subjective one and needs to be designed and implemented according to the required objective/application in monitoring water. The quantification of heavy metal content in water can be accomplished using the voltammogram generated. Extracted features can be correlated to the heavy metal content; they can be predicted as well using an algorithm like a PLS or neural networks. The extracted features may not always be PCA-based features. The weblate or slantlate-based feature extraction using voltammograms are not uncommon [8]. Detection and quantification of arsenic in contaminated water will be presented in Chapter 7 under miscellaneous application of e-tongues.

Recycled water monitoring in WWTPs is done by measuring parameters such as BOD, COD, ammonia, orthophosphate, etc. in recycled water samples in order to assess their acceptability in view of irrigation and public health. This also can be done by e-tongue instrumentation using cyclic voltammetry [11,13]. The e-tongue features may be correlated with the analytically estimated parameters (like COD, BOD, ammonia, orthophosphate, sulfate, acetic acid, and alkalinity) in WWTP plants using PLS algorithms.

If the problem is related to natural and drinking water quality monitoring, that is classification and authentication of various categories of water, this can be accomplished with ET instrumentation using pulse voltammetry (as described in this chapter). The extracted features can also qualify various categories of water using the proposed machine-learning algorithms.

References

1. R. Martinez-Manez, J. Soto, E. Garcia-Breijo, L. Gil, J. Ibanez, E. Llobet (2005). An electronic tongue design for the qualitative analysis of natural waters. *Sensors and Actuators B*, vol. 104, 302–307.
2. E. Garcia-Breijo, J. Atkinson, L. Gil-Sanchez, R. Mascot, J. Ibanez, J. Garrigues, M. Glanc, N. Laguarda-Miro, C. Olguin (2011). A comparison study of pattern recognition algorithms implemented on a microcontroller for use in an electronic tongue for monitoring drinking waters. *Sensors and Actuators A: Physical*, vol. 172, 570–582.
3. L. Moreno, A. Merlos, N. Abramova, C. Jimenez, A. Bratov (2006). Multi-sensor array used as an electronic tongue for mineral water analysis. *Sensors and Actuators B*, vol. 116, 130–134.
4. F. Winquist, P. Wide, I. Lundstrom (1997). An electronic tongue based on voltammetry. *Analytica Chimica Acta*, vol. 357, 21–31.
5. C. Krantz-Rulcker, M. Stenberg, F. Winquist, I. Lundstrom (2001). Electronic tongue for environmental monitoring based on sensor arrays and pattern recognition: A review. *Analytica Chimica Acta*, vol. 426, 217–226.

6. J. Olsson, P. Ivarsson, F. Winquist (2008). Determination of detergents in washing machine wastewater with a VET. *Talanta*, vol. 76, 91–95.
7. P. Kundu, P. Pancharya, M. Kundu (2011). Classification & authentication of unknown water sample using machine learning algorithms. *ISA Transactions*, vol. 50, 487–495.
8. P. Kundu, A. Chatterjee, P. Panchariya (2011). Electronic tongue system for water sample authentication: A slantlet-transform-based approach. *IEEE Transactions on Instrumentation and Measurement*, vol. 60(6), 1959–1966.
9. J. Soto, I. Campos, R. Martinez-Manez (2013). Monitoring wastewater treatment using voltammetric electronic tongues. *Real-Time Water Quality Monitoring, SSMI* 4, S.C. Mukhopadhyay and A. Mason (Eds.). Berlin: Springer-Verlag Berlin Heidelberg, pp. 65–103.
10. I. Campos, R. Masot, M. Alcaniz, L. Gil, J. Soto, J.L. Vivancos, E. Garcia, R.H. Labrador, J. Barat, R. Martinez-Manez (2010). Accurate concentration determination of anions nitrate, nitrite and chloride in minced meat using a VET. *Sensors and Actuators B: Chemical*, vol. 149, 71–78.
11. C. DiNatale, A. Macagnano, F. Davide, A. D'Amico, A. Legin, Y. Vlasov, A. Rudnitskaya, B. Selezenev (1997). Multicomponent analysis on polluted waters by means of an electronic tongue. *Sensors and Actuators B*, vol. 44, 423–428.
12. A. Gutes, F. Cespedes, M. Del Valle, D. Louthander, C. Krantz-Rulcker, F. Winquis (2006). A flow injection VET applied to paper mill industrial waters. *Sensors and Actuators B*, vol. 115, 390–395.
13. I. Campos, M. Alcaniz, D. Aguado, R. Barat, J. Ferrer, L. Gil, M. Marrakchi, R. Martinez-Manez, J. Soto, J. Vivancos (2012). A VET as tool for water quality monitoring in wastewater treatment plants. *Water Research*, vol. 46, 2605–2614.
14. C. Rosen (2001). A chemometric approach to process monitoring and control with applications to wastewater treatment operation. PhD thesis, Department of Industrial Electrical Engineering and Automation, Lund University, Sweden.
15. M. Kundu, S.K. Agir (2016). Detection and quantification of arsenic in water using electronic tongue. *In the Proceedings of IEEE Sponsored International Conference on Control, Measurement and Instrumentation (CMI-16)*, 8–10 January 2016, Kolkata, India.
16. J.H.T. Luong, E. Lam, K.B. Male (2014). Recent advances in electrochemical detection of arsenic in drinking and ground waters. *Analytical Methods* , vol. 6, 6157–6169.
17. K.G. Walsh, P. Salaun, M.K. Uroic, J. Feldmann, J.M. McArthur, C.M.G. van den Berg (2011). Voltammetric determination of arsenic in high iron and manganese groundwaters. *Talanta*, vol. 85, 1404–1411.
18. M.R. Rahman, T. Okajima, T. Ohsaka (2010). Selective detection of As(III) at the Au(111)-like polycrystalline gold electrode. *Analytical Chemistry*, vol. 82, 9169–9176.
19. L. Bu, T. Gu, Y. Ma, C. Chen, Y. Tan, Q. Xie, S. Yao (2015). Enhanced cathodic preconcentration of As(0) at Au and Pt electrodes for anodic stripping voltammetry analysis of As(III) and As(V). *The Journal of Physical Chemistry C* , vol. 119, 11400–11409.
20. R.T. Kachoosangi, R.G. Compton (2013). Voltammetric determination of Chromium(VI) using a gold film modified carbon composite electrode. *Sensors and Actuators B*, vol. 178, 555–562.
21. J.P. Metters, R.O. Kadara, C.E. Banks (2012). Electroanalytical sensing of chromium(III) and (VI) utilising gold screen printed macro electrodes. *Analyst*, vol. 137, 896–902.
22. S. Abbasi, A. Bahiraei (2012). Ultra trace quantification of chromium (VI) in food and water samples by highly sensitive catalytic adsorptive stripping voltammetry with rubeanic acid. *Food Chemistry*, vol. 133, 1075–1080.
23. S. Abbasi, A. Bahiraei, F. Abbasai (2011). A highly sensitive method for simultaneous determination of ultra-trace levels of copper and cadmium in food and water samples with luminol as a chelating agent by adsorptive stripping voltammetry. *Food Chemistry*, vol. 129, 1274–1280.
24. C. Soderstrom, H. Boren, F. Winquist, C. Krantz-Rülcker (2003). Use of an electronic tongue to analyze mold growth in liquid media. *International Journal of Food Microbiology*, vol. 83, 253–261.
25. M. Kundu, P. Kundu (2013). The e-tongue-based classification and authentication of mineral water samples using cross-correlation-based PCA and Sammon's nonlinear mapping. *Journal of Chemometrics*, vol. 27(11), 379–393

chapter seven

Miscellaneous application of chemometrics

This chapter details two of the finest and diversified applications of chemometrics: one in patient care monitoring and another in monitoring arsenic content in contaminated water. Patient care monitoring is a vast subject area; therefore, this chapter will focus only on ECG monitoring (in terms of electrocardiogram synthesis/modeling and analysis). Arsenic poisoning from drinking water has been the worst natural debacle (accentuated by human activity) in recent times. Application of chemometrics in arsenic quantification (in contaminated water) is presented here.

7.1 Electrocardiogram as patient care monitor

Continuous measurement of patient parameters such as heart rate and rhythm, respiratory rate, blood pressure, blood-oxygen saturation, and many other parameters has become a common feature in the care of critically ill patients. When accurate and immediate decision-making is crucial for effective patient care, electronic monitors frequently are used to collect and display physiological data. Increasingly, such data are collected using noninvasive sensors from less seriously ill patients in a hospital's medical-surgical units, labor and delivery suites, nursing homes, or patients' own homes to detect unexpected life-threatening conditions or to record routine, but required, data efficiently. A patient monitor watches for and warns against life-threatening events in patients, critically ill or otherwise. In this context, telemedicine/telehealth needs to be mentioned. Videoconferencing, transmission of still images, e-health and patient portals, continuing medical education, and nursing call centers are all inclusive of a telemedicine or telehealth concept, and the induction of ICT (information and communication technology) made it possible. Electrocardiography (ECG), photoplethysmography (PPG), ultrasound, CT scan (computed tomography scan), and ECHO (echocardiogram) are the technologies installed at rural and primary health service posts that can be connected to the specialty hospitals/healthcare network using even simple webcams and integrated services digital network (ISDN) telephone lines. Health care of people in remote locations (especially in villages of developing countries) is a staggering issue because of high population density, the lack of infrastructural facilities, and experienced doctors. In such circumstances, interactive patient care and forward type tools, developed using different biomedical signals, may help the local medical practitioners in complementing their clinical knowledge and judgments. With the networking facilities available, treatment also can be conducted from any location, whether distant or convenient, urban or rural consulting with with experienced physicians. Telemedicine is the integrated technology platform with which a patient can be treated and monitored through a communication link by a remotely located physician. This section of this book presents synthesis and automated pattern classification of human electrocardiograms using chemometrics as a part of patient care monitoring effort, which may also act as a tutor for medical professionals.

7.1.1 Heart and cardiovascular system

The heart is considered to be one of the three most important body organs, along with the kidneys and the central nervous system, insofar as life support is concerned. The human heart is a four-chambered, hollow, and flexible organ that collects impure blood from the other organs, purifies it, and then circulates oxygenated blood to the whole body. It is placed in the thoracic chamber, anterior to the vertebral column and posterior to the sternum, slightly offset to the left. Depending on the age, it weighs between 250 and 350 g. The human heart consists of two pairs of atria and ventricles, longitudinally connected. The heart along with the network of veins (which carry deoxygenated blood) and arteries (which carry oxygenated blood) forms the cardiovascular system to supply blood to the whole body. The word cardiac is derived from Greek word *kardia*, meaning "related to the heart." Details of the human heart physiology can be found in [1,2]. The atria receive the blood from the body, and ventricles supply the blood. The pair of atria and ventricles work in tandem, that is they contract and expand together to collect and supply blood, respectively. In between these two operations, there is an important function of purification, which is performed by sending the impure (deoxygenated) blood to the lungs. To achieve this, the right ventricle sends the impure blood to the lungs and receives the pure blood at the left atria. A complete cardiac cycle consists of collection of blood from body organs, purifying it, and sending it back to the whole body. The pumping action of the heart supports the following circulatory systems:

- Pulmonary circulation: This supports blood circulation to the lungs for purification.
- Systemic circulation: This supports blood circulation to the entire body except the lungs.
- Coronary circulation: Coronary circulation takes care of circulating oxygenated blood to the heart cells.

A healthy adult human heart contracts (or expands) 72–80 times per minute. Figure 7.1 [3] shows the schematic of blood flow directions between various heart chambers.

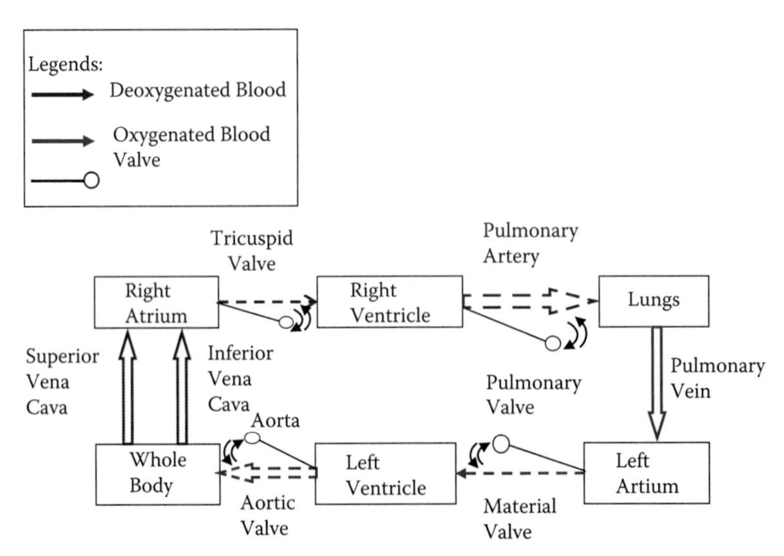

Figure 7.1 Schematic blood flow direction between heart chambers. (Adapted from R. Gupta et al., *ECG Acquisition and Automated Remote Processing*, Springer, India, 2014.)

7.1.2 Events in electrocardiogram (ECG) and ECG morphology

The mechanical action of the heart is due to generation of small electrical impulses and their propagation through the heart surface. Electrocardiography (ECG) is a popular noninvasive technique for preliminary level investigation on cardiovascular assessment, which represents a time-averaged representation of these electrical potentials picked up by placing electrodes on body surface. As a convention, an activity is recorded as positive (or negative) when the resulting electrical impulses move toward (or away from) the electrode. Electrocardiography (i.e., the science and practice of making and interpreting recordings of cardiac electrical activity) can be divided into morphology and arrhythmology. The electrocardiogram (also ECG), introduced into clinical practice more than 100 years ago by Dutch doctor and physiologist Willem Einthoven, comprises a linear recording of cardiac electrical activity as a function of time. The electrical activity of heart muscles is initiated at the sinoatrial (SA) node located at the upper region of the right atrium. A specialized group of cells named pacemaker cells at this node spontaneously depolarize and repolarize at a rate of 60–100 times/min. The electrical impulses gradually spread over the atria at a speed of 4 m/s, causing them to contract together. This produces a noticeable deflection on the ECG record, which is named a P wave. After flowing through the atria, the electrical impulses reach the atrioventricular (AV) node, located at the lower end of the right atria. Here, the conduction is delayed, at a speed of 0.1 m/s. The AV node is the only route through which these electrical impulses can reach the ventricles, since the rest of atrial myocardium is separated from the ventricles by a nonconducting ring of fibrous tissue. The conduction at the AV node provides a gap between the atrial conduction and ventricular conduction and is reflected as an equipotential line, which is called a PR segment in the ECG record. From the AV node, the electrical impulses enter the bundle of cardiac muscle fibers (His bundle), which is bifurcated into left bundle and right bundle branches. As the impulses flow through the bundle branches, the contraction (depolarization) of the ventricles starts together. This generates a combination of sharp downward and upward deflection in the ECG record, named the QRS complex. The right bundle branch conducts the pathway of conduction to the right ventricle, while the left bundle branch is divided into anterior and posterior fascicles that conduct the wave to the left ventricle. The conduction is gradually distributed in Purkinje fibers, which are spread out to the left and right atria. Since the systemic circulation dominates in overall contraction of the ventricles in terms of muscle activity, the QRS complex mainly represents ventricular contraction due to the left ventricle. During ventricular contraction, the atria also expand (repolarize). However, the generated electrical activity due to this is very feeble and suppressed by the ventricular activity. There is a small time gap between the ventricular contraction and expansion, where no electrical activity is recorded in the ECG record. This is represented by the equipotential ST segment (the section on the ECG between the end of the S wave and the beginning of the T wave), after which the ventricles start expanding (repolarization) together. The ventricular repolarization is represented by a T wave in the ECG record. In certain electrodes, a U wave is recorded as small afterpotentials after the T wave, representing slow repolarization of interventricular septum or slow repolarization of ventricles. The cardiac events and corresponding ECG waves are summarized in Table 7.1 and Figure 7.2. Though the depolarization of SA and AV nodes is important, it does not make any detectable waves in an ECG.

Each beat of the heart can be observed as a series of deflections away from the baseline on the ECG. A single sinus (normal) cycle of the ECG, corresponding to one heartbeat, is traditionally labeled with the letters P, Q, R, S and T on each of its turning points; a functional diagram of one heartbeat is shown by Figure 7.3.

Table 7.1 ECG waves and their corresponding events

Wave or segment	Event name
P wave	Atrial depolarization
PR interval	Start of atrial depolarization to start of ventricular depolarization
QRS complex	Ventricular depolarization
ST segment	Pause in ventricular electrical activity before repolarization
T wave	Ventricular repolarization
QT interval	Total time taken by the ventricular depolarization and repolarization
U wave function	Slow ventricular repolarization

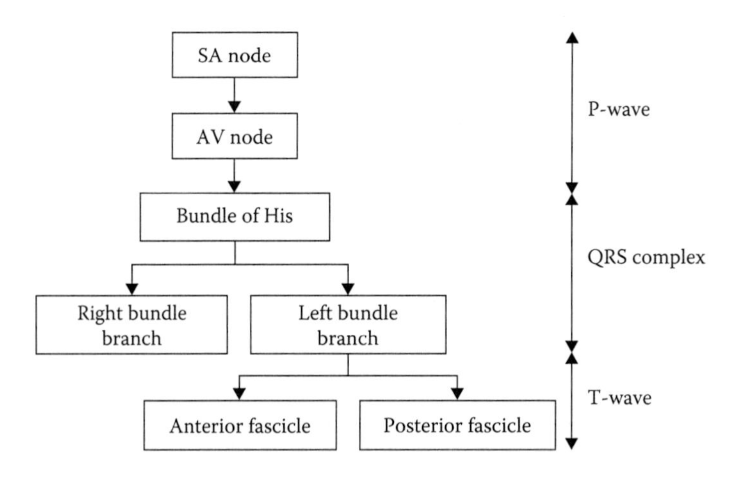

Figure 7.2 Pathways of electrical signals in heart.

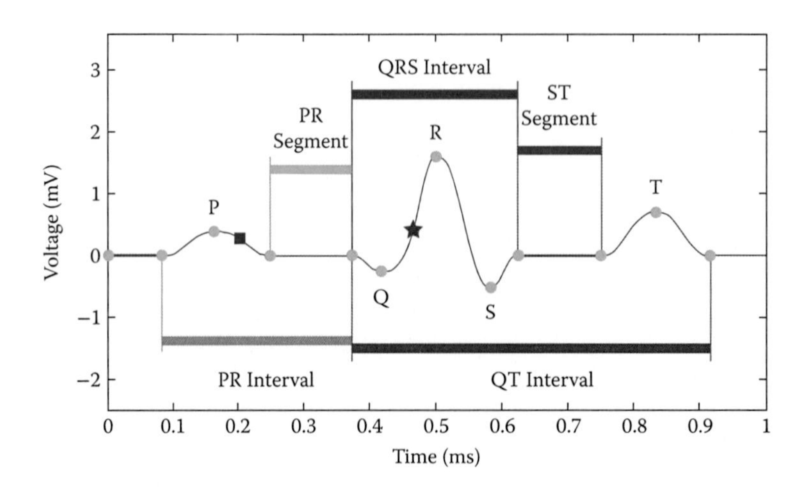

Figure 7.3 ECG morphology.

7.1.3 ECG lead system

The ECG lead system describes internationally accepted standards for placing electrodes on external body surface to acquire an ECG signal. This provides a framework for easy reproducibility, comparison, and nomenclatures for the physicians and medical research fraternity to analyze and interpret ECG records. The first work on ECG lead standardization was performed by W. Einthoven (awardee of Nobel Prize in Medicine for his work in 1924). Einthoven postulated that the human body is a perfect sphere with homogeneous density, with the human heart located at the center. He considered three leads placed on the right arm, left arm, and left leg of a person standing with laterally stretched hands, which forms an equilateral triangle with the vertices on a circle. Einthoven's bipolar leads, also called standard leads, are defined as

Lead I = VLA-VRA
Lead II = VLL-VRA
Lead III = VLL-VLA

where VLA is potential of the left arm, VRA is potential of the right arm, and VLL is potential of the left leg. The lead positions and connections for the standard leads, or limb leads, developed by Einthoven are shown in Figure 7.4.

Later on, Frank Wilson defined and standardized unipolar leads with three unipolar limb leads and six precordial chest leads, measured with respect to a reference terminal outside the body, named the Wilson central terminal (WCT). WCT is realized as a resistive network using 5-kΩ resistances and given as

$$\text{VWCT} = \frac{1}{3}(\text{VLA} + \text{VRA} + \text{VLL}) \tag{7.1}$$

The precordial chest leads are defined as

Lead VL = VLA-VWCT
Lead VR = VRA-VWCT
Lead VF = VLL-VWCT

Later on, E. Goldberger modified the precordial leads to develop "Augmented Limb Leads," aVR, aVL, and aVF, respectively, by opening the exploring lead connections with

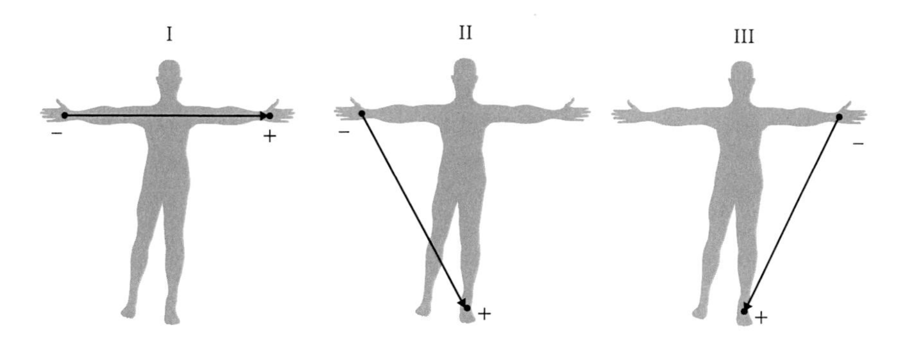

Figure 7.4 Standard lead positions of Einthoven.

Table 7.2 Chest lead positions

Lead name	Position
V1	Right fourth intercostal space
V2	Left fourth intercostal space
V3	Halfway between V2 and V4
V4	Left fifth intercostal space
V5	Horizontal to V4, anterior auxiliary plane
V6	Horizontal to V5, mid-auxiliary plane

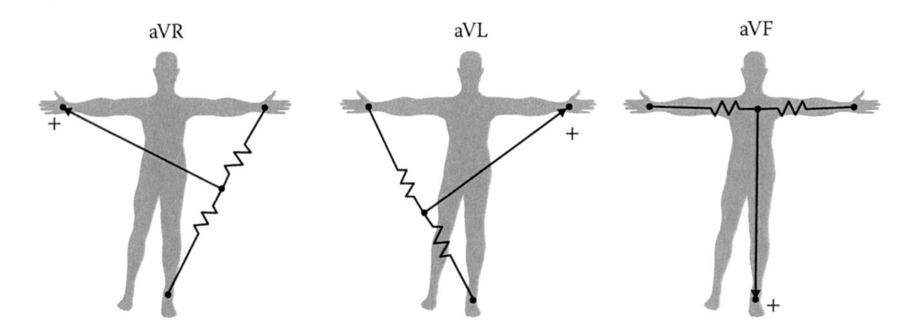

Figure 7.5 Augmented lead positions and connections.

the WCT. The lead positions are given in Table 7.2 and Figure 7.5. The chest leads v1, v2, v3, v4, v5, and v6 measure the cardiac potentials at specified intercostal spaces of the chest with respect to the WCT. These six leads define the ECG in a horizontal plane. The six frontal plane leads (I, II, III) and six horizontal plane leads constitute a 12-lead ECG system, which is by far the most common and accepted method of mapping an ECG signal from a patient. More details of ECG lead systems can be found in [4].

7.1.4 ECG recording

An ECG record in a clinical setup is generated through an electrocardiograph machine, with electrodes connected to the patient body. It incorporates an amplifier, filter, patient isolation, and mechanical printing arrangement. The fundamental component to the electrocardiograph is the amplifier, which is responsible for taking the voltage difference between leads and amplifying the signal. ECG voltages measured across the body are on the order of hundreds of microvolts up to 1 millivolt (the small square on a standard ECG is 100 microvolts). This low voltage necessitates a low noise circuit and the instrumentation amplifier is the key. Early electrocardiographs were constructed with analog electronics and the signal could drive a motor to print the signal on paper. A thermally sensitive, graduated strip chart paper would be driven at a speed of 25 mm/s (mostly recommended) and it would generate the impression by a moving stylus controlled by amplifier output. Each smallest square on the paper record along the x-axis (time) represents 40 ms, and the same along the y-axis (millivolt amplitudes) represents 0.1 mV. Old or traditional recordings facilitate sequential recording; that is, one lead plot can be obtained at a time by a mechanical switching arrangement. Modern electrocardiographs use analog-to-digital converters to convert an analog signal to a digital one befitting digital electronics, which permits digital recordings of ECGs and the use of computers. Modern electrocardiograph machines allow simultaneous and multilead recording.

Table 7.3 Common ECG features useful in clinical judgment

Clinical signature	Typical values (ms)	Nominal limits (ms)
P width	110 ms	20 ms
T width	180 ms	40 ms
PR interval	120 ms	20 ms
QRS interval	100 ms	20 ms
QT interval	400 ms	40 ms
P amplitude	0.15 mV	005 mV
T amplitude	0.3 mV	0.2 mV
QRS amplitude	1.2 mV	0.5 mV

7.1.5 Clinical signature encased in ECG morphology

Figure 7.3 and Table 7.3 shows the useful ECG features and wave durations commonly used for clinical diagnosis. Among the common clinical signatures, the distance between two consecutive R peaks, named RR duration, signifies heart rhythms and is used for heart rate computation. The ventricular activity (depolarization and repolarization) is recorded in the QT segment, which is the region of interest to the cardiologists for diagnosis of major cardiac diseases. Since the heart rate of a normal patient may change during continuous recording, a corrected QT interval named QT_c is sometimes used to compensate for heart rate variation, given by Bazett's formula (Equation 7.2) or Fridericia's formula (Equation 7.3):

$$QT_c = \frac{QT \text{ interval}}{\sqrt{R-R \text{ duration}}} \tag{7.2}$$

$$QT_c = \frac{QT \text{ interval}}{\sqrt{R-R \text{ interval}}} \tag{7.3}$$

For clinical diagnosis, the medical experts have identified some quantitative and qualitative signatures. Some of the common clinical signatures expressed in a quantitative manner are given in Table 7.3 against their nominal range for a healthy adult. However, for a complete feature extraction, as many as 18 clinical signatures are reported in the literature. However, the actual clinical diagnosis procedure is much more complex, and many other parameters like age, sex, hereditary symptoms, food habits, and demographic factors are also taken into consideration.

7.1.6 Abnormality in ECG morphology

If the PR interval is greater than 0.2 s, then we call it a first-degree block. All the waves will still be present; there will just be a gap between the P wave and QRS complex. A first-degree heart block is not in itself very important: it can be a sign of coronary artery disease, acute rheumatic carditis, digoxin toxicity, or electrolyte disturbance. A second-degree heart block reveals an intermittent absence of QRS complexes, which indicates a blockage somewhere between the AV nodes and the ventricles. There are three types of this particular phenomenon:

1. **Mobitz type 2 phenomenon:** There is a regular rhythm, and a fairly constant PR interval, but a consistent absence of QRS (basically for every QRS, there are two or three P waves).

2. **Wenckebach phenomenon** (Mobitz type 1): Progressive lengthening of the PR interval followed by an absence of the QRS, then a shortened PR interval and normal QRS, and the cycle repeats. The cycle is variable in length.
3. R-R interval shortens with the lengthening of the PR interval.

In the bundle branch block (BBB) there is a normal PR interval and lengthened QRS duration. In the right bundle branch block (RBBB), there is a second R wave (R1) and the QRS complex will also be wide—greater than 12 ms. RBBB indicates right-sided heart disease. The left bundle branch block (LBBB) is exactly opposite to that of RBBB.

Myocardial infarction (MI), commonly known as a heart attack, occurs when blood flow stops to a part of the heart causing damage to the heart muscle. An inferior myocardial infarction (IMI) is a problem with the heart where cells along the inferior wall of the heart die in response to oxygen deprivation. This most commonly occurs as a result of a blockage in the right coronary artery, cutting off the supply of blood to this area of the heart. In anterior myocardial infarction (AMI)), the anterior wall of heart is involved. This is caused often by occlusion of the left anterior descending coronary artery. It can be categorized as infarction involving the anterior wall of the left ventricle, and produces indicative electrocardiographic changes in the anterior chest leads and often in limb leads. Myocardial infarction is classified into two types: ST-segment elevation myocardial infarction (STEMI) and non-STEMI.

7.1.7 Computerized ECG signal analysis

Computerized analysis and interpretation of ECG can contribute towards fast, consistent, and reproducible results which can save great deal of time for the physicians rather than manually checking paper-based ECG records. In addition, digitized transmission and storage of ECG can save memory space as well as facilitate remote monitoring and treatment of cardiac patients. Accurate detection of ECG fiducial points, P, QRS, and T, and their respective onset and offset points for computing the wave durations along with the heights of wave peaks are the prime purposes of any ECG analysis software. For clinical diagnosis, the cardiologists need at least three to four cardiac cycles of a 12-lead ECG record for visual analysis. For rhythm analysis, however, long-duration (sometimes 24–36 h) ECG records are needed. In principle, a cardiac cycle can be divided into low-frequency (P and T waves, and equipotential segments) and high-frequency (QRS complex) regions. In principle, the detection of a QRS complex is easier using statistical templates, consisting of slope and amplitude measures. In the case of ECG signals, however, the presence of different artifacts poses a challenge to the accurate estimation of the position of the fiducial points. Moreover, the shape and magnitude of the ECG vary widely across populations of different continents and are influenced by food habits and demographic and hereditary factors. Therefore, the different algorithms proposed over the years for ECG signal analysis have claimed to achieve comparable accuracy, if not better than the others, without promising a 100% perfection level. In spite of best practices and precautions during acquisition, ECG signals are corrupted by different types of noise, power line interference (PLI), baseline wander (BW), and muscle tremors (EMG noise) and are affecting the ECG wave shapes. Detection of a baseline is of paramount importance in the case of ECG feature extraction, since all the amplitude features (wave peak heights) are measured with respect to baseline voltage. The scope of computerized ECG analysis covers R-peak detection for rhythm analysis and feature extraction for disease identification. A detailed discussion on computerized ECG signal analysis is available in [3].

7.1.8 Electrocardiogram and nonstationarity

The ECG database comprises a collection of time-series data sets signifying a sequence of measurements recorded by one electrode during one heartbeat. Each heartbeat has an assigned classification of normal or abnormal. All abnormal heartbeats are representative of a cardiac pathology known as supraventricular premature beat.

The ECG signal is a nonstationary one (with bursts like QRS features contributing a localized high-frequency component) that can be ascertained by simple and preliminary measures like covariance, skewness, kurtosis, and autocorrelation function (ACF) of the ECG signal. Skewness is a measure of symmetry, or more precisely, the lack of symmetry in data distribution. A distribution, or data set, is symmetric if it looks the same to the left and right of the center point. Kurtosis is a measure of whether the data are heavy-tailed or light-tailed relative to a normal distribution. That is, data sets with high kurtosis tend to have heavy tails, or outliers. Data sets with low kurtosis tend to have light tails, or a lack of outliers. Many classical statistical tests and intervals depend on normality assumptions. Significant skewness and kurtosis clearly indicate that the data are not normal. If a data set exhibits significant skewness or kurtosis, some transformation can be applied to make the data normal. Another approach is to use techniques based on distributions other than normal. Figures 7.6 through 7.9 present the kurtosis, covariance, skewness, and ACF present in the MI inferior ECG data

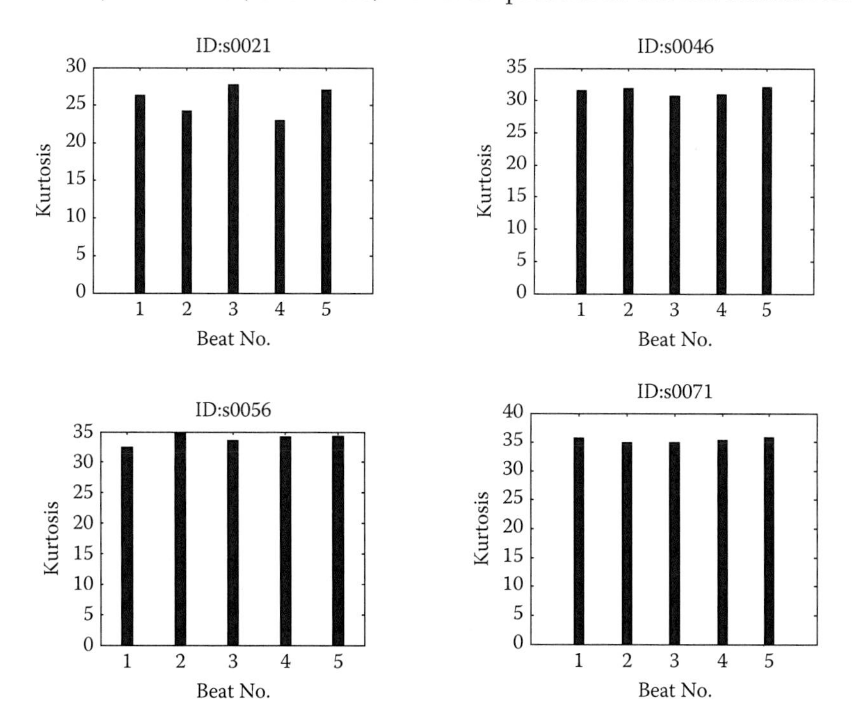

Figure 7.6 Beat-wise kurtosis.

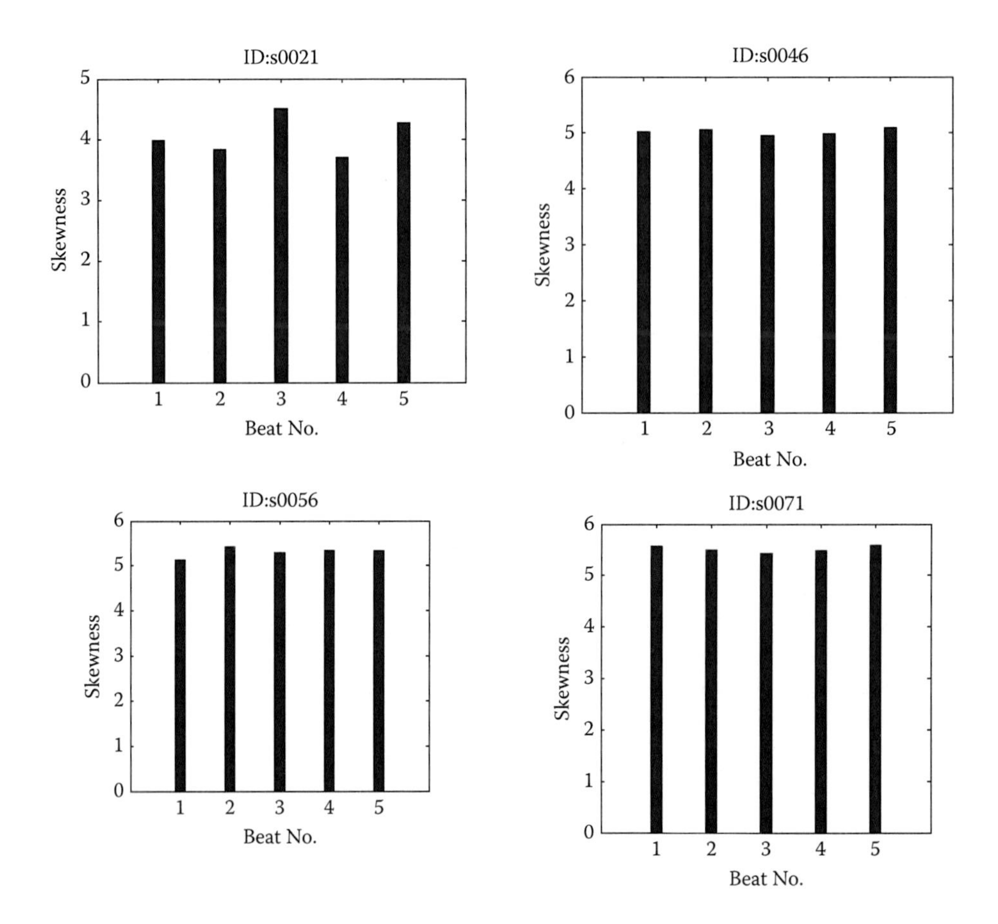

Figure 7.7 Beat-wise skewness.

revealing the nonstationary aspect present in the ECG signal. The programs required to determine the previously mentioned time series features of ECG data follow.

Program 7.1 To find covariance

```
clc
clear all
K=[];

[file] = uigetfile('*.txt');
x = dlmread([file]);
x=load(file,'-ascii');
n=size(x);

X1=[];
for i=1:n(1,1)
    if(x(i,1)==-5)
        break;
    end
    X1=[X1;x(i,1)];
end
```

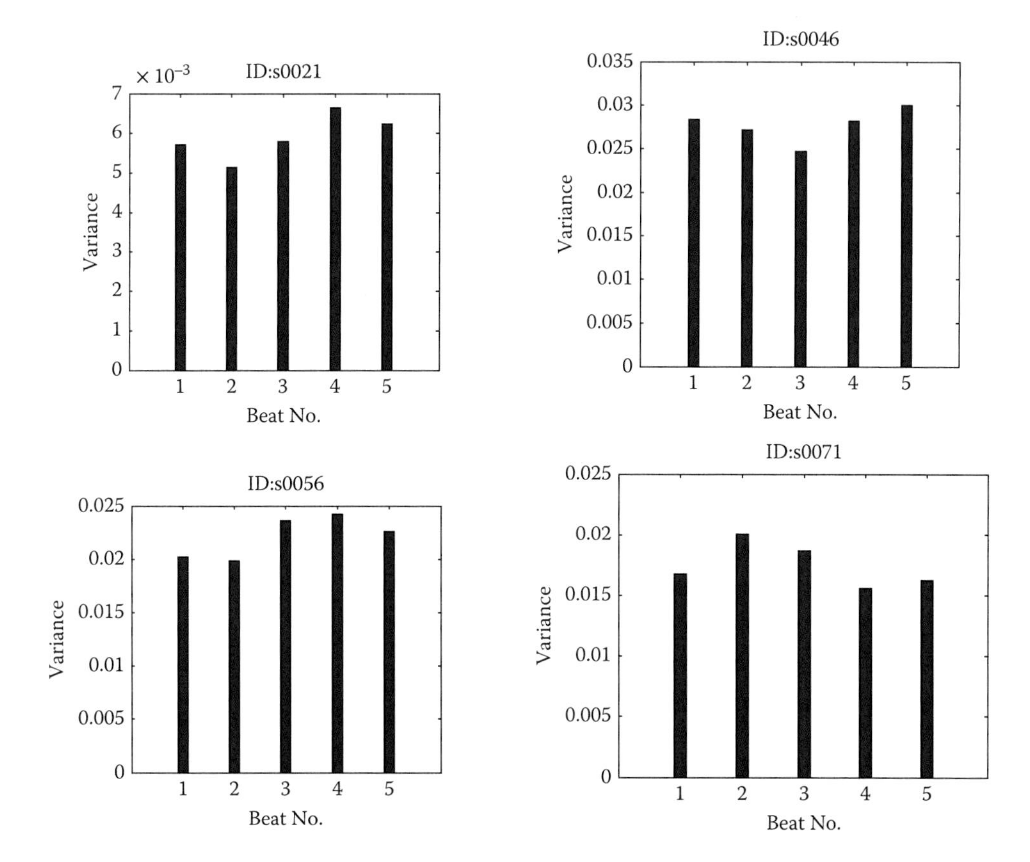

Figure 7.8 Beat-wise covariance.

```
X=zeropad(x);

for i=1:5
Y=X(:,i)

K=[K; cov(Y)];
end
K

bar(K,0.2)
xlabel('Beat No.');ylabel('Variance')

text=strcat('ID:',file(1:5))

title(text)
```

Program 7.2 To find kurtosis

```
clc
clear all
K=[];
```

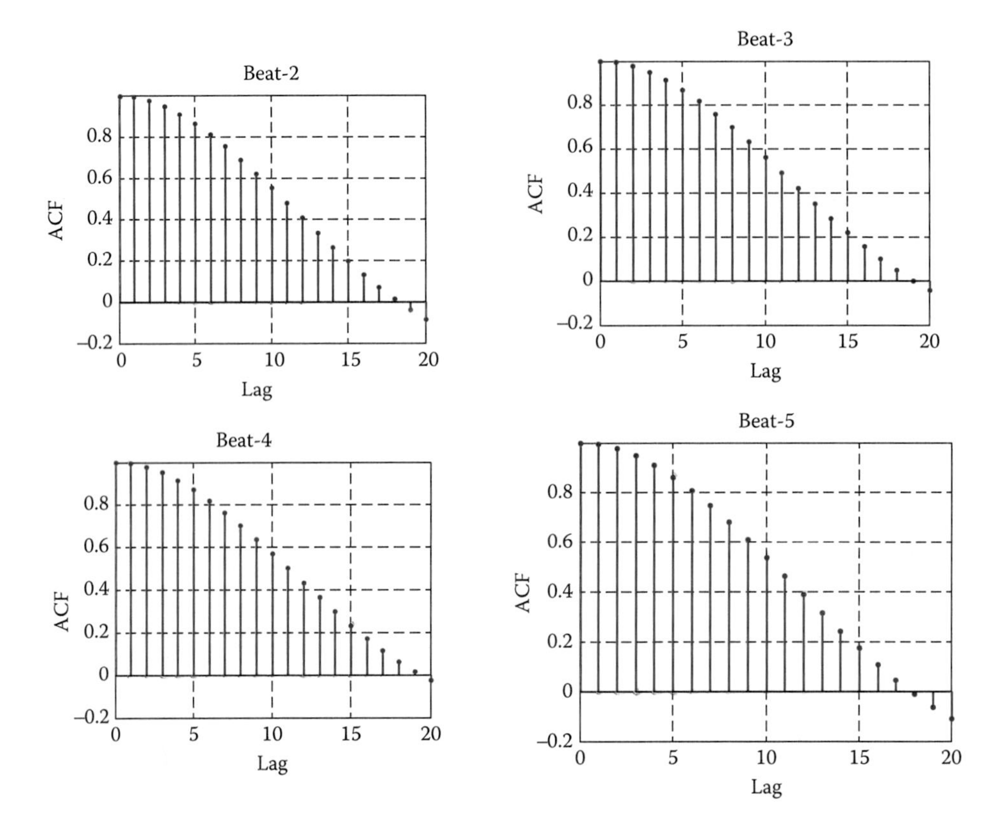

Figure 7.9 Beat-wise ACF.

```
[file] = uigetfile('*.txt');
x = dlmread([file]);
x=load(file,'-ascii');


X=zeropad(x);

for i=1:5
Y=X(:,i)

K=[K; kurtosis(Y)];
end
K

bar(K,0.2)
xlabel('Beat No.');ylabel('Kurtosis')
text=strcat('ID:',file(1:5))

title(text)
```

Program 7.3 To find skewness

```
clc
clear all
```

```
K=[];

[file] = uigetfile('*.txt');
x = dlmread([file]);
x=load(file,'-ascii');
n=size(x);

X1=[];
for i=1:n(1,1)
   if(x(i,1)==-5)
       break;
   end
   X1=[X1;x(i,1)];
end


X=zeropad(x);

for i=1:5
Y=X(:,i)

K=[K; skewness(Y)];
end
K

bar(K,0.2)
xlabel('Beat No.');ylabel('skewness')

text=strcat('ID:',file(1:5))

title(text)
```

Program 7.4 To find autocorrelation

```
clc
clear all
K=[];

[file] = uigetfile('*.txt');
x = dlmread([file]);
x=load(file,'-ascii');
n=size(x);

X1=[];
for i=1:n(1,1)
   if(x(i,1)==-5)
       break;
   end
   X1=[X1;x(i,1)];
end


X=zeropad(x);
```

```
text=strcat('ID:',file(1:5))
figure(1)
Y=X(:,1)
autocorr(Y, [], []);
ylabel('ACF')
title('Beat-1')


figure(2)
Y=X(:,2)
autocorr(Y, [], []);
ylabel('ACF')
title('Beat-2')

figure(3)
Y=X(:,3)
autocorr(Y, [], []);
ylabel('ACF')
title('Beat-3')

figure(4)
Y=X(:,4)
autocorr(Y, [], []);
ylabel('ACF')
title('Beat-4')

figure(5)
Y=X(:,5)
autocorr(Y, [], []);
ylabel('ACF')
title('Beat-5')
```

The preprocessed data "s0021are.txt_ld_I_full_beat" required for testing the above programs are available on the book webpage as an e-resource.

7.2 Morphological modeling of ECG

In this section, a morphological modeling method for a single-lead ECG is presented using two different approaches (*viz.*, Fourier and Gaussian models [5]). Single-lead ECG data are preprocessed to remove unwanted noise and segmented in three zones, P-R, Q-R-S, and S-T. The individual segments are then modeled to extract model coefficients. The residual of each segment, computed as the difference between the original and reconstructed samples, is also modeled using the Fourier model. The algorithms are validated with lead V2 from normal and anterior myocardial infarction (AMI) data. An ECG cycle (also called a beat) is obtained as a sequence of P, QRS, and T waves (with occasional U waves), representing atrial and ventricular depolarization and depolarization sequences. The clinical features and pathological inferences are related to the wave durations and intervals of the constituent ECG waves [6,7]. Over the years, computerized analysis and interpretation of ECG has remained as one of the prominent areas of medical research. The time-dependent autoregressive (AR)/autoregressive moving average (ARMA) model represents the general class of nonstationery signals. However, ECG signals possess some distinct features because of their pseudoperiodicity, such as different features of the constituent signals (P, QRS, and T) representing the actions of various parts of the heart; hence, they are difficult to model. Most of the early models of ECGs were nonparametric in nature [8–9]. Automated ECG processing can be divided into two broad subcategories: wave delineation and pattern recognition [10,11], and compression of ECG for bulk data storage by compression techniques [12]. ECG modeling or synthesis is useful in applications like beat classification, compression for bulk storage, and generation of synthetic ECG waves. There are two broad approaches used for ECG modeling. One is a data-fitting model, which has been mostly used for ECG compression. The second one is direct generation of ECGs by combining differential equations with known and controlled abnormalities. An ECG beat modeling and classification method is described in [13]. After preprocessing, the QRS segment is modeled using a Hermite function, followed by clustering using a self-organizing map [13]. A model for generating a synthetic ECG signal using three coupled differential equations is described in [14]. A trajectory in three-dimentional state-spaces with x, y, z coordinates generates the individual ECG cycles. Each revolution of this circle corresponds to one cardiac beat (R-R interval). Different morphological features (abnormalities) were introduced by varying the model parameters [14]. Use of data flow graphs for ECG generation is described in [15], where the ECG wave is fragmented into its constituent waves P, QRS, ST, T and U, and each of them is described in discrete form. The different waves and segments were generated using different delay elements, and sampling intervals. A data-driven model using a Gaussian function is described in [16]. The entire beat, without any fragmentation, was generated using a number of Gaussians, based on the number of local maxima. The model range was determined using two methods, namely, zero crossing by maximum local maxima, and bank methods using one local maximum and two surrounding maxima. In another approach, all fiducial points and wave segments are detected using a single Gaussian mesa function (GMF) to model a complete beat [17]. Five characteristic waves (P, Q, R, S, and T) of ECG are described by a GMF with a sixth one describing the biphasic nature of the waves. The modeled ECGs were also labeled as normal or abnormal. Fourier analysis for ECG modeling is described in [18] to find the maximum harmonic content. A Hopfield neural network (HNN) is applied and proposed for ECG signal modeling and noise reduction in [19]. The HNN is a recurrent neural network that stores the information in a dynamic stable pattern. This algorithm retrieves a pattern stored in memory in response to the presentation of an incomplete or noisy version of that pattern. Computer simulation results show that this method can successfully model the ECG signal and remove high-frequency noise. In this chapter, the synthesis of a single-lead ECG is proposed using two approaches, Gaussian model (GM) and Fourier model (FM), and their performance is compared using normal and abnormal (anterior myocardial infarction (AMI) ECG data from the Physionet database [20]. While most of the researchers have used MIT BIH arrhythmia data (data from the Massachusetts Institute of Technology and Boston's Beth Israel Hospital) for model validation, PTB Diagnostic ECG data have been used here in order to test the performance with diagnostic ECGs. Analysis and synthesis of an ECG wave consists of the following steps:

- Preprocessing of ECD data array
- R-peak detection

- Baseline point detection
- Baseline modulation correction
- Fiducial point determination
- Calculation of wave duration, intervals, and amplitude

7.2.1 Preprocessing of ECG

The proposed model extracts a single beat from a time-aligned ECG obtained from a single lead. The beat is segmented into three zones. The beat boundary is considered from one TP segment to the following one. The first segment is taken as the beat start to a Q-on index, the second one from Q-onset to S offset, and the third one is S offset to the end of the beat. Each of the segments is subjected to preprocessing followed by a Gaussian model (GM) and a Fourier model (FM) fit. Single-lead ECG data is used from the PTB Diagnostic ECG database under Physionet with a sampling frequency of 1 kHz. The single-lead ECG data is preprocessed for elimination of baseline wander and other artifacts using a discrete wavelet transformation (DWT) followed by PCA-based (principal component analysis) filtering. Other methods of denoising techniques include digital filtering techniques, the source separation method, neural networks, and the empirical decomposition method.

- The raw ECG signals are subjected to DWT using Daubechie's 6 (db6) up to the 10th level of decomposition so that the baseline wander frequency near 0.05 Hz can be eliminated.
- The wavelet coefficients are reconstructed discarding approximate coefficient A10 (representing baseline wander and DC level) and detail coefficients D1 and D2 (representing high-frequency components: muscle noise and electrosurgical noise). The resulting array say c is obtained.
- In the next stage, beat extraction is done. For this a reference fiducial point, an R-peak index, is considered (since the QRS complex is the most significant part of an ECG wave). The tentative QRS regions are identified in the wavelet decomposed structure (in the D4 + D5 array) using a thresholding-based approach. Figure 7.10a shows the original ECG, Figure 7.10b shows the ECG with baseline and muscle noise eliminated, and Figure 7.10c shows the D4 + D5 array used for QRS zone detection, the detailed procedure of which can be found in [3].
- The R-peaks are detected in the denoised array c as the maximum absolute amplitude in a window of max ± 70 ms, where max denotes the local maxima in array c. The identified R-peaks and baseline points are shown in Figure 7.10b. The beat start and ending points are detected between two successive R-peaks by dividing each PR interval in 2:1 ratio.
- In the next stage, time aligned ECG beats are obtained using 10 consecutive beats from array c. The extracted beats are cut and aligned with respect to R-peaks. When the beat lengths are found unequal, the shorter beats are padded at the respective tails. The aforementioned processes lead to the formation of multivariate data *t_beat*, which can be expressed as follows:

$$t\_beat = [c_1, c_2, c_3 .... c_n] \tag{7.4}$$

where $n = 10$ for the current case and the kth beat can be represented as follows:

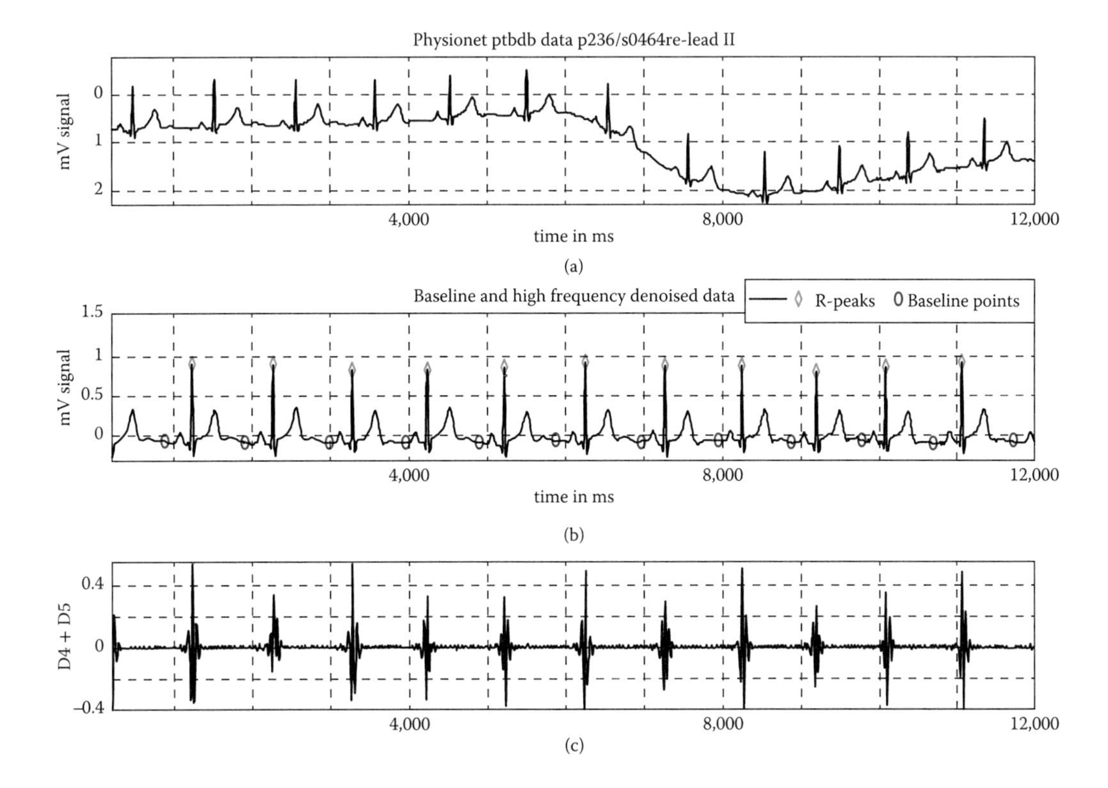

Figure 7.10 Preprocessed results with QRS and baseline points detected from ECG.

$$c_k = \left[c_{1k},\, c_{2k},\, c_{3k} \ldots . c_{lk}\right]^T \tag{7.5}$$

where l is the length of the longest beat. The *t_beat* array will have the R-peak positions at *r*_in index in all columns and a beat having length $m \le 1$, and will have $(1-m)$ total number of elements padded up at the start and end tails.

- The time-aligned beat matrix *t_beat* of dimension $l \times n$ is subjected to PC decomposition after mean adjustment from each column beat c_k (which is nothing but a PCA-based filtering). The principal components are obtained by applying linear orthogonal transform $\psi = [\psi_1\ \psi_2\ \psi_3\ \ldots .\ \psi_l]$ to t_*beat*, $PC = \psi^T.\text{t\_}beat$, where $PC = [PC_1\ PC_2\ PC_3 \ldots \ldots PC_l]^T$ are the principal components which are mutually uncorrelated. The first PC is obtained as a scaler product of $PC_1 = \psi_1.\text{t\_}beat$, which possesses the maximum variance in the eigenvector. The first PC is retained in the single beat reconstruction, which is normalized on a scale of 0–1 and used in the subsequent stages.

7.2.2 Beat segmentation for extracting zones

Each beat is segmented into three zones: beat start to Q-onset (denoted as P-R), Q-onset to S offset (denoted as Q-R-S), and S offset to the end of the beat (denoted as S-T), which implies the first part is basically the left part of the QRS, the second zone contains the entire QRS, and the last zone is the right-hand side of the QRS. QRS morphologies vary with ECG and lead type. Q and S points are searched in a window of r_in ± 70 ms around the R-peak, denoted by r_in. Considering the QRS shape symmetric w.r.t. the R-peak, a similar procedure is followed for Q-onset and S-offset detection. The beat extraction and beat segmentation algorithms are as follows:

(A) Procedure: Beat extraction from raw ECG:
Input variables: input array: y(raw ECG array)
Output variables: bt (beat arrays), blp: baseline point index
$Slp\_15 = \sum y(k) - y(k-15)/15$ (15-point average slope)

1. Execute discrete wavelet transform (DWT) decomposition using db6 up to tenth level.
2. Dk = k^{th} level detail coefficients.
3. $X = D4 + D5$.
4. Use threshold $\geq 15\%$ of max (X) to identify tentative maximum X amplitude, store in qrs.
5. For each window $(qrs-25)$ to $(qrs+25)$, identify maximum and minimum amplitude indices from y such that slope inversions across them are identified.
6. max (y) = maximum amplitude index of y; min (y) = minimum amplitude index of y.
7. Calculate $abs(\text{span}) = \{\max(y) - \min(y)\}$.
8. If $\max(y) \geq 10\%$ of span, max (y) = R-peak, /* nature of R-peak */. Else min (y) = R-peak.
9. Divide $\max(y)$ to $\max(y+1)$ index in 2:1 ratio, to get index z.
10. For each window $(z-30)$ to $(z+30)$, calculate slp_15.
11. blp = index with min (slp_15).
12. $bt\ (k) = blp\ (k)$ to $blp\ (k+1)$ /* extract the beat between two successive baseline points */.

(B) Q-peak and Q-on point detection:
Variables:
Input array: y (raw ECG array), r_pk (R–peak index),

slp_a (10–point average slope)
Output: q_in (Q–peak index), q_on (QRS onset index)
Variables: $slp\_a = |y(i) - y(i-10)|/10$; 10-point average slope
$Slp(k) = |y(k) - y(k-10)|$; 10-point slope
/* procedure Q, Q onset detection*/

1. Calculate slp_a in window $(r\_pk - 35)$ *to* r_pk
2. Look for slope sign inversions within $(r\_in - 35)$
   /* Case a: if there are slope inversions */
   a. Identify each index of slope sign inversions in $(r\_in - 35)$ *to* r_in window, Say q_t array
   b. Consider first slope inversion index as q_in, with the restriction that its amplitude is lower than the others in q_t.
   c. Search Q-onset in the window $(q\_in - 35)$ to q_in
   d. Calculate $Slp10(k)$ at each index k
   e. If $slp(k) \leq 0.5 \times slp\_a$, consider k *as* q_on.
   /* Case b: if there is no slope inversion within $(r\_pk - 35)$*/

Search within the window (r_pk-70) to (r_pk-35).
Calculate $slp(k)$ at each index k.
If at index k, either slope sign inversion occurs or $slp(k) \leq 0.5 \times slp\_a$, $q\_on = k$

3. If no condition in case (a) or case (b) matches, consider q_on as lowest amplitude index in $(r\_in-70 : r\_in)$ window.

(c) T peak and T-wave extraction:
Input variable: r_in array (R peak / QRS index); blp (baseline point index)
Output variables: t (T-peak index); t_on (T-onset index), t_off (T-offset index)
Variables: t_w array (T-wave); r_off (QRS offset)
$st\_slp = |y(r\_off+20) - y(r\_off)|/20$; average 20-point slope of ST segment
$t\_slp = |y(k+20) - y(k)|/20$; average 20-point slope at k^{th} index on T-wave
$blp\_slp = |y(blp+10) - y(blp-10)|/20$; average 20-point slope around baseline point
/* procedure T, T onset and T offset detection*/

1. Divide the r_in to (r_in+1) zone into 2:1 ratio.
2. Take the wider window as t_w.
3. Consider index with absolute maximum amplitude in t_w as t (T-index).

/* Find T-onset in the window (r_off) to T */

4. Calculate t_slp from T and proceed towards r_off
5. If $t\_slp(k) \leq 30\%$ of st_slp, $t\_on = k$.

/* Find T-offset in the window T to *blp**/

6. Calculate t_slp from T and proceed towards blp index.
7. If $t\_slp(k) \leq 30\%$ of blp_slp, $t\_off = k$.

All the extracted segments were normalized on a 0–1 scale before being processed in subsequent stages, as shown in Figure 7.11.

7.2.3 *Modeling and reconstruction of ECG waves*

7.2.3.1 *Fourier model*

It is considered that the single-beat ECG is a periodic signal, and hence, the segmented zones P-R, Q-R-S, or S-T can be individually represented by the Fourier harmonic components given by

$$x(t) = C_0 + \sum_{i=1}^{n} A_i \sin(i\omega_0 t) + B_i \cos(i\omega_0 t) \tag{7.6}$$

where, $x(t)$ = instantaneous value of zone potential, C_0 = average value of zone potential signal, ω_0 = angular frequency of fundamental component $2\pi/T$, T = time period of zone potential wave, n = total number of data points, $A_i = \frac{2}{T}\int_0^T x(t)\sin(i\omega_0 t)\,dt$, and $B_i = \frac{2}{T}\int_0^T x(t)\cos(i\omega_0 t)\,dt$.

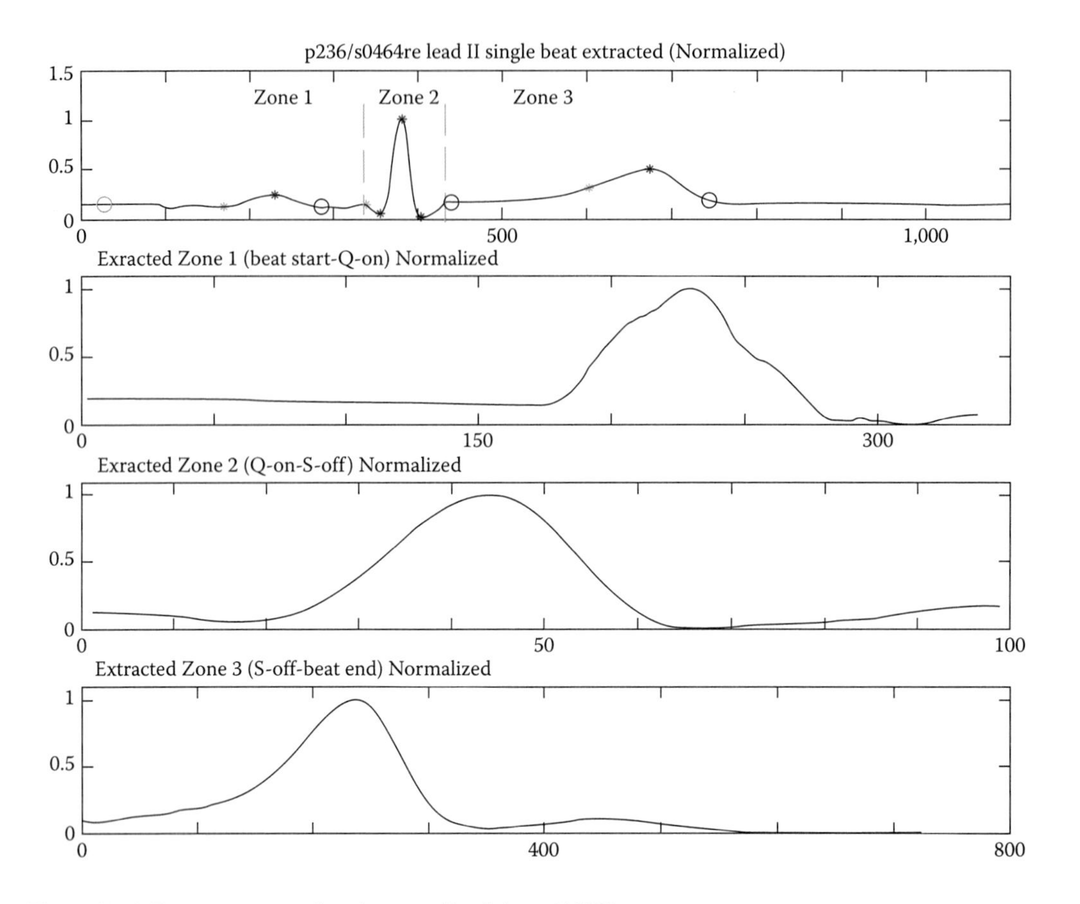

Figure 7.11 Zone segmented and normalized from ECG beat.

Assuming, $A_n = C_n \cos\theta_n$ and $B_n = C_n \sin\theta_n$, Equation 7.6 may be written as

$$x(t) = C_0 + \sum_{i=1}^{n} C_n \sin(n\omega_0 t + \theta_n) \tag{7.7}$$

where $C_n = \sqrt{A_n^2 + B_n^2}$, $\theta_n = \tan^{-1}\left(\dfrac{A_n}{B_n}\right)$.

In Fourier modeling, the model parameters considered are ω_0, C_0, and θ_n. The model coefficients are extracted using the Curve Fitting Toolbox (*cftool*) of MATLAB® for each of the individual zones. For computing Fourier coefficients, a fourth-order model is considered. The residue (E) is calculated between the actual zonal sample array (preprocessed) and reconstructed data using Fourier coefficients. The residue signal is again modeled using a fourth-order Fourier model.

7.2.3.2 Gaussian model

In this category, the segmented ECG zones P-R, Q-R-S, and S-T are modeled as a combination of a set of independent Gaussian functions called GC:

$$GC(x) = \sum_{i=1}^{N} a_i e^{-\left(\frac{x-\mu_i}{2\sigma_i}\right)^2} \tag{7.8}$$

where μ_i and σ_i are the mean and standard deviation of ith component of a Gaussian mixture (combination). The model coefficients are computed using the *cftool* of MATLAB® with the fitting option for a second-order Gaussian model. The model parameters (e.g., a_i, μ_i, and σ_i) are determined and individual wave zones are reconstructed over the specific time range of the corresponding part. The residue (E) between the original and reconstructed data set is again computed by a fourth-order Fourier model.

7.2.4 Reconstruction of single-beat ECG waveform

The individual data segments (i.e., P-R, Q-R-S, and ST) are computed using corresponding Fourier or Gaussian model parameters. The errors are also computed using error model parameters. The final normalized data sets corresponding to each segment are computed after summing the segment data sets and error data sets. Each normalized segment data set is then converted into absolute form using Equation 7.9:

$$y_i = \frac{(y_{ci} - \text{bias}_k)}{\text{gain}_k} \tag{7.9}$$

where y_i is final data point, y_{ci} is the ith final normalized data point, and bias_k and gain_k are bias and gain value of the kth segment respectively. Finally, the data sets for all segments are concatenated to form the data sets of a single ECG beat. The plot of time against amplitude indicates the reconstructed single-beat ECG waveform. The modeling and reconstruction performance parameters are ascertained by computing percentage root mean squared difference normalized (PRDN), mean square error (MSE), signal to noise ratio (SNR) and maximum error ($E_{\max}$). Among these, PRDN, MSE, and SNR provide global error measures and $E_{\max}$ provides local error.

$$\text{PRDN} = 100 \times \sqrt{\frac{\sum_{k=1}^{N} [x(k) - \hat{x}(k)]^2}{\sum_{k=1}^{N} [x(k) - \bar{x}]^2}} \tag{7.10}$$

$$\text{MSE} = \frac{\sqrt{[x(k) - \hat{\bar{x}}(k)]^2}}{N} \tag{7.11}$$

$$\text{SNR} = 10\log \frac{\sum_{k=1}^{N} [x(k) - \bar{x}]^2}{\sum_{k=1}^{N} [x(k) - \hat{x}(k)]^2} \tag{7.12}$$

$$E_{\max} = \max[x(k) - \hat{x}(k)] \tag{7.13}$$

where x = original ECG sample, $\hat{x}$ = reconstructed sample, N = total number of data points, $\bar{x}$ = mean of the original data, and E_{max} = maximum reconstruction error.

7.2.5 Performance evaluation of ECG synthesizer

Normal and abnormal ECG (AMI) data from the PTB Diagnostic ECG database (ptbdb) have been used for evaluating the performance of the ECG synthesizer. This database contains 12-lead ECG normal and abnormal patients. AMI symptoms are diagnosed in leads v1–v6 of ECG records at 1 kHz sampling, with ST elevation, prominent QS, and sometimes T-wave inversion. For this work, lead v2 records from 25 AMI and 40 healthy control records under ptbdb have been used. Table 7.4 shows the reconstruction results with Fourier and Gaussian models. For 25 AMI data sets, the average PRDN and SNR are found to be 9.33 and 20.71 respectively with a Gaussian model. The same quantities for a Fourier model were 5.43 and 22.16 respectively. For 40 normal (healthy control) data sets, the average PRDN and SNR were 1.22 and 34.21 with GM and 3.23 and 25.91 with FM, respectively. From Table 7.4, one can also conclude that average error figures are lower with a Fourier model. The mean square error (MSE) using both models is almost insignificant, whereas the maximum absolute error is less for the Fourier model, as evidenced from Table 7.4. Therefore, it can be concluded that the model performance varies with abnormality, and hence, the orders of GM and FM need to be varied to get the desired performance. Figure 7.12 shows the reconstruction plot against the original signal along with the error plot. Figure 7.12a and Figure 7.12b show plots with P169/s0329lre (normal) data for a FM and a GM, respectively. The reconstruction error plot shows better results with the Fourier model. The same for AMI data p081/s0266lre is shown in Figure 7.13a and b. A close scrutiny of the figures reveals that the FM performs better than the GM. This lower efficiency is due to the lower order used for the Gaussian model. It has been found that choosing a third- or fourth-order GM produces a reconstruction that, in most cases, is better than the second-order model. However, increasing the order number also enhances the number of coefficients. Tables 7.5 and 7.6 show the coefficients for a fourth-order FM and a second-order GM respectively. It is observed that a total of 54 coefficients need to be stored per beat of ECG using an FM, whereas the same for GM is 45. Thus, the GM is more memory efficient, but shows less reconstruction efficiency compared to the FM. All the reconstructed records were checked by a cardiologist, who confirmed that clinical signatures were preserved in the reconstructed/synthesized record. A study is also performed to check the performance without residual encoding. Table 7.7 shows the PRDN and SNR figures for a normal ECG computed without the residual model. If the corresponding items are compared in Table 7.4, it is observed that in all cases, the PRDN increases and SNR decreases, showing a fall in performance.

The model parameters with a suitable encoding technique can be used for ECG compression. The current section focuses on modeling a single beat. For studies involving beat-to-beat variability in ECG, like in arrhythmia or QT variability in cardiomyopathy, multiple beats are required.

The preprocessed AMI data set used for both model applications (*s0021are.txt_ld_I_full_beat*) is available on the book webpage as an e-resource. Programs 7.5 and 7.6 present code for the Fourier and Gaussian models, respectively.

Table 7.4 Performance of ECG waveform reconstruction by Fourier and Gaussian model parameters

	Gaussian model				Fourier model			
ID	MSE	PRDN	SNR	E_{max}	MSE	PRDN	SNR	E_{max}
P010/s0036(AMI)	0.0003	13.14	17.6	0.0915	0.0001	6.72	23.4	0.0270
P026/s0095(AMI)	0.0002	9.40	20.5	0.0746	0	4.20	27.5	0.0384
P044/s0146(AMI)	0.0002	10.85	19.2	0.02777	0.0001	7.04	23.0	0.0447
P081/s0266(AMI)	0.0004	14.49	16.7	0.1134	0.0001	5.69	24.8	0.0367
P081/s0346(AMI)	0.0001	6.93	23.1	0.0483	0.0003	11.59	18.7	0.0721
P044/s0159(AMI)	0.0001	8.69	21.2	0.0333	0	5.28	25.5	0.0208
P169/s0329(Normal)	0	1.78	35.0	0.0054	0	2.68	31.4	0.0010
P174/s0324(Normal)	0	1.83	34.7	0.0137	0.0004	1.58	16.1	0.0115
P237/s0465(Normal)	0	1.31	37.7	0.0053	0	2.84	30.9	0.0012
P247/s0480(Normal)	0	2.47	32.2	0.0219	0.00026	14.7	16.7	0.0078
P265/s0501(Normal)	0	4.01	27.9	0.0223	0	6.49	23.8	0.0028

(*N: healthy control)

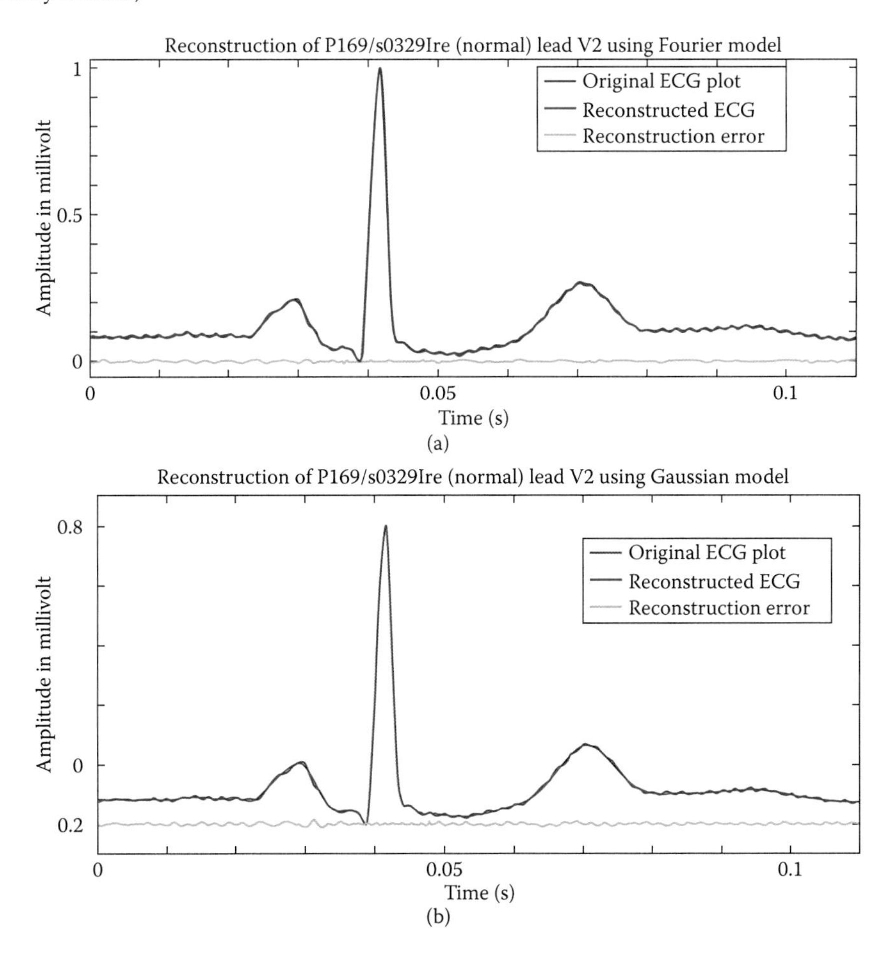

Figure 7.12 Reconstruction signal plot with original along with error (normal data): (a) Fourier model, (b) Gaussian model.

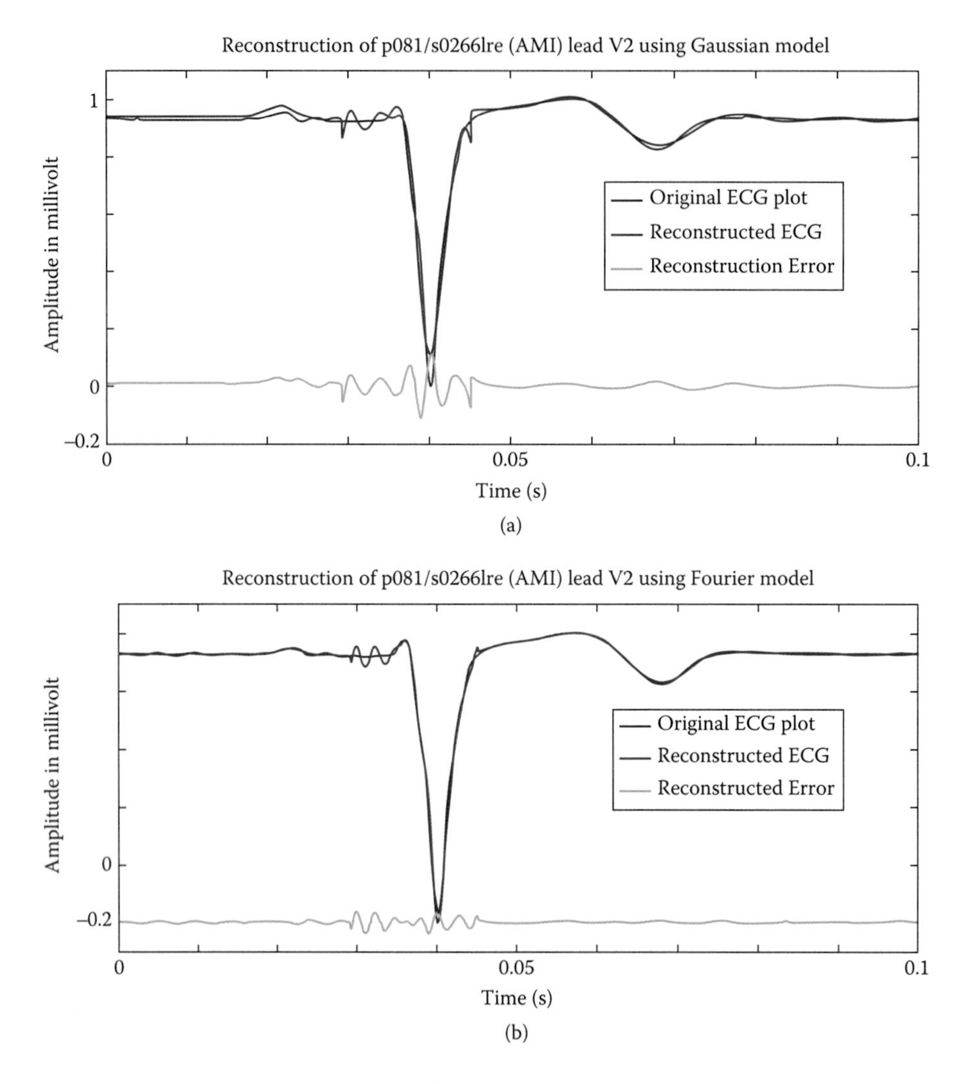

Figure 7.13 Reconstruction signal plot with original along with error (AMI data): (a) Gaussian model, (b) Fourier model.

Program 7.5 Fourier model for single-beat ECG

```
clc
clear all

[file1] = uigetfile('*.txt');
X = dlmread([file1]);

X=load(file1,'-ascii')
% file=Fullbeat(all three preprocessed segments in one)
% pre-processed AMI data 's0021are.txt_ld_I_full_beat'
X2=X(:,1);
n1=size(X)
X1=[];
for i=1:n1(1,1)
```

Table 7.5 Model coefficients: Fourier

Patient ID with lead no.	Zone	**Original signal model parameters**				**Residue model parameters**			
		n	**C_n**	**n (in rad)**	**(rad/ sec)**	**n**	**C_n**	**n**	**(rad/sec)**
p081_s0266lre (AMI)	PR	1	0.1323	1.1624	212.1	1	0.0075	0.7597	374.6
		2	0.1460	−0.4787		2	0.0132	0.4511	
		3	0.1155	1.2429		3	0.0626	−1.3457	
		4	0.0544	−0.5002		4	0.0625	1.5487	
	QRS	1	0.3500	1.5278	425.4	1	0.0012	0.2423	687.5
		2	0.2009	−0.2139		2	0.0040	−0.0554	
		3	0.0676	1.0496		3	0.0384	0.6924	
		4	0.0320	−0.5471		4	0.0263	1.1880	
	ST	1	0.1646	0.3085	107.3	1	0.0020	1.3503	176.6
		2	0.2205	−1.1858		2	0.0026	1.5140	
		3	0.1454	0.8223		3	0.0575	−0.8451	
		4	0.0753	0.2602		4	0.0194	0.7322	

Table 7.6 Model coefficients: Gaussian

Patient ID with lead no.	Zone	Original signal model parameters				Residue model parameters			
		n	a_n	b_n	c_n	n	C_n	n	(rad/sec)
p026/s0095lre (AMI)	PR	1	0.7741	0.01439	0.002608	1	0.0168	0.3505	334.7
		2	0.4018	0.005825	0.01154	2	0.0166	0.6283	
						3	0.0264	0.0180	
						4	0.0384	−1.5469	
	QRS	1	0.5861	0.002427	0.002719	1	0.1883	0.0090	457.3
		2	0.956e13	−20.16	3.591	2	0.1915	1.5058	
						3	0.0370	−0.8891	
						4	0.0644	0.5658	
	ST	1	0.5555	0.02198	0.005816	1	0.0099	−0.6127	131.8
		2	0.762	0.04069	0.02535	2	0.0194	0.4185	
						3	0.0227	0.7729	
						4	0.0104	−0.7391	

Table 7.7 Reconstruction results without residual model

Patient ID and record no.	Gaussian model		Fourier model	
	PRDN	*SNR*	*PRDN*	*SNR*
P169/s0329lre (N)	9.85	20.1	2.88	30.8
P174/s0324lre (N)	13.3	17.5	10.8	19.3
P237/s0465re (N)	48.6	6.27	4.02	27.9
P245/s0480re (N)	15.6	16.1	13.6	17.4
P265/s0501re (N)	35.8	8.93	11.7	18.6

```
 if (X2(i,1)==-5)
     break
 end
 X1=[X1;X2(i,1)];

end
size(X1)
X=X1;
%pause

file1
file1(1:18) % data file nomenclature
fname1=strcat(file1(1:18),'zone1.txt')
fname2=strcat(file1(1:18),'zone2.txt')
fname3=strcat(file1(1:18),'zone3.txt')

x1=load(fname1,'-ascii');

n1=size(x1)

y1=[];x1(:,1);

for i=1:n1(1,1)

  if (x1(i,1)==-5)
      break
  else
      y1=[y1;x1(i,1)];
  end

end
n1=size(y1);
gain1=y1(1,1);
bias1=y1(2,1)
y1=y1(3:n1(1,1),:);
n1=size(y1);
t1=0:1:n1(1,1)-1;
t1=0.001*t1;

% If we provide a pause here and get out of the program with % ctrl+c and
give a cftool prompt in the command window,
% zone 1 of the full % beat preprocessed diseased AMI data will % be
fitted to a 4th order Fourier model and generate the
% following 10 model parameters listed below. Re-run %the
% program.


          a0 =       0.503   ;
         a(1)  =     -0.3535   ;
         b(1)  =    -0.02874   ;
         a(2)  =     -0.0584   ;
         b(2)  =      0.1511   ;
         a(3)  =     0.09179   ;
         b(3)  =     0.02266   ;
         a(4)  =     0.02639   ;
```

```
       b(4) =      -0.06559   ;
       w =         42.66;

       x=t1;

        C1=[];theta1=[];

       for i=1:4
         C1=[C1;sqrt(a(i)^2+b(i)^2)];
         theta1=[theta1;atan(b(i)/a(i))];

               end

yc1 =a0 + a(1)*cos(x*w) + b(1)*sin(x*w) + a(2)*cos(2*x*w) +
b(2)*sin(2*x*w) + a(3)*cos(3*x*w) + b(3)*sin(3*x*w) + a(4)*cos(4*x*w) +
b(4)*sin(4*x*w)

% depending on the performance matrix the order is selected.

  ya1=(yc1-bias1)/gain1 ;

      C1 =[a0 ; C1];
      theta1=[0;theta1];

x2=load(fname2,'-ascii');
n2=size(x2)

y2=[];x2(:,1);

for i=1:n2(1,1)

  if (x2(i,1)==-5)
      break
  else
      y2=[y2;x2(i,1)];
  end

end
n2=size(y2);
gain2=y2(1,1);
bias2=y2(2,1)

y2=y2(3:n2(1,1),:);
n2=size(y2);
t2=0:1:n2(1,1)-1;
t2=0.001*t2 ;

%If we provide a pause here and get out of the program with %ctrl+c % and
give a cftool prompt in the command window,%zone 2 of the full beat
preprocessed diseased AMI data will %be fitted to a 4th order Fourier
model and generate the %following 10 model parameters listed below.
Re-run the %program.
         a0 =       0.3491   ;
       a(1) =      -0.3365   ;
       b(1) =      0.08011   ;
       a(2) =       0.2018   ;
```

```
        b(2) =      0.05654 ;
        a(3) =     -0.03565  ;
        b(3) =     -0.01436  ;
        a(4) =      0.02777  ;
        b(4) =      0.01371  ;
        w =        83.34;

              x=t2;

               C2=[];theta2=[];

              for i=1:4
          C2=[C2;sqrt(a(i)^2+b(i)^2)];
          theta2=[theta2;atan(b(i)/a(i))];

        end
yc2 =a0 + a(1)*cos(x*w) + b(1)*sin(x*w) + a(2)*cos(2*x*w) +
b(2)*sin(2*x*w) + a(3)*cos(3*x*w) + b(3)*sin(3*x*w) + a(4)*cos(4*x*w) +
b(4)*sin(4*x*w)

 ya2=(yc2-bias2)/gain2 ;


          C2 =[a0 ; C2];
       theta2=[0;theta2];

      x3=load(fname3,'-ascii');
n3=size(x3)


y3=[];x3(:,1);

for i=1:n3(1,1)

      if (x3(i,1)==-5)
       break
  else
       y3=[y3;x3(i,1)];
  end

 end
n3=size(y3);
gain3=y3(1,1)
bias3=y3(2,1)


y3=y3(3:n3(1,1),:);
n3=size(y3);
t3=0:1:n3(1,1)-1;
t3=0.001*t3 ;

%If we provide a pause here and get out of the program with % ctrl+c %
and give a cftool prompt in the command window, % zone 3 of the full beat
```

```
preprocessed diseased AMI data will % be fitted to a 4th order Fourier
model and generate the
% following 10 model parameters listed below. Re-run the
% program.

       a0 =      -3.552;
       a(1) =      -2.063  ;
       b(1) =        8.12  ;
       a(2) =         7.1  ;
       b(2) =       1.944  ;
       a(3) =      0.5128  ;
       b(3) =      -4.041  ;
       a(4) =      -1.308  ;
       b(4) =      0.3502  ;
       w =       7.936;

      x=t3;

        C3=[];theta3=[];

       for i=1:4
         C3=[C3;sqrt(a(i)^2+b(i)^2)];
         theta3=[theta3;atan(b(i)/a(i))];

       end

 yc3 =a0 + a(1)*cos(x*w) + b(1)*sin(x*w) + a(2)*cos(2*x*w) +
b(2)*sin(2*x*w) + a(3)*cos(3*x*w) + b(3)*sin(3*x*w) + a(4)*cos(4*x*w) +
b(4)*sin(4*x*w)

ya3=(yc3-bias3)/gain3 ; % converted into absolute form after
% reconstruction

      C3=[a0 ; C3];
      theta3=[0 ;theta3];

      Y=[ya1';ya2';ya3'];
      N=size(Y)

     size(X)

     % pause

T=0:1:N(1,1)-1;
T=0.001*T;
T=T';
X=X(:,1);
Y=Y(1:N(1,1)-2 ,:);
T=T(1:N(1,1)-2,:);
E=X-Y;

text1='Original'
text2='Reconstructed'
text3='Error'
```

```
plot(T,Y,'r',T,X,'b',T,E,'m','LineWidth',2)% X=experimental,
y=constructed, E=error
grid
xlabel('Time(s)'); ylabel('Amplitude')
title('Reconstruction of ecg waveform using Fourier Model')
gtext(text1)
gtext(text2)
gtext(text3)
C=[C1 ; C2 ; C3];
Theta=[theta1;theta2;theta3];
Z=[C Theta]

% performance matrices of reconstruction process
MSE=sum(E.^2);
MSE=MSE/N(1,1);
MSE=sqrt(MSE);

Emax=max(E);

R=sum(E.^2);
M=mean(X,1);
L=sum((X-M).^2);
PRDN=sqrt(sum(E.^2)/L);

SNR=10*log(L/sum(E.^2));

disp('PRDN      MSE     SNR     Emax')
[PRDN MSE SNR Emax]
```

Program 7.6 Gaussian model for single-beat ECG

```
clc
clear all

[file1] = uigetfile('*.txt');
X = dlmread([file1]);
X=load(file1,'-ascii')

% file1=Fullbeat(all three preprocessed segments are in one) pre-
processed AMI data 's0021are.txt_ld_I_full_beat'

n1=size(X)
X2=X(:,1);
n1=size(X)
X1=[];
for i=1:n1(1,1)
 if (X2(i,1)==-5)
     break
 end
 X1=[X1;X2(i,1)];

    end
size(X1)
```

```
X=X1;

file1
file1(1:18)%% data file nomenclature
fname1=strcat(file1(1:18),'zone1.txt')
fname2=strcat(file1(1:18),'zone2.txt')
fname3=strcat(file1(1:18),'zone3.txt')

x1=load(fname1,'-ascii');

n1=size(x1)


y1=[];x1(:,1);

for i=1:n1(1,1)

  if (x1(i,1)==-5)
      break
  else
      y1=[y1;x1(i,1)];
  end

end
n1=size(y1);
gain1=y1(1,1);
bias1=y1(2,1)

y1=y1(3:n1(1,1),:);
n1=size(y1);
t1=0:1:n1(1,1)-1;
t1=0.001*t1 ;

% If we provide a pause here and get out of the programme
% with ctrl+c  and give a cftool prompt in the command window, %zone 1 of
the full beat preprocessed diseased AMI data %will be fitted to a 2nd
order Gaussian  model and generate the %following six model parameters
listed below. Re-run the
% program.


        a1 =      0.6215   ;
        b1 =     0.09937   ;
        c1 =     0.01487 ;
        a2 =      0.6916   ;
        b2 =     0.06161 ;
        c2 =     0.04913   ;

 A1=[a1 b1 c1 ; a2 b2 c2];

  x=t1;
    yc1 =  a1*exp(-((x-b1)/c1).^2) + a2*exp(-((x-b2)/c2).^2);
  ya1=(yc1-bias1)/gain1 ;

x2=load(fname2,'-ascii');
```

```
n2=size(x2)

y2=[];x2(:,1);

for i=1:n2(1,1)

  if (x2(i,1)==-5)
      break
  else
      y2=[y2;x2(i,1)];
  end

end
n2=size(y2);
gain2=y2(1,1);
bias2=y2(2,1)

y2=y2(3:n2(1,1),:);
n2=size(y2);
t2=0:1:n2(1,1)-1;
t2=0.001*t2 ;


%If we provide a pause here and get out of the program with %ctrl+c % and
give a cftool prompt in the command window, %zone 2 of the full  beat
preprocessed diseased AMI data will %be fitted to a 2nd order Gaussian
model and generate the following %6 model parameters listed below. Re-run
the
% program.


          a1 =       0.7956   ;
          b1 =      0.03789   ;
          c1 =      0.01035   ;
          a2 =    1.546e+013    ;
          b2 =       -9.642   ;
          c2 =        1.707   ;

  A2=[a1 b1 c1 ; a2 b2 c2];

   x=t2;
   yc2 =  a1*exp(-((x-b1)/c1).^2) + a2*exp(-((x-b2)/c2).^2);

  ya2=(yc2-bias2)/gain2 ;

x3=load(fname3,'-ascii');
n3=size(x3)

y3=[];x3(:,1);

for i=1:n3(1,1)

  if (x3(i,1)==-5)
      break
  else
```

```
      y3=[y3;x3(i,1)];
  end

end
n3=size(y3);
gain3=y3(1,1);
bias3=y3(2,1)

y3=y3(3:n3(1,1),:);
n3=size(y3);
t3=0:1:n3(1,1)-1;
t3=0.001*t3 ;

%If we provide a pause here and get out of the program with %ctrl+c  and
give a cftool prompt in the command window, %zone 3 of the full beat
preprocessed diseased AMI data will %be fitted to a 2nd order Gaussian
model and generate the following %6 model parameters listed below. Re-run
the
% program.

        a1 =      0.9294   ;
       b1 =     0.05427   ;
       c1 =     0.08331   ;
       a2 =      0.7392   ;
       b2 =      0.2685   ;
       c2 =     0.06166   ;

  A3=[a1 b1 c1 ; a2 b2 c2];
  x=t3;
  yc3 =  a1*exp(-((x-b1)/c1).^2) + a2*exp(-((x-b2)/c2).^2);

ya3=(yc3-bias3)/gain3 ;% converted into absolute form after
reconstruction

   Y=[ya1';ya2';ya3'];
   N=size(Y)

T=0:1:N(1,1)-1;
T=0.001*T;
T=T';
X=X(:,1);
Y=Y(1:N(1,1)-2 ,:);
T=T(1:N(1,1)-2,:);
E=X-Y;

text1='Original'
text2='Reconstructed'
text3='Error'

plot(T,Y,'r',T,X,'b',T,E,'m','LineWidth',2)
grid
xlabel('Time(s)'); ylabel('Amplitude')
title('Reconstruction of ecg waveform using Gauss Model')

gtext(text1)
```

```
gtext(text2)
gtext(text3)

A=[A1 ; A2 ; A3]
% performance matrices for reconstruction process

MSE=sum(E.^2);
MSE=MSE/N(1,1);
MSE=sqrt(MSE);

Emax=maxZ(E);

R=sum(E.^2);
M=mean(X,1);
L=sum((X-M).^2);
PRDN=sqrt(sum(E.^2)/L);

SNR=10*log(L/sum(E.^2));

disp('PRDN      MSE     SNR     Emax')
[PRDN MSE SNR Emax]
```

7.3 Automated ECG pattern classifier

Conventional automated ECG classifiers employing soft computing tools may suffer from inconsistencies due to less reliable clinical feature extraction procedures. In this section, the dissimilarity factor (D) based on Karhumen-Loeve (KL) expansion is deployed for ECG classification among normal and inferior myocardial infarction (IMI) patients [21]. Time-aligned ECG beats are obtained through filtering and wavelet decomposition processes, followed by PCA-based beat enhancement, which is basically multivariate time series data. PCA has wide applications in medical signal processing including ECG enhancement, classification, and compression [22–24]. The T wave and QRS segments of IMI data sets from Leads II, III and aVF are extracted and compared with corresponding segments of healthy patients using Physionet ptbdb data deploying dissimilarity and showing the promise of descriptive statistical tools as an alternative for medical signal analysis, without using direct clinical feature extraction.

7.3.1 Preprocessing of the ECG

The preprocessing of single-lead ECG data is done in three stages. The raw data used in the study has a sampling frequency of 1 kHz.

1. **Noise filtering**: In the first stage, a fourth-order low-pass Butterworth filter is used to select the clinical bandwidth of the ECG (0–90 Hz) and discard a major component of muscle noise from the data. The resulting filtered array is obtained as *b*[].
2. **Application of DWT transform on the denoised signal**: The resulting *b* array is decomposed using a discrete wavelet transform (using Daubechie's 6 (db6) wavelet and method illustrated in [25]). The resulting detail (Di) and approximation

coefficients (Ai) are reconstructed discarding D1 and D2 (representing high frequency) and discarding A10 (representing low frequency) to obtain an array *c*[].

3. **PCA-based beat enhancement/filtering**: The array *c*[] is subjected to PCA-based filtering after a beat-alignment process to further enhance (denoise) the signal. At first, the tentative locations of QRS regions are identified from the D2+D3 array, which contains the QRS frequency from the discrete wavelet (DWT) decomposed structure. Adapting a thresholding-based approach, the exact location of the R peak (or QS peak) is identified. The distance between two successive QRS peaks measured from baseline points are called ECG beats. The beat extraction algorithm outline is shown below:

/*Beat extraction procedure*/
I. $XX_2 = D2 + D3$.
II. Use threshold $\geq k \times \max(XX_2)$ to identify tentative QRS zones in array z.
III. For each z element repeat steps IV–XI.
IV. $qrs(k) = \max \text{ or} \min$ position around $z(k) \pm 70$ ms window. /* $qrs(k)$ defines QRS peaks */
V. $rr(j) = qrs(j+1) - qrs(j)$ /* RR interval */
VI. segment[] = index after dividing each $rr(j)$ interval in 2:1
VII. Repeat steps VII–XI for each segment $[k]$.
VIII. $slp\_15(j) =$ 15-sample average slope at j
IX. Compute slp_15 in segment $[k]$ ±50 = slp_a
X. $blp(k) = \min(slp\_a)$
XI. $beat(j) = blp(j) \ to \ blp(j+1)$
/*---*/
Following this procedure, a total of m beats are extracted.

7.3.2 ECG enhancement through beat-aligned PCA

Before the PC decomposition, it is fairly mandatory that the ECG beats are to be time aligned, with reference to a fiducial point, and marked. Since the lengths of the beat segments may be unequal, beat by beat modification is necessary. The resulting k^{th} beat array can be represented by the following column vector:

$$beat_k = [c_1 \ c_2 \ c_3 \ \ldots\ldots c_l] \tag{7.14}$$

where, c_1, c_2, c_3 are the ECG samples of the kth beat array, and l is the length of the beat k. For PC decomposition of kth beat array, the following steps are undertaken:

- The beat matrices are time aligned with reference to their respective R-peak index and arranged in a matrix whose column lengths are equal:

$$X = [beat_1 \ beat_2 \ beat_3 \ \ldots beat_m \ ], \text{with } m = 10 \tag{7.15}$$

To obtain X, the beat vectors are modified to include the fiducial point at the pth row, such that in each column k, $g\_t(k) = p\_l - lbt(k)$ and $g\_\_b(k) = p\_r - rbt(k)$ are the first and end elements of column k, to be prefixed and suffixed respectively, p_l and p_r:

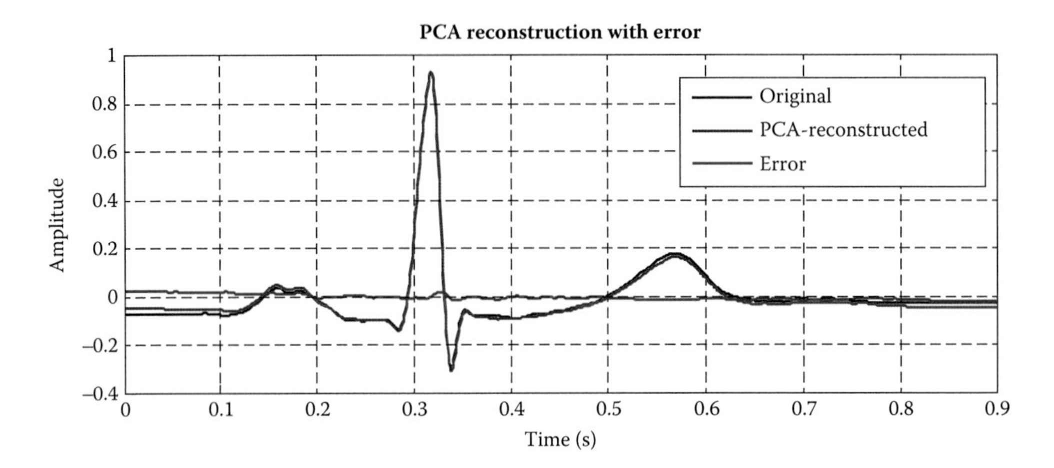

Figure 7.14 PCA reconstructed ECG beat

maximum length of left-side and right-side beat respectively; $lbt(k)$ and $rbt(k)$ are the lengths of the left-side and right-side beats, respectively, for column k.

- The matrix X with dimension $m\times(p\_l+p\_r)$ is subjected to PC decomposition after mean adjustment of each beat k. The covariance matrix obtained as $C_x = \text{cov}\ (X)$ yields the eigenvectors α_j and eigenvalues (λ_j) defined by $C_x \times \alpha_j = \lambda_j \times \alpha_j$, where $j = 1,2....t$. The PCs are linear transformation of the beats with transformation coefficients given by the eigenvectors α_j. The signal is reconstructed retaining the first PC, which provides maximum variance in the eigenvectors.
- Using n beats of the single-lead ECG data set, the following matrix is formed:

$$Y_{m\times n} = \left[beat\_d_1 \;\; beat\_d_2 \;\; beat\_d_3 \ldots beat\_d_k .. beat\_d_n\right] \tag{7.16}$$

where $beat\_d_k$ is a column vector obtained from m number of beats through PCA denoising (considering $n = 5$). The performance of PCA-based reconstruction can be expressed in terms of PRD and PRDN. Figure 7.14 shows the time-aligned PCA constructed ECG beats done using PCA-based reconstruction program 7.7. The pre-processed (time-aligned) data sets (beat_p240_s0468re (normal), beat_p284_s0551lre (abnormal)) needed to test the program are available on the book webpage as an e-resource.

7.3.3 *Extraction of QRS and T wave from inferior myocardial infarction data*

As discussed earlier in subsection 7.1.6, myocardial infarction (MI) refers to a condition where blood circulation to a group of cardiac cells is affected, resulting in the formation of dead cells. When this affects the bottom layer of cardiac cells, the result is an inferior MI (IMI). IMI is manifested as the occurrence of prominent Q waves, and inverted T waves in leads II, III, and aVF, as shown in Figure 7.15. Thirty-five ECG records from the Physionet database were selected and clinically verified with a cardiologist for the pathological signatures as mentioned. These clinically identified records were only used in the present case study. From each column $beat\_d$ vector, QRS zones and T waves are

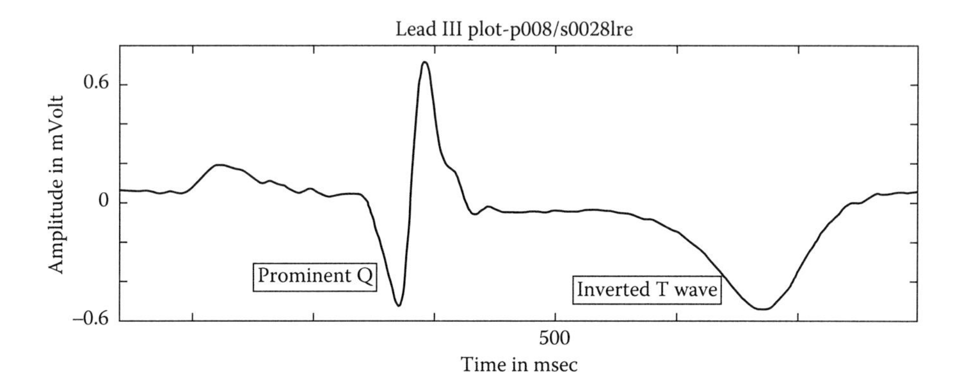

Figure 7.15 A typical IMI lead III plot showing pathological symptoms.

extracted. R-peak locations are already known in each column of Y (Equation 7.16) at the $p\_l+p\_r=r\_in$ (say). The QRS zone is extracted as (r_in-50) to (r_in+40) samples in the z array. For each beat, $beat\_d$, the QRS zone is stored as array qrs_z.

The T peak is extracted in the ST-region as (r_in+600) to the end of the $beat\_d$ array by computing the absolute 20-sample upside and downside slope at every sample position k, using the following equations:

$$spl\_l(k)=\left[z(k)-z(k-20)\right] \tag{7.17}$$

$$spl\_r(k)=\left[z(k)-z(k+20)\right] \tag{7.18}$$

where spl_l and spl_r refer to the absolute upside and downside slope respectively. The sample with maximum aggregate slope is considered the T-peak index t_in. The T-wave is extracted as (t_in-80) to (t_in+80) sample positions in array z and taken in array t_z. Finally, all columns in qrs_z and t_z are normalized in the range 0–1.

Program 7.7 Reconstruction of ECG beats using PCA

```
% This program finds PCs and re-converts to get original data
% Starts with time-aligned ECG beats. Store the encoded PCA,
% eigenvalues and beat-ends. To test the program, ECG data set %required
are beat_p240_s0468re (normal) and %beat_p284_s0551lre %(abnormal)

clc
clear all

[file1] = uigetfile('*.txt');
x = dlmread([file1]);
%'data.txt');  %;

n=fix(length(x(1,:))/10);

 kk3=1;
 k9=1; k8=1;k7=1;
```

```
%for ld=1:3

ld = input ('Enter the PC, 1, 2 or 3 for reconstruction:  ')
if ld==1
    col=10;
elseif ld==2
    col=9;
else
    col=8;
end

prd=[];
prdn=[];
for kk=1:1
  close all
 kk2=1;
X_k=x(:,kk:kk+9);                    % x(1:2000,2:5);

[COEFF,SCORE] = princomp(X_k);

[eig_vec,eig_val]=eig(cov(X_k));
pca=(eig_vec)'*(X_k)';
pca=pca';
A=(eig_vec)*pca';
B=A';                                 % original data

% modifying the eig matrix
for k1=1:10
    for k2=1:10
    null_e(k1,k2)=0;
    end

end

temp_e=null_e;
temp_e(:,10-(10-col))=eig_vec(:,col);
m_eig= temp_e;

%modifying the PCA matrix

for k1=1:length(A(1,:))
    for k2=1:10
    null_p(k1,k2)=0;
    end

end

temp_p=null_p;
temp_p(:,10-(10-col))=pca(:,col);
m_pca=temp_p;

rec_data= m_eig*(m_pca)';

clear temp_e temp_p
% error calculation
```

```
sum_sqerr=0;
org_samp_9=0;
sum_mu=0;

for k1=1:10                          % Column index
    mu(k1)=mean(B(:,k1));
    for k2=1:length(A(1,:))          % k2 = row index
    err(k1,k2)=rec_data(k1,k2)-A(k1,k2);
    sqr_err= err(k1,k2)^2;
    sum_sqerr =(sum_sqerr+sqr_err);
    org_samp_9=A(k1,k2)^2+org_samp_9;
    sum_mu=(A(k1,k2)-mu(k1))^2+sum_mu;
    end

end
    prd_pc=sqrt(sum_sqerr/org_samp_9);
    prdn_pc=sqrt(sum_sqerr/sum_mu);
    clear sqr_err sum_sqerr sum_mu
end

n=size(rec_data);
 t=0:1:n(1,2)-1;
 t=0.001*t;
 t=t';

    plot(t,A(1,:),'Color','k','LineWidth',2)
    %plot(t,A','r')

    title('original data')
    hold on
 %subplot(4,1,ld+1),

 plot(t,rec_data(1,:),'Color','b','LineWidth',2)
 %plot(Z(:,1), Z(:,2),'b')

 hold on
 %subplot(4,1,ld+1),

 plot(t,err(1,:),'Color','r','LineWidth',2)

 title('reconstruction with error retaining required PCA')
 hold on
 %end
 ylabel('Amplitude') ;xlabel('Time(S)')

  grid



  Z=[ rec_data(1,:)' A(1,:)'];
  %Z=[Z; prd_pc prdn_pc];
  fname=file1(6:15);

  fname=strcat(fname,'_pca_recon_',int2str(ld),'.txt');
```

```
save(fname,'Z','-ascii')

PRD=prd_pc
PRDN=prdn_pc
```

7.3.4 Dissimilarity-based classification/authentication of ECG

The theory of the dissimilarity factor (D) between two multivariate data sets X_1 and X_2 is deduced from KL expansion (as already described in Chapter 2).

The dissimilarity factor (D) between two multivariate data sets is as follows:

$$D = \frac{4}{m}\sum_{j=1}^{m}\left(\lambda_j - 0.5\right)^2 \tag{7.19}$$

where λ_j is the eigenvalue of the jth population, and $m = 3$ since, for each data, three subsets are considered. D varies between 0 and 1. When the data sets are similar, D is near zero and it is a value near to 1 when the data sets are dissimilar. The dissimilarity, D, is computed separately for QRS and T waves using five beats of normal and of IMI data (the MATLAB® code for calculating dissimilarity is provided in Chapter 2). The proposed dissimilarity-based algorithm is tested with ptbdb data under Physionet. A total of 15 normal and 35 IMI subjects are used for testing. For computing the dissimilarity factor (D), the following steps are executed:

- The QRS and T waves from five beats are extracted from normal and IMI patients. In consultation with a cardiologist, 35 IMI data are selected and the corresponding clinical symptoms are marked.
- Comparison is done among normal and IMI data using matrix *qrs_z* for QRS segments and matrix *t_z* for T waves. In the current study, X_1 is formed using three columns of QRS_z, each from lead II, III and aVF of normal data.
- Similarly, X_2 is formed using IMI data. For T waves, separate sets of X_1 and X_2 are formed.

The index D_{qrs} in Table 7.8 and 7.9 reflects the dissimilarity factor (D) computed between QRS segments using a few normal and IMI ECG data. Similarly D_t represents the dissimilarity factor (D) computed between normal and abnormal T segments. At first, D_{qrs} and D_t between 15 normal ECG records are computed to verify their similarity. Table 7.8 shows the dissimilarity between the normal records, using p240/s0468lre and p243/s0472lre as representative normal data. A composite dissimilarity factor, denoted by D_c, is computed by averaging D_{qrs} and D_t over the 15 normal data records used. D_c is found

Table 7.8 Dissimilarity factor calculated between normal ECG data

Ref Data/ *Patient ID (Normal)*	p240/s0468re		p243/s0472re	
	D_{qrs}	D_t	D_{qrs}	D_t
p243/s0472re	0.41	0.21	0.00	0.00
p156/s0299lre	0.12	0.25	0.43	0.11
p155/s0301lre	0.41	0.43	0.38	0.34
p166/s0275re	0.33	0.31	0.45	0.23
P240/s0468re	0.00	0.00	0.37	0.41

Table 7.9 Dissimilarity factor calculated between normal and IMI data

Reference data (normal)	p236_s0464re		p240_s0468re	
Patient ID and record number from Physionet	D_{qrs}	D_t	D_{qrs}	D_t
p093/s0396lre	0.6095	0.6853	0.6299	0.4482
p093/s0378lre	0.6254	0.6439	0.6369	0.5389
p100/s0399lre	0.5539	0.6986	0.6361	0.5947
p057/s0198lre	0.5078	0.7164	0.5733	0.5739
p066/s0280lre	0.4667	0.8099	0.6047	0.4422
p070/s0235lre	0.6324	0.5946	0.6624	0.5879
p075/s0242lre	0.6037	0.7996	0.7115	0.4781
p075/s0246lre	0.6652	0.6547	0.7373	0.3466
p100/s0401lre	0.5952	0.6981	0.6576	0.4343
p100/s0407lre	0.5480	0.7025	0.6284	0.4815

to be 0.39 revealing the similarity among them. The threshold limit of 0.5 has been considered both for D_{qrs} and D_t to identify the dissimilarity between two data sets (dissimilarity value below 0.5 qualifies two data sets as similar).

In the next stage D_{qrs} and D_t are computed between three normal data and 35 IMI data. Representative results are provided in Table 7.9. In Table 7.9, p236_s0464re and p240_s0468re are the two normal ECG records (out of three) and p093/s0396lre, p093/s0378lre, p100/s0399lre, p057/s0198lre, p066/s0280lre, p070/s0235lre, p075/s0242lre, p075/s0246lre, p100/s0401lre, and p100/s0407lre are representative IMI records (among 35). For this, overall average D_c is found to be 0.65 for the majority of the results. Careful observation of Table 7.9 reveals that in some discrete cases, the D_{qrs} D_t values are low, below 0.5, which is the lower threshold limit of concluding that two data sets (which are in consideration) are dissimilar. The majority of the data in Table 7.9 shows high D_c values (in the range of 0.58 to

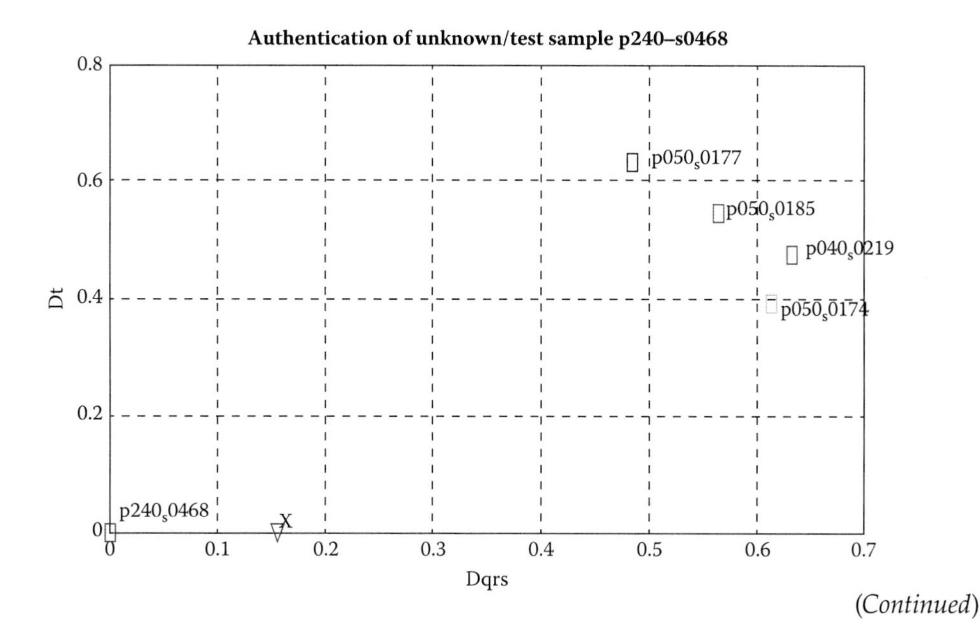

(Continued)

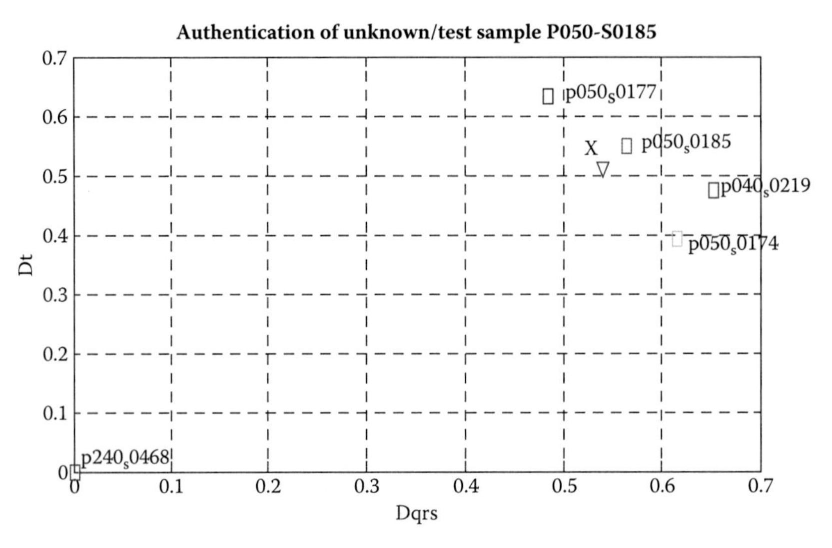
Authentication of unknown/test sample P050-S0185
Dt
Dqrs
p050s0177
p050s0185
p040s0219
p050s0174
p240s0468
0 0.1 0.2 0.3 0.4 0.5 0.6 0.7

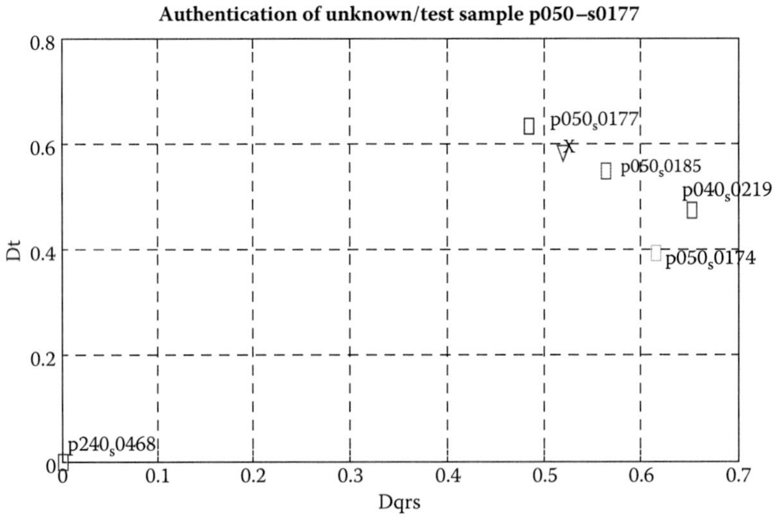
Authentication of unknown/test sample p050–s0177
Dt
Dqrs
p050s0177
p050s0185
p040s0219
p050s0174
p240s0468

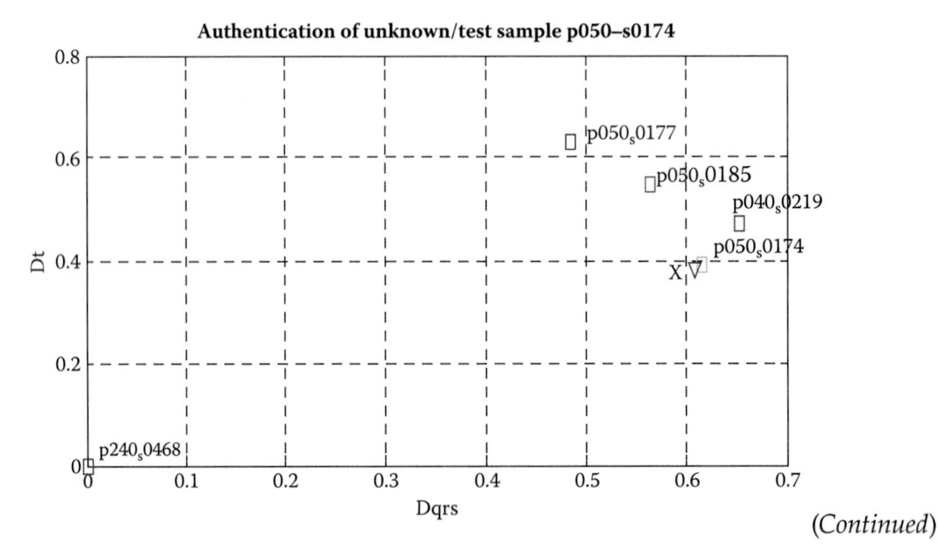
Authentication of unknown/test sample p050–s0174
Dt
Dqrs
p050s0177
p050s0185
p040s0219
p050s0174
p240s0468

(Continued)

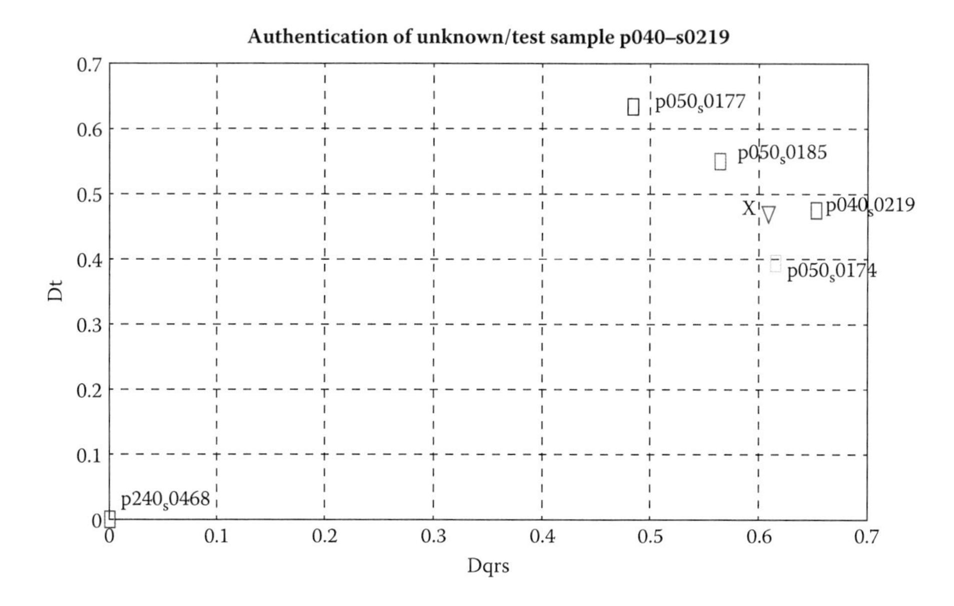

Figure 7.16 Authentication of unknown ECG record based on dissimilarity.

Table 7.10 Distance matrix for identifying unknown ECG patterns with respect to known ECG patterns

Reference/test data sets	p240_s0468	p050_s0185	p050_s0177	p050_s0174	p040_s0219
p240_s0468	**1.42E-01**	6.90E-01	7.18E-01	6.14E-01	6.95E-01
p050_s0185	7.38E-01	**4.96E-02**	1.40E-01	1.34E-01	1.17E-01
p050_s0177	7.83E-01	7.10E-02	**4.54E-02**	2.28E-01	1.88E-01
p050_s0174	7.29E-01	1.68E-01	2.77E-01	**4.11E-03**	9.10E-02
p040_s0219	7.75E-01	1.01E-01	2.16E-01	7.24E-02	**3.28E-02**

0.73 in most cases), which justifies the ability of the algorithm to detect the degree of clinical abnormality compared to reference (normal) data sets in ECG waveforms.

A dissimilarity-based pattern authenticator/classifier is designed using P240-S0468 (normal ECG; as reference database) and four other IMI data sets namely, P050-S0185, P050-S0177, P050-S0174, and P050-S0219. Every time, each of the above data sets is adulterated with a random bias to become test data and is compared with respect to its dissimilarity with the five others, hence a (5 × 6) dissimilarity matrix is created. Keeping in view the fact that ECG data varies from patient to patient, even representing a specific category of cardiac normality or abnormality, the test data samples are created as simulated patients with varying ECGs. Figure 7.16 shows a representative dissimilarity-based authentication as well as classification performance. The ordinate of Figure 7.16 is D_{qrs} and abscissa is D_t. Each of the preprocessed reference data (normal and IMI data) along with the simulated test data (belonging to any of the five categories) are plotted in $D_{qrs} - D_t$ plane in Figure 7.16. The unknown test data will show the minimum distance from the category it belongs to, as is revealed in Figure 7.16 and Table 7.10. The reference data are laid along the rows of Table 7.10 and simulated test data appear along the columns, hence, the resulting minimum entry values are along the diagonals of Table 7.10. Program 7.8 is developed as an unknown ECG authenticator/classifier, and is presented below.

Program 7.8 Authentication/classification of unknown ECG patterns

```
%% data set required p240_s0468, p050_s0168, p050_s0177, %p050_s0174,
p050_s0219. This program requires function %DISSIM.m which is already
%presented in chapter 2.
clc
clear all

format long;
fname11='p240_s0468_QRS.txt';
fname12='p240_s0468_T.txt'

fname21='p050_s0185_QRS.txt'
fname22='p050_s0185_T.txt'

fname31='p050_s0177_QRS.txt'
fname32='p050_s0177_T.txt'

fname41='p050_s0174_QRS.txt'
fname42='p050_s0174_T.txt'

fname51='p040_s0219_QRS.txt'
fname52='p040_s0219_T.txt'

text=['p240_s0468';'p050_s0185';'p050_s0177';'p050_s0174';'p040_s0219';'X
']

x11=load(fname11,'-ascii');
x12=load(fname12,'-ascii');

x21=load(fname21,'-ascii');
x22=load(fname22,'-ascii');

x31=load(fname31,'-ascii');
x32=load(fname32,'-ascii');

x41=load(fname41,'-ascii');
x42=load(fname42,'-ascii');

x51=load(fname51,'-ascii');
x52=load(fname52,'-ascii');

n11=size(x11);
n12=size(x12);

n21=size(x21);
n22=size(x22);

n31=size(x31);
n32=size(x32);

n41=size(x41);
n42=size(x42);
```

```
n51=size(x51);
n52=size(x52);

b=input('Enter bias:')

r11=randn(n11(1,1),n11(1,2));
x11u=x11+b*r11;

r12=randn(n12(1,1),n12(1,2));
x12u=x12+b*r12;

r21=randn(n21(1,1),n21(1,2));
x21u=x21+b*r21;

r22=randn(n22(1,1),n22(1,2));
x22u=x22+b*r22;

r31=randn(n31(1,1),n31(1,2));
x31u=x31+b*r31;

r32=randn(n32(1,1),n32(1,2));
x32u=x32+b*r32;

r41=randn(n41(1,1),n41(1,2));
x41u=x41+b*r41;

r42=randn(n42(1,1),n42(1,2));
x42u=x42+b*r42;

r51=randn(n51(1,1),n51(1,2));
x51u=x51+b*r51;

r52=randn(n52(1,1),n52(1,2));
x52u=x52+b*r52;

D=[];
D1=[];
D2=[];
for u=1:5

if(u==1)
    Y1=x11u, Y2=x12u;
    N1=size(x11) ; N2=size(x12);

    T1=0:1:N1(1,1)-1;
    T1=0.001*T1;
    T1=T1';

    T2=0:1:N2(1,1)-1;
    T2=0.001*T2;
    T2=T2';


end
```

```
if(u==2)
    Y1=x21u, Y2=x22u;
    N1=size(x21) ; N2=size(x22);

    T1=0:1:N1(1,1)-1;
    T1=0.001*T1;
    T1=T1';

    T2=0:1:N2(1,1)-1;
    T2=0.001*T2;
    T2=T2';
end

if(u==3)
    Y1=x31u, Y2=x32u;
    N1=size(x31) ; N2=size(x32);

    T1=0:1:N1(1,1)-1;
    T1=0.001*T1;
    T1=T1';

    T2=0:1:N2(1,1)-1;
    T2=0.001*T2;
    T2=T2';
end

if(u==4)
     Y1=x41u, Y2=x42u;
    N1=size(x41) ; N2=size(x42);

    T1=0:1:N1(1,1)-1;
    T1=0.001*T1;
    T1=T1';

    T2=0:1:N2(1,1)-1;
    T2=0.001*T2;
    T2=T2';
end

if(u==5)
    Y1=x51u, Y2=x52u;
    N1=size(x51) ; N2=size(x52);

    T1=0:1:N1(1,1)-1;
    T1=0.001*T1;
    T1=T1';

    T2=0:1:N2(1,1)-1;
    T2=0.001*T2;
    T2=T2';
end
```

```
D11=DISSIM(x11,x11);
D12=DISSIM(x12,x12);

D21=DISSIM(x11,x21);
D22=DISSIM(x12,x22);

D31=DISSIM(x11,x31);
D32=DISSIM(x12,x32);

D41=DISSIM(x11,x41);
D42=DISSIM(x12,x42);

D51=DISSIM(x11,x51);
D52=DISSIM(x12,x52);

D61=DISSIM(x11,Y1);
D62=DISSIM(x12,Y2);

D= [D11 D12 D21 D22 D31 D32 D41 D42 D51 D52 D61 D62]
D2=[D2;D];

%figure(u)

plot(D(1,1),D(1,2),'bs',D(1,3),D(1,4),'ms',D(1,5),D(1,6),'ks',D(1,7),D(1,
8),'cs',D(1,9),D(1,10),'rs',D(1,11),D(1,12),'rv','LineWidth',2)

grid

xlabel('Dqrs'); ylabel('Dt')
gtext(text)

D=[D11 D12 ; D21 D22 ; D31 D32;D41 D42;D51 D52;D61 D62];


d=[];

for i=1:5
d1=[D(i,1) D(i,2); D(6,1) D(6,2)];

d=[d pdist(d1)]
end
D1=[D1;d]
pause

end

save('distance.txt','D1', '-ascii','-tabs')
save('dissimilarity.txt','D2','-ascii','-tabs')
```

The required data sets, p240_s0468 (normal), and P050-S0185, p050_s0177, p050_s0174, and p050_s0219 (IMI data), are available on the book webpage as an e-resource.

7.4 Arsenic quantification in contaminated water using chemometrics

This section aims to focus on quantification of arsenic in water (to a 5-ppb level) using simple electronic tongue instrumentation and chemometric processing tools. Anodic stripping voltammetry (ASV) is used to detect arsenic. Subsequently, a low cost and indigenous possible arsenic meter scheme suitable for bulk monitoring is proposed. It seems to be necessary to provide a brief introduction to the severity of arsenic pollution and the current state of the art of its detection, as well as future possibilities for technical enhancement with respect to its detection and quantification.

Arsenic occurs naturally in the earth's crust, in sea water, and in the human body. Arsenic predominantly exists in +V (arsenate), +III (arsenite), and –III (arsine) oxidation states. It cannot be transformed into a nontoxic material, and there is always a need to monitor its threshold level. The source of arsenic is geological: human activities such as deep mining and the application of pesticides in agriculture can increase its levels in the environment. The mobilization of arsenic from geological deposits into ground and drinking water resources are possibly the greatest threat to human health. Arsenic poisoning from drinking water has been the worst natural debacle in recent times. Prolonged exposure to inorganic soluble arsenic would result in the following sorts of human health disorders: discoloration of the skin, stomach pain, nausea, vomiting, diarrhea, numbness in hands and feet, partial paralysis, and blindness. Arsenic is also carcinogenic and may lead to cancer of the bladder, lungs, skin, kidney, nasal passages, liver, and prostate. The toxicity of trivalent arsenic is due to its high affinity for the sulfhydryl groups of biomolecules such as glutathione (GSH), lipoic acid, and the cysteinyl residues of many enzymes. The formation of *As (III)*–sulfur bonds inhibits the activities of enzymes. Besides their toxicity to humans, arsenic +V and +III are highly toxic to plants because they decouple phosphorylation and restrain phosphate uptake. Bangladesh, India, Vietnam, and Cambodia are severely affected by arsenic poisoning in drinking water and groundwater. The concentration of arsenic is usually less than 2 μg/L in seawater. The levels of arsenic in unpolluted surface water and groundwater vary typically from 1–10 μg/L. The World Health Organization (WHO) as well as the US Environmental Protection Agency (EPA) has set the arsenic standard for drinking water at 10 ppb, and drinking water systems have to comply with this standard to protect the public health [26]. Different analytical techniques are used for the detection of inorganic As such as atomic absorption spectroscopy (AAS), hydride generation-AAS, inductively coupled plasma (ICP), neutron activation analysis, capillary electrophoresis (CE), laser-induced breakdown spectroscopy (LIBS), atomic force microscopy (AFM), surface-enhanced Raman spectroscopy (SERS), electrochemical method, and colorimetry. Such techniques provide limits of detection (LOD) well below the WHO arsenic guideline (10 ppb), but are lab based, time consuming, and require costly instrumentation for routine monitoring of bulk samples and highly trained personnel. A few of these techniques have been proven to be hazardous to the technician's health due to generation of arsine gas. There is an urgent need to develop portable, uncomplicated, consistent, sensitive, and inexpensive equipment for field measurements.

Low-cost electrochemical methods including polarography, cathodic stripping voltammetry (CSV), and anodic stripping voltammetry (ASV) provide an attractive alternative because of their portability and simple instrumentation. All of these techniques are applicable for the determination of electroactive *As (III)*. An electronic tongue utilizes electrochemical analysis methods, which consist of a sensor/sensor array coupled with *chemometric* processing techniques that can be utilized in quantification of arsenic in contaminated water. ASV is an electroanalytical method that detects ion in a solution by the potential at which they oxidize and strip away from the surface of the electrode. The electrode is first plated and the analyte is deposited cathodically on the electrode. The analyte is then stripped (oxidized) from the electrode during a potential scan in the anodic direction. Thus, the name anodic strips voltammetry. The stripping current is proportional to the amount of metal analyte present in the solution. Different electrodes such as platinum, gold, mercury, modified glassy carbon (GC), and boron-doped diamond (BDD), in combination with a variety of acids such as HCl, H_2SO4, HClO4 and HNO3 as supporting electrolytes, have been used for the As (III) detection using ASV. The stripping step can be linear, staircase, square wave, or pulse. From the perspective of current e-tongue (ET) research, its usage, and trepidation of severe arsenic pollution in ground and drinking water, the detection of *As* in simulated contaminated water was carried out. The proposed arsenic meter is an assembly of ET-based signal generation and chemometric processing techniques.

7.4.1 *E-tongue-based experimentation*

The experiment consists of a potentiostat (make: Gamry, USA; model: Reference-600), which is interfaced to a PC through USB, and is operated through data acquisition software (DAS). The DAS uses a PHE200™ physical electrochemistry software package. The potentiostat is commonly interfaced to an electrochemical cell containing three electrodes, which are a reference electrode (Ag/AgCl wire dipped in 3 M KCl solution was used as the reference electrode). The tip of the Ag/AgCl electrode was fitted with a porous Vycor frit, working electrode (a gold wire with a length of 5 mm and diameter of 0.6 mm was used as the working electrode), and counter electrode (a platinum wire of 5 mm length and 0.46 mm diameter was used as the counter electrode). The potentiostat is connected to a PC or a microcontroller-based system. The reference electrode establishes a constant reference potential in the electrochemical cell, against which the working electrode potential may be determined. The current flows into the solution via the working electrode and leaves the cell via the counter electrode.

A 75 mg/L stock solution of *As (III)* was prepared by dissolving 19.81 mg of reagent grade As_2O_3 in ultrapure water with 1 M HCl as the supporting electrolyte. Successive dilutions were made as required. The concentrations of arsenic solution considered for the experimentation were 5, 7.5, 10, 25, 50, 75, 100, 250, 500, 750 ppb. 1 M HCl was prepared by dilution of 35–37% of hydrochloric acid solution and ultrapure water (Milli-Q Ultrapure Water Purification System). All the solutions were prepared and stored at room temperature. As a standard experimental procedure, the gold electrode was kept in a solution of 50 mM KOH and 25% H_2O_2 for 10 minutes followed by:

1. Potential sweep from 200 mV to –1200 mV in 50 mM KOH.
2. Rinsed well with Milli-Q water.
3. Sonicated for 10 minutes.
4. Dried and polished using a 0.05 µ polishing-grade alumina on a polishing pad.
5. Finally the electrode is rinsed to remove the traces of adhering alumina particles.

25 mL of the simulated arsenic sample to be analyzed was transferred into the cell; the arsenic was allowed to be deposited for 180 s. During the deposition, the solution was constantly stirred using a magnetic stirrer. The stirring action was stopped at the end of the deposition period and after an idle period of 5 s; the voltage was swept in the range from −400 mV to 600 mV with a pulse size of 25 mV and frequency of 50 Hz (optimized ASV values). The resulting voltammogram was recorded. DAS conducted the above steps. Deposition potentials below −400 mV resulted in frequent generation of hydrogen bubbles at the working electrode, and therefore hampered the arsenic preconcentration step. For each of the 10 varying concentrations of arsenic solution, three samples were prepared; thus, thirty total samples were considered for experimentation. Figure 7.17 shows the detected current versus sampling instant for various concentrations of arsenic solutions. The maximum and minimum detected currents were 345 μA and 102 μA for 750 and 5 ppb solutions, respectively [27].

7.4.2 *Arsenic quantifier design*

Principal component analysis is used to develop a subspace model using the experimental data. The data belonging to the different concentrations of arsenic-contaminated samples find themselves segregated in different clusters possessing individual cluster centers. The unknown sample finds its place in an appropriate cluster following the nearest neighborhood principle.

The data matrix resulted from the collected experimental time series signatures is 603 × 10. The mean of three samples are taken for each category of arsenic solutions, hence, a data matrix of 201 × 10 is considered for PCA decomposition. The resulting scores are projected along two principal component directions as shown in Figure 7.18, which reveals

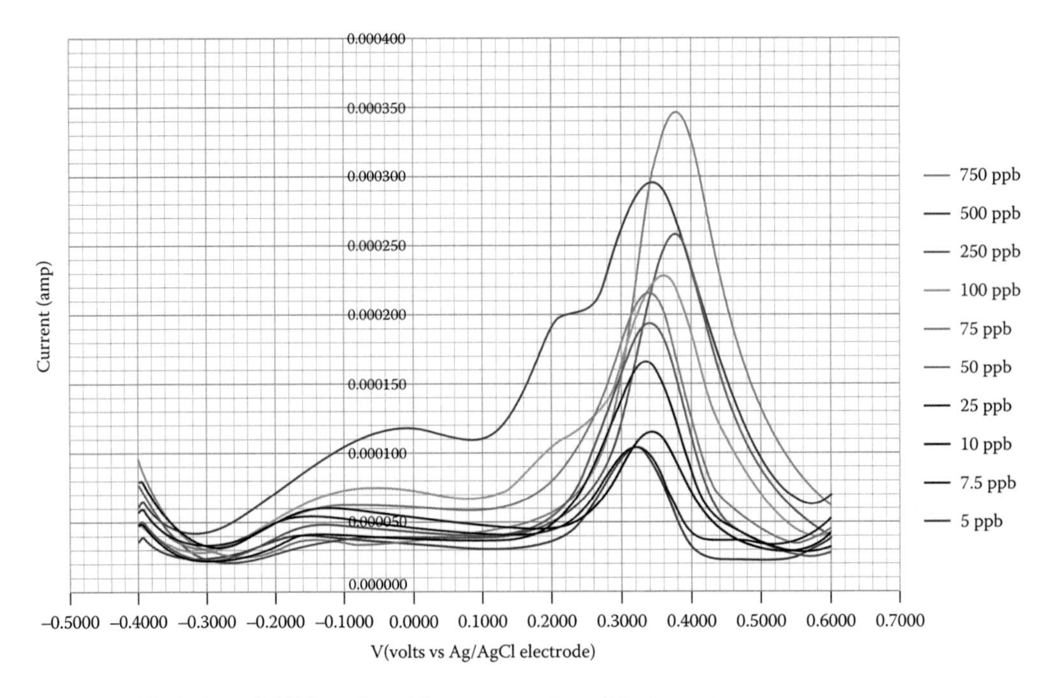

Figure 7.17 Variation of ASV peaks with concentration of As in water.

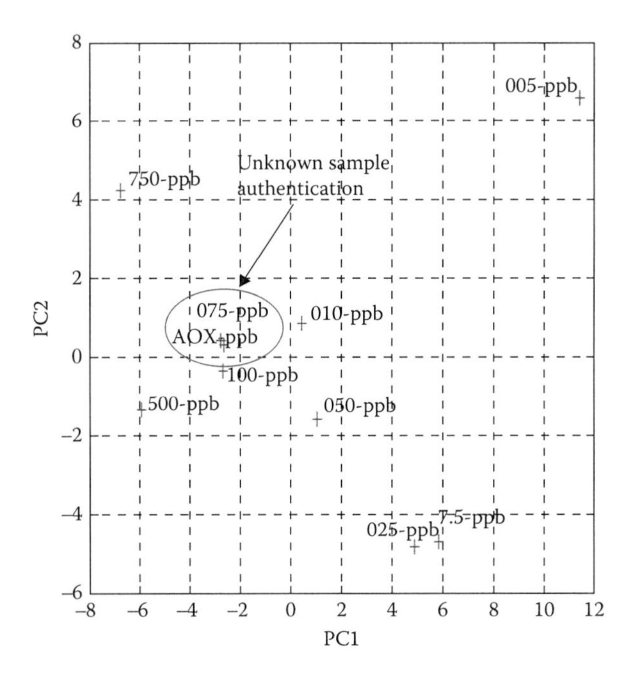

Figure 7.18 Unknown arsenic sample (75 ppb) authentication.

the fair amount of separability among the data pertaining to different concentrations in ppb level and the authentication of unknown sample. The unknown arsenic contaminated solution is simulated by incorporating 5% noise to any of the 10 categories of data (mean of the three samples). The Euclidean distance of each of the data points from the origin is calculated. Then the distances are correlated with arsenic concentrations using *cftool* (MATLAB®). The resulting model is as follows:

$f(x) = ax^b + c$, goodness of fit: R-square: 0.9692
Coefficients (with 95% confidence bounds):

a = 5.463e-013 (-8.87e-012, 9.963e-012)
b = 16.14 (8.148, 24.14)
c = 35.87 (–13.44, 85.19)

The quantity of any unknown solution can be also found using the developed empirical model. For developing a good quantifier, the range of concentration should not be that large and arsenic concentration ranging from 100 ppb down to five ppb level is crucial to quantify and monitor. An exhaustive data collection within that range will ensure successful quantifier development.

The developed quantifier is intended to be a part of a cheap and indigenous arsenic meter development. An arsenic quantifier suitable for bulk monitoring consists of a data acquisition (DAQ) card, PC-based signal generation and reception via the DAQ card using LABVIEW software, and dedicated hardware for signal reception, amplification, and transduction in conjunction with the developed quantifier or quantization software (the general schematic of an ET-based quantifier is provided in Chapter 5; Figure 5.14). A more economic and compact version of the quantifier uses a custom-designed microcontroller unit (MCU) board responsible for ASV signal generation, current signal reception, and flow chart scaling, data logging, and communication to a PC/laptop through a serial communication unit. Detailed hardware design is not within the scope of this book.

References

1. A.M. Katz (2011). *Physiology of the Heart*. Philadelphia, PA: Lippincott Williams & Wilkins.
2. P.A. Iaizio (Ed.) (2009). *Handbook of Cardiac Anatomy, Physiology and Devices*. New York, NY: Springer.
3. R. Gupta, M. Mitra, J. Bera (2014). *ECG Acquisition and Automated Remote Processing*. India: Springer.
4. R. Plonsey, D.G. Fleming (1969). *Bioelectric Phenomena*. New York, NY: McGraw Hill.
5. P. Kundu, R. Gupta (2015). Electrocardiogram synthesis using Gaussian and Fourier models. *IEEE International Conference on Research in Computational Intelligence and Communication Networks* (ICRCICN), IEEEXplore, 20–22 November 2015.
6. F. Kusumoto (2009). *ECG Interpretation from Pathology to Clinical Interpretation*. New York, NY: Springer.
7. F. Morris, W.J. Brady, J. Camm (Eds.) (2008). *ABC of Clinical Electrocardiography*. Boston, MA: Blackwell Pub Co.
8. M. Graham (1976). Latent component in electrocardiogram. *IEEE Transactions on Biomedical Engineering*, vol. BME-23(3), 220–224.
9. S.M.S. Jalaleddnne, G.G. Hutchens, R.D. Strattan and W.A. Coberly (1990). ECG data compression Technique-a unified approach. *IEEE Transactions on Biomedical Engineering*, vol. BME-37(4), 329–342.
10. C. Li, C. Zheng, C. Tai (1995). Detection of ECG characteristic points using wavelet transform. *IEEE Transactions on Biomedical Engineering*, vol. 42, 21–28.
11. W. Yu, L. Jhang (2011). ECG classification using ICA features and support vector machines. In *Proceedings of International Conference on Neural Information Processing ICONIP 2011*, vol. 7062, Shanghai, China, pp. 146–154.
12. S.M.S. Jalaleddine, C.G. Hutchens, R.B Strattan, W.A. Coberly (1990). ECG data compression techniques-a unified approach. *IEEE Transactions on Biomedical Engineering*, vol. 37(4), 329–342.
13. M. Lagerholm, C. Peterson, G. Bracccini, L. Edenbrandt, L. Sornmo (2000). Clustering ECG complexes using hetmite functions and self organizing maps. *IEEE Transactions on Biomedical Engineering*, vol. 47(7), 838–848.
14. P.E. Mcsharry, G.D. Clifford, L. Tarassenko, L.A. Smith (2003). A dynamic model for generating synthetic electrocardiogram signals. *IEEE Transactions on Biomedical Engineering*, vol. 50(3), 289–294.
15. Z. Li, M. Ma (2005). ECG modeling with DFG. In *Proceedings of the 2005 IEEE Engineering in Medicine and Biology 27th Annual Conference*, Shanghai, China, pp. 2691–2694.
16. S. Parvaneh, M. Pashna (2007). Electrocardiogram synthesis using a Gaussian combination model (GCM). *Computers in Cardiology*, vol. 34, 621–624.
17. R. Dubios, P.M. Blanche, B. Quenet, G. Dreyfus (2007). Automatic ECG wave extraction in long-term recordings using gaussian mesa function models and nonlinear probability estimators. *Computer Methods and Programs in Biomedicine*, vol. 88(3), 217–233.
18. S.C. Bera, R. Sarkar (2010). Fourier anlysis of normal ECG signal to find its maximum harmonic content by signal reconstruction. *Sensors and Transducers*, vol. 123(12), 106–117.
19. F. Bagheri, N. Ghafarnia, F. Bahrami (2013). Electrocardiogram (ECG) signal modeling and noise reduction using hopfield neural networks. *ETASR – Engineering, Technology and Applied Science Research*, vol. 3(1), 345–348.
20. Physionet database: http://www.physionet.org
21. R. Gupta, P. Kundu (2016). Dissimilarity factor based classification of inferior microcardial infraction ECG. *IEEE First International Conference on Control, Measurement and Instrumentation (CMI-16)*, 8–10 January 2016.

22. F. Castells, P. Laguna, L. Sornmo, A. Bollmann, J.M.M. Roig (2007). Principal component analysis in ECG signal processing. *EURASIP Journal on Advances in Signal Processing*, 1–21.
23. M.P.S. Chawla (2011). PCA and ICA processing methods for removal of artifacts and noise in electrocardiograms—A survey and comparison. *Applied Soft Computing*, vol. 11(2), 2216–2226.
24. M. Kotas (2006). Application of projection pursuit based robust principal component analysis to ECG enhancement. *Biomedical Signal Processing and Control*, vol. 1(4), 289–298.
25. S. Banerjee, R. Gupta, M. Mitra (2012). Delineation of ECG characteristic features using multiresolution wavelet analysis method. *Measurement*, vol. 45(3), 474–487.
26. USEPA 816-F-01-004 (January 2001), Arsenic and Clarifications to Compliance and New Source Monitoring Rule: A Quick Reference Guide, Office of Water [Online]. Available: http://water.epa.gov/drink/info/arsenic/upload/2005_11_10_arsenic_quickguide.pdf
27. M. Kundu, S.K. Agir (2016). Detection and quantification of arsenic in water using electronic tongue. In *The Proceedings of IEEE Sponsored International Conference on Control, Measurement and Instrumentation (CMI-16)*, 8–10 January 2016, Kolkata, India.

Index

PGSTL